Modern Digital Technologies for the Built Environment of the Future

Modern Digital Technologies for the Built Environment of the Future

Guest Editor

Andrzej Szymon Borkowski

MDPI

Basel • Beijing • Wuhan • Barcelona • Belgrade • Novi Sad • Cluj • Manchester

Guest Editor
Andrzej Szymon Borkowski
Faculty of Geodesy and
Cartography
Warsaw University of
Technology
Warsaw
Poland

Editorial Office
MDPI AG
Grosspeteranlage 5
4052 Basel, Switzerland

This is a reprint of the Special Issue, published open access by the journal *Infrastructures* (ISSN 2412-3811), freely accessible at: https://www.mdpi.com/journal/infrastructures/special_issues/O2V83QIPH9.

For citation purposes, cite each article independently as indicated on the article page online and as indicated below:

Lastname, A.A.; Lastname, B.B. Article Title. *Journal Name* **Year**, *Volume Number*, Page Range.

ISBN 978-3-7258-7935-9 (Hbk)
ISBN 978-3-7258-7936-6 (PDF)
https://doi.org/10.3390/books978-3-7258-7936-6

Contents

About the Editor

Andrzej Szymon Borkowski

Andrzej Szymon Borkowski is an Associate Professor at the Faculty of Geodesy and Cartography, Warsaw University of Technology, where he habilitated in technical sciences. He graduated from the same Faculty in the field of spatial planning and has been affiliated with the Department of Spatial Planning and Environmental Sciences since 2014. His research interests focus on Building Information Modelling (BIM), BIM–GIS integration, formal ontology applied to the built environment, computer vision and large language models in construction, digital twins, and geospatial methods for spatial planning. He is the author of more than 120 scientific publications, including articles in high ranking international journals, as well as scripts and tutorials used in academic teaching. He leads a dedicated research laboratory on ResearchGate and cooperates with industrial partners on applied research and development projects. He is an Authorized Autodesk Certified Instructor and an accredited ECDL CAD examiner, and has long standing training experience with design software from Autodesk, Adobe, and Esri. He has served on many organizing committees and research teams, and has been a Guest Editor of Special Issues in international journals.

Preface

The digitization of the built environment has moved from an emerging trend to a core expectation of the architecture, engineering, construction, and operations sectors. Within the last decade, methods such as Building Information Modeling, Geographic Information Systems, computer vision, machine learning, and Virtual Reality have evolved from isolated research experiments into integrated components of professional workflows. At the same time, the concept of the digital twin is reshaping how the lifecycle of buildings and infrastructure is understood, from early design, through construction, to long-term operation and maintenance.

This Reprint was conceived as a platform to bring together contributions that reflect both the diversity and the maturity of this transformation. The invitation to the original Special Issue emphasized building, technical, transmission, as well as green and blue infrastructure, and encouraged authors to submit work grounded in real case studies or rigorous literature synthesis. The collected articles meet that aim. They span early conceptual design with immersive technologies, information requirement management according to the ISO 19650 standard, artificial intelligence applied to imagery and text in construction, structural health monitoring of concrete bridges, parametric and epistemic approaches to the representation of existing buildings, blockchain-based sustainable procurement, and integrated decision making for conservation-oriented projects.

The purpose of this Reprint is twofold. The first goal is to offer a coherent reference for researchers exploring the frontier between construction informatics, artificial intelligence, and digital twin engineering. The second goal is to provide practitioners, teachers, and graduate students with a tangible map of current methods, their limitations, and their open research questions. Each contribution can be read on its own, yet together, they sketch a broader picture of where the field is heading and which problems deserve further investigation.

I wish to thank all the authors for their valuable work, the reviewers for their careful and constructive feedback, and the editorial team of *Infrastructures* for their continuous support throughout the production of the Special Issue and this Reprint. I hope that the ideas collected here will stimulate further interdisciplinary research and support the ongoing digital transformation of the built environment.

Andrzej Szymon Borkowski
Guest Editor

infrastructures

Editorial

Editorial for the Special Issue "Modern Digital Technologies for the Built Environment of the Future"

Andrzej Szymon Borkowski

Faculty of Geodesy and Cartography, Warsaw University of Technology, Politechniki Square 1, 00-661 Warsaw, Poland; andrzej.borkowski@pw.edu.pl

1. Introduction

The construction sector has long been perceived as one of the least digitized branches of the economy. That perception is changing rapidly. Over the last decade, Building Information Modeling (BIM) has moved from a specialist technique into a widely accepted standard, Geographic Information Systems (GIS) have become an essential component of infrastructure planning, and artificial intelligence (AI) has been incorporated into the daily practice of design, inspection, and facility management. Digital Twins (DT), once limited to manufacturing and aerospace, now represent a realistic horizon for buildings and civil infrastructure as well. This Special Issue of *Infrastructures*, titled "Modern Digital Technologies for the Built Environment of the Future", was conceived against this background. Its aim was to collect contributions that reflect both the breadth and depth of the ongoing transformation in this field, covering building, technical, transmission, and green or blue infrastructure elements, and bringing together studies on topics ranging from early conceptual design to long-term asset management.

The Special Issue closed with eight accepted contributions, including seven original research articles and one systematic review. The authors come from research centers from Europe, North America, Asia, and Australia, which confirms the international relevance of the topic. Taken together, the collected papers form a coherent panorama of current methods, open problems, and emerging research directions in construction informatics.

2. Overview of the Contributions

This Special Issue [1] opens with the work of Mitera-Kiełbasa and Zima 1, who address the challenge of automating the preparation of Exchange Information Requirements (EIR) for construction projects under the ISO 19650 standard [2]. Using Word2Vec for text vectorization and Support Vector Machines (SVMs) for classification, the authors achieve an average F1 score of 0.7, demonstrating that natural language processing can meaningfully support information management in BIM-based project delivery. The authors also highlight the current limits of text generation methods in this domain and indicate large language models (LLMs) as a promising direction for further work.

Christenson 2 focuses on the epistemological role of digital modeling. Based on a case study of the Federal Archive Building in New York City, the author argues that parametric analysis should be regarded not only as a tool for geometric representation but also as an instrument for generating architectural hypotheses. This paper invites a rethink of the purpose of digital models of existing buildings, and lays groundwork for later contributions in this Special Issue that revisit the representation of heritage and existing structures.

Renganathan and colleagues 3 examine the early conceptual design phase, in which traditional BIM workflows are often too rigid. By integrating Virtual Reality (VR) with BIM

 https://doi.org/10.3390/infrastructures11050150

in a structured framework, validated through a pilot study, the authors show measurable improvements in spatial cognition, emotional engagement, and iterative decision making. Their contribution illustrates how immersive technologies can complement data-driven methods and support more human-centered design practice.

Singh and co-authors 4 shift the focus to sustainable construction procurement and the adoption of blockchain technology. Combining a systematic literature review, expert consultation, Fuzzy DEMATEL, k means clustering, and social network analysis, the study identifies decentralization, data security, quality, and project cost as key drivers, while system stability, overall project performance, and customer satisfaction emerge as central nodes in the adoption network. The authors demonstrate how hybrid artificial intelligence-based methods can inform strategic decisions about technology adoption in this sector.

Borkowski, Kochański, and Rukat 5 provide a thorough review of the evolution of convolutional and recurrent neural networks in the context of BIM, organized along three axes: computer vision coupled with BIM models; sequence and time series modeling for cost, energy, and risk prediction; and the integration of deep learning outputs with the semantics and topology of Industry Foundation Classes (IFC) models. The paper also introduces Bimetria, a practical tool that combines convolutional networks with optical character recognition to generate estimated three-dimensional models from two-dimensional drawings. This contribution positions mature convolutional and recurrent architectures alongside emerging attention- and graph-based models, and raises fundamental questions about benchmarks, synthetic data, and explainability in the AECOO (Architecture, Engineering, Construction, Owner, Operator) domain.

Hanson, Liu, and Christenson 6 reflect on existing building representation by proposing a method that promotes epistemic resilience rather than epistemic closure. Implemented in Autodesk Revit, their parametric reference frame foregrounds reciprocal relationships between orthographic views and deliberately sacrifices efficiency in favor of interpretive depth. Their study expands the role of BIM beyond documentation and situates it as an instrument of architectural knowledge production, with implications for software development, pedagogy, and critical representation.

Kwast-Kotlarek and Szóstak 7 present an integrated RACI (Responsible, Accountable, Consulted, Informed), Analytic Hierarchy Process (AHP), and BIM methodology for projects with high functional complexity and conservation constraints. The approach is validated on a higher education conservation project with a BIM execution plan and scan to BIM procedures. The authors demonstrate that coupling a responsibility matrix with AHP-based decision making and BIM data reduces accountability gaps, improves interdisciplinary coordination, and supports transparent, data-driven design under the design build model.

The Special Issue closes with a systematic review by Karkan and colleagues 8, who critically evaluated 77 studies on deep learning-based visual inspection of concrete bridges using two-dimensional images. Their review maps classification, object detection, and segmentation models, and identifies transfer learning, data augmentation, and careful dataset composition as key success factors. The authors also outline future directions, including tighter integration with BIM, edge computing for real-time monitoring, and the development of richer annotated datasets for better generalizability.

3. Common Threads and Future Directions

Read as a whole, the contributions in this Special Issue point toward several converging trends. First, BIM is increasingly understood not as a single technology but as a semantic and geometric backbone onto which artificial intelligence, Virtual Reality, blockchain, and structured decision making methods can be grafted. Second, there is growing awareness

https://doi.org/10.3390/infrastructures11050150

that the value of digital models lies not only in geometric accuracy but also in their capacity to support knowledge production, critical inquiry, and transparent governance. Third, interoperability, standardization, and the quality of domain-specific datasets remain persistent bottlenecks that no single study can resolve on its own.

On behalf of the Guest Editorship, I would like to thank all the authors for their valuable contributions, the reviewers for their rigorous and constructive feedback, and the *Infrastructures* Editorial Office for their continuous support throughout preparations for this Special Issue and its Reprint. I hope that the collected papers will inspire further interdisciplinary research and contribute to a more mature and responsible digital transformation of the built environment.

Conflicts of Interest: The author declares no conflicts of interest.

List of Contributions:

1. Mitera-Kiełbasa, E.; Zima, K. Automated Classification of Exchange Information Requirements for Construction Projects Using Word2Vec and SVM. *Infrastructures* **2024**, *9*, 194. https://doi.org/10.3390/infrastructures9110194.
2. Christenson, M. Parametric Analysis as a Tool for Hypothesis Generation: A Case Study of the Federal Archive Building in New York City. *Infrastructures* **2025**, *10*, 71. https://doi.org/10.3390/infrastructures10040071.
3. Renganathan, B.; Shanthi Priya, R.; Kumar, G.R.; Thiruvengadam, J.; Senthil, R. Intuitive and Experiential Approaches to Enhance Conceptual Design in Architecture Using Building Information Modeling and Virtual Reality. *Infrastructures* **2025**, *10*, 127. https://doi.org/10.3390/infrastructures10060127.
4. Singh, A.K.; Mohandes, S.R.; Shakor, P.; Cheung, C.; Arashpour, M.; Kidd, C.; Kumar, V.R.P. Blockchain Technology Adoption for Sustainable Construction Procurement Management: A Multi-Pronged Artificial Intelligence-Based Approach. *Infrastructures* **2025**, *10*, 207. https://doi.org/10.3390/infrastructures10080207.
5. Borkowski, A.S.; Kochański, Ł.; Rukat, K. Evolution of Convolutional and Recurrent Artificial Neural Networks in the Context of BIM: Deep Insight and New Tool, Bimetria. *Infrastructures* **2026**, *11*, 6. https://doi.org/10.3390/infrastructures11010006.
6. Hanson, C.; Liu, X.; Christenson, M. Reframing BIM: Toward Epistemic Resilience in Existing-Building Representation. *Infrastructures* **2026**, *11*, 40. https://doi.org/10.3390/infrastructures11020040.
7. Kwast-Kotlarek, U.; Szóstak, M. RACI–AHP–BIM Methodology in Projects with High Functional Complexity and Conservation Constraints. *Infrastructures* **2026**, *11*, 105. https://doi.org/10.3390/infrastructures11030105.
8. Lotfi Karkan, N.; Shakeri, E.; Sadeghi, N.; Banihashemi, S. Smart Surveillance of Structural Health: A Systematic Review of Deep Learning-Based Visual Inspection of Concrete Bridges Using 2D Images. *Infrastructures* **2025**, *10*, 338. https://doi.org/10.3390/infrastructures10120338.

References

1. Special Issue "Modern Digital Technologies for the Built Environment of the Future" Infrastructures Journal. Available online: https://www.mdpi.com/journal/infrastructures/special_issues/0LRF202A18 (accessed on 21 April 2026).
2. *ISO 19650-1:2018*; Organization and Digitization of Information about Buildings and Civil Engineering Works, Including Building Information Modeling (BIM)—Information Management Using Building Information Modeling—Part 1: Concepts and Principles. International Organization for Standardization: Geneva, Switzerland, 2018.

https://doi.org/10.3390/infrastructures11050150

infrastructures

MDPI

Article

Automated Classification of Exchange Information Requirements for Construction Projects Using Word2Vec and SVM

Ewelina Mitera-Kiełbasa * and Krzysztof Zima

Division of Management in Civil Engineering, Faculty of Civil Engineering, Cracow University of Technology, 31-155 Kraków, Poland; krzysztof.zima@pk.edu.pl

* Correspondence: e.mitera@pk.edu.pl

Abstract: This study addresses the challenge of automating the creation of Exchange Information Requirements (EIRs) for construction projects using Building Information Modelling (BIM) and Digital Twins, as specified in the ISO 19650 standard. This paper focuses on automating the classification of EIR paragraphs according to the ISO 19650 standard's categories, aiming to improve information management in construction projects. It addresses a gap in applying AI to enhance BIM project management, where barriers often include technological limitations, a shortage of specialists, and limited understanding of the methodology. The proposed method uses Word2Vec for text vectorisation and Support Vector Machines (SVMs) with an RBF kernel for text classification, and it attempts to apply Word2Vec with cosine similarity for text generation. The model achieved an average F1 score of 0.7, with predicted categories for provided sentences and similar matches for selected phrases. While the text classification results were promising, further refinement is required for the text generation component. This study concludes that integrating AI tools such as Word2Vec and SVM offers a feasible solution for enhancing EIR creation. However, further development of text generation, particularly using advanced techniques such as GPT, is recommended. These findings contribute to improving managing complex construction projects and advancing digitalization in the AECO sector.

Keywords: EIR; Word2Vec; SVM; cosine similarity; AI; BIM; Digital Twin; AECO; text classification; text generation

1. Introduction

The Architecture, Engineering, Construction, and Operation (AECO) sector is experiencing a significant digital transformation in alignment with the European Commission's *Industry 5.0* proposal, which promotes a digital and green transition [1]. There is a growing emphasis on the implementation of Building Information Modelling (BIM) and Digital Twins to enhance the efficiency of both public and private construction projects. These methodologies support the European Union's goal of fostering digitalization within the industry, and some countries (35% of European countries [2]) have committed to their use for public construction projects to varying degrees.

Building Information Modelling (BIM) facilitates the creation of a shared digital representation of built assets, which streamlines design, construction, and operational processes [3]. In the literature, BIM refers to both the methodology and the models themselves [4]. However, it is more accurate to speak of „models" in the plural, as in practice, each discipline typically operates using its own model. This methodology outlines the process for creating, storing, and exchanging information on building assets, and aims at interoperability. The Digital Twin (DT) extends this by integrating real-time data [5] from interconnected devices, such as sensors and Internet of Things (IoT) applications, enabling advanced simulations and AI-driven analyses [6]. DT plays a crucial role in intelligent and smart construction [7]. BIM is supported by the ISO 19650 international standard, which standardizes information management and facilitates experience sharing across projects. Despite these advantages, managing a construction project using BIM remains complex, requiring specialized knowledge and strategic planning for the entire life cycle of a built asset. Technological barriers, a lack of expertise, and insufficient understanding of the methodology are often cited as challenges to BIM implementation [8,9]. The ISO 19650 standard underscores the importance of engaging expert knowledge in such processes [10].

This study addresses these challenges by proposing an automated, personalized approach to creating Exchange Information Requirements (EIRs), which are crucial for outlining the client's (project owner's, general contractor's, or designer's) requirements for the contractor (designer or subcontractor). Here, automation focuses on classifying EIR paragraphs according to the ISO 19650 standard's managerial, commercial, and technical categories, contributing to more efficient information management in BIM/Digital Twin projects. AI integration offers a viable solution to these challenges, streamlining the process and facilitating the adoption of these methodologies. This is particularly relevant to the European Union's digital transition in the AECO sector.

While existing research highlights the potential of AI in the construction sector, gaps remain in its application to automated information management. Gohel et al. reviewed AI's role in construction management [11], and Rangasamy et al. examined the integration of BIM and AI in prefabricated construction [12]. Some studies have proposed improvements through the use of classification algorithms, such as that of Ma et al., which uses rule-based methods for diagnosing faults in urban rail systems [13]. Zheng et al. developed a domain-specific language model for the AEC sector, improving text classification and entity recognition [14]. Dolhopolov et al. utilized AI in analyzing Digital Twins of buildings and construction sites [15]. Advanced classification models have been introduced outside the AECO sector, such as Bartal et al.'s ADA model for narrative classification to effectively detect post-traumatic stress disorder following childbirth [16] and Huang et al.'s FinBERT language model, which is adapted to the financial sector [17]. However, despite these advancements, the automation of information requirements (IRs) in the BIM/Digital Twin context has received insufficient attention.

Further studies, such as those by Piazzi et al., Goonetillake et al., and Tomczak et al., have focused on improving the specification and exchange of information within BIM [18–20]. Piazzi et al. focused on graphical Exchange Information Requirements, using advanced algorithms and rule-based systems to enhance data exchange efficiency and accuracy [18]. Goonetillake et al. developed a prototype tool for embedding digital Exchange Information Requirements, enhancing the automation and accuracy of capturing as-built information through process maps and sample information requirements [19]. Tomczak et al. presented a comprehensive review of methods used to specify information requirements in digital construction projects, comparing standardized and non-standardized approaches to improve efficiency and communication in BIM environments [20].

Text generation research, which attempts to create text sequences based on existing documents, dates back to the 1950s. Luhn described an algorithm for automatically summarizing documents [21], and today, techniques such as abstractive summarisation and deep learning models are employed [22,23]. However, in the early stages of such research, simpler methods like semantic similarity searches can be used [24,25].

The objective of this research was to develop the automatic categorization of data from a database of published Exchange Information Requirements (EIRs) in both public and private tenders and bidding procedures conducted using BIM or Digital Twin methodologies. This categorization aligns with the ISO 19650 international standard, with the potential future integration of a chatbot. This study addresses the challenge of automating EIR creation, streamlining the information management process in BIM/Digital Twins, and responding to the need for an easier implementation of these methodologies. Furthermore, it aligns with the European Union's digital transition goals within the AECO sector. By addressing this gap, this study not only advances the application of AI in construction project management but also contributes to the practical implementation of standardized methodologies, such as the ISO 19650 standard, thereby facilitating the smoother adoption of BIM and Digital Twins across the sector.

2. Materials and Methods

This paper reports on the development of an automated model for proposing Exchange Information Requirements (EIRs) tailored to user preferences and a specific construction project, implemented in BIM/as a Digital Twin, with focus text classification into the ISO 19650 standard. This section highlights the use of Word2Vec, Support Vector Machines (SVMs), and the Radial Basis Function (RBF) kernel for text classification in accordance with the ISO 19650 standard. A future paper will also discuss the application of the Term Frequency–Inverse Document Frequency (TF-IDF) statistical method for the numerical representation of text, as well as Bidirectional Encoder Representations from Transformers (BERT) for vectorization and classification, based on transformer architectures. Figure 1 illustrates the overall methodology, with the section relevant to this paper highlighted in red.

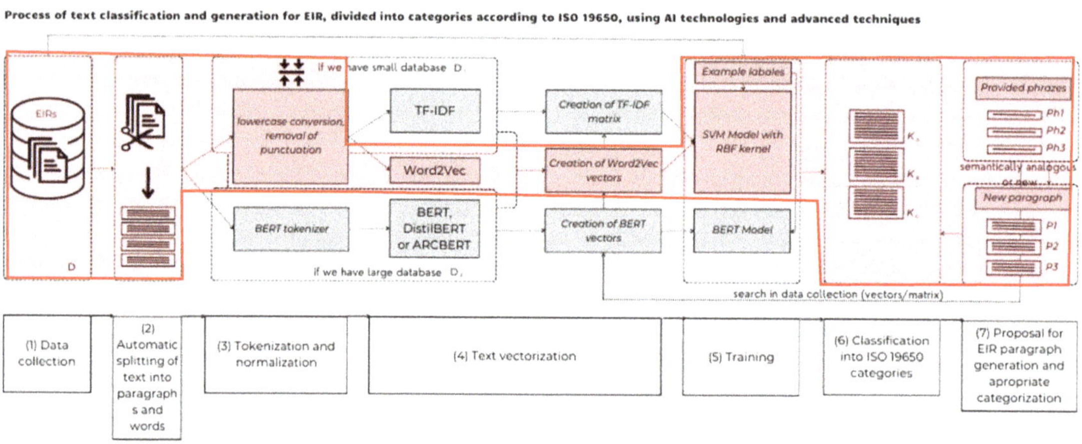

Figure 1. Process of text classification and generation for the EIR proposal presented (in red using Word2Vec and SVM).

2.1. Step (1): Creation of Data Collection

The dataset required for training the model comprises Exchange Information Requirement (EIR) documents from both public and private tenders. The user will also be able to upload their own documents, for instance, from previous projects, allowing for the refinement of established patterns. Several countries publish EIR templates, which provide an appropriate foundation for this process. It is important to note that these documents do not necessarily need to be categorized according to the ISO 19650 standard, as the model is designed to automatically assign paragraphs to the relevant categories.

2.2. Steps (2) and (3): Text Preprocessing

The collected data must be adequately prepared for text vectorization, which requires text preprocessing. Since the aim is to use fragments of EIRs for a proposal form, tailored to user preferences, the text is first divided into paragraphs (step 2), followed by normalization and tokenization, as it is from tokens that feature vectors are created. The text undergoes normalization, meaning letters are converted to lowercase, and special characters and whitespace, such as spaces, are removed. The text is subsequently tokenized, which involves splitting the text into individual words. For example, the sentence "The contractor shall compile the BEP." would be tokenized using a Natural Language Toolkit (NLTK)-based script as: ["The", "contractor", "shall", "compile", "the", "BEP", "."]. This structured approach ensures that the data are prepared for efficient processing in the subsequent stages of classification and categorization.

2.3. Step (4): Text Vectorization Using Word2Vec

To achieve text categorization, the text must first be converted into numerical values, specifically creating numerical feature vectors for paragraphs. In this case, the feature vector for a paragraph is the arithmetic mean of the feature vectors of the tokens within that paragraph, generated during the preprocessing phase. The Word2Vec method was selected for this, using static word embedding, meaning static vector representation. Some methods require larger datasets for training, while others can function effectively with smaller datasets. Therefore, it was decided to explore the implementation of text vectorization using TF-IDF, which performs well with smaller datasets (in this case, EIRs), and BERT, which yields better results with larger datasets (as shown in Figure 1). This paper focuses on the use of Word2Vec, which can be applied to large datasets, but due to its pre-trained nature on a vast corpus, it should also perform adequately with smaller datasets.

Word2Vec falls under unsupervised learning methods, which do not require labelled data for training (labelling is applied during classification). While it uses neural networks, it employs shallow networks, with only one hidden layer of neurons.

One key advantage of this method is that the vectors generated possess semantic relationships, allowing for the identification of similar words. For example, "LOD" (Level of Detail or Level of Development in BIM models, as specified by the PAS 1192 standard and BIM Forum) and "LOIN" (Level of Information Need used in the ISO 19650 standard) can be compared by calculating the cosine of the angle between their vectors. Moreover, Word2Vec is pre-trained on a large corpus, enabling it to perform well even with limited training resources.

Two architectures can be employed: the Continuous Bag of Words (CBoW) architecture for smaller datasets, and the Skip-Gram architecture for larger ones [26], with the code in [27]. The difference between these architectures is illustrated in Figure 2. CBoW predicts a target word based on its surrounding context, while Skip-Gram predicts surrounding words based on a target word [26,28].

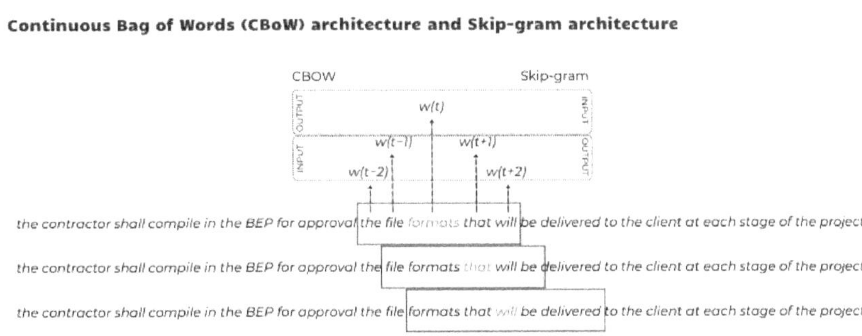

Figure 2. Differences between Continuous Bag of Words and Skip-Gram architectures.

The mathematical formulation of the objective function for the Skip-Gram model is shown in Equation (1) [28], which maximizes the average log probability, where $\{w_1, w_2, \ldots, w_T\}$ is the sequence of training words, c is the size of the training context, and w_T is the central word. The softmax function $p(w_O|w_I)$ is used, where v_w and v'_w are the "input" and "output" vector representations of w, and W is the number of words in the vocabulary.

$$SG = \frac{1}{T}\sum_{t=1}^{T} \sum_{-c\leq j\leq c, j\neq 0} \log\, p(w_{t+j}|w_t), \text{ where } p(w_O|\, w_I) = \frac{\exp\left({v'_{wO}}^{T} v_{wI}\right)}{\sum_{w=1}^{W} \exp\left({v'_w}^{T} v_{wI}\right)} \tag{1}$$

Mikolov noted that calculating the denominator of this probability can be computationally problematic, considering that it accounts for all words in the vocabulary. Thus, hierarchical softmax or the approximation of the softmax function by Noise-Contrastive Estimation (NCE), which can be simplified by negative sampling (NEG), was proposed.

Word embedding for the word "format" for a sample EIR fragment (i.e., creating domain-specific models) is illustrated in Figure 3. With different assumptions, such as a smaller vector size, the results will differ. The *gensim* library allows Word2Vec models to be trained on a custom dataset.

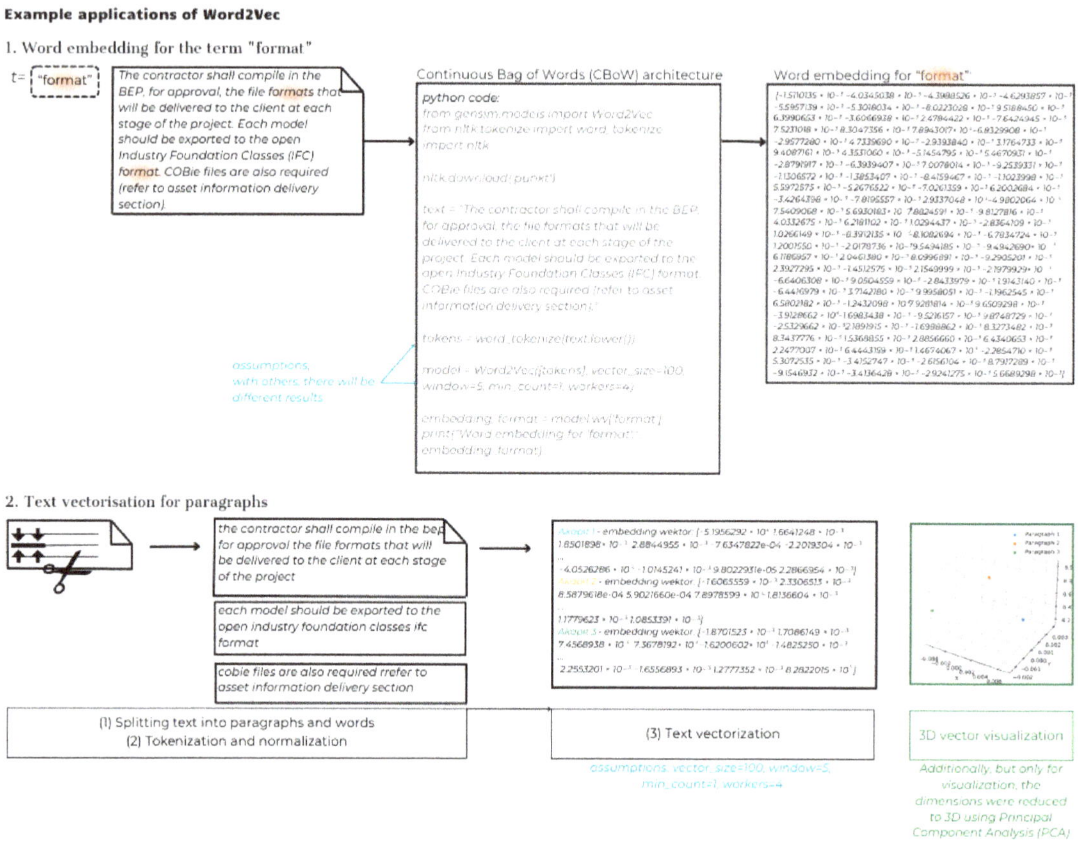

Figure 3. Example applications of Word2Vec for text vectorization.

Text vectorization for paragraphs begins with splitting the text into paragraphs (schematically illustrated in Figure 3), followed by tokenization and normalization, and then text vectorization for each paragraph separately, where the average embedding vector for all words in the paragraph is calculated.

Improvements to the Word2Vec method include Global Vectors for Word Representation (GloVe) and FastText, which are suitable topics for future research.

2.4. Steps (5) and (6): Text Classification into the ISO 19650 Standard Categories

After preparing the text, the core of this study can be addressed: the classification of text into ISO 19650 categories. Before delving into the classification algorithms, it is essential to provide an overview of the Exchange Information Requirements (EIRs) as outlined in the ISO 19650 and explain the associated categories of information.

2.4.1. Exchange Information Requirements According to the ISO 19650 Standard

The ISO 19650 standard is an international standard for managing construction projects using BIM. According to this standard, an Exchange Information Requirements (EIR) document is a "specification for what, when, how, and for whom information is to be produced in relation to an appointment (agreed instruction for the provision of information concerning works, goods, or services)" [3]. This document is crucial in real-estate development process, as proper preparation aims to prevent dissatisfaction with the project outcome, as it clearly outlines the client's expectations [29]. In cases where specialized knowledge is lacking, it is advisable to consult experts in the field.

The contractor—whether it is a design firm tasked with producing comprehensive design documentation, a general contractor responsible for construction in a Design–Bid–Build system, or a subcontractor—has to submit a BIM Execution Plan for approval. This plan should meet the EIRs and provide more detailed project-specific information. It is essential to consider the entire life cycle of the building asset when specifying expectations, as a well-designed model can prove beneficial in later stages, such as during a building's operation, facilitating quicker access to equipment warranties or user manuals.

The ISO 19650 standard categorizes EIR information into three main groups: managerial, commercial, and technical. In the non-mandatory Polish BIM guidelines, managerial aspects include standards, roles and responsibilities, data security, coordination and clash detection, collaboration organization, meetings and model reviews, Health and Safety (H&S) management, and project management. Commercial aspects cover data exchange/delivery schedule, strategic goals, defining expectations regarding BIM scope, and competency assessment, while technical aspects involve technical or software platforms, data exchange formats, coordinates, levels of detail, and training [30].

For example, "roles and responsibilities" within managerial aspects might include a table that outlines who is responsible for what tasks in the project, ensuring clarity in accountability.

"Strategic goals" in the commercial domain might detail not only project-specific objectives but also the project owner's broader organizational goals. It is important to note that information requirements go beyond just EIRs. The ISO 19650 standard outlines six main types of information requirements: Organizational Information Requirements, Project Information Requirements, Asset Information Requirements, Exchange Information Requirements, and the deliverables associated with them: the Asset Information Model and the Project Information Model. This highlights the complexity of the issues that must be addressed to develop appropriate guidelines.

One example of technical requirements includes the Levels of Information Need, which are divided into geometrical, alphanumerical, and document-based information. Geometrical information pertains to dimensions, such as the size of windows or spans, while alphanumerical data specify materials, such as the concrete class to be used. It is crucial to ensure that the information provided is necessary for the projects, as excessive detail can lead to waste, such as overly detailed geometrical representations or unnecessary model parameters that are irrelevant for analysis or operation [31].

2.4.2. Support Vector Machine

For text classification into the ISO 19650 categories, the Support Vector Machine (SVM) method was applied. Text classification using BERT was also proposed for future studies, as outlined in Figure 1. However, this transformer-based model is more effective when applied to large datasets, which is why an alternative for smaller datasets was suggested. One key consideration was selecting a method capable of handling classification across three categories that may not be linearly separable in a multidimensional space (specifically, a 100-dimensional space in our case). Therefore, the Radial Basis Function (RBF) kernel was proposed for SVM, which is particularly suited to such scenarios, which will be explained further in the paper.

SVM is a supervised learning method used for classification, regression, and anomaly detection tasks. It operates by identifying a hyperplane or a set of hyperplanes in a multidimensional space to separate different classes. SVM is particularly effective in complex scenarios where data are not linearly separable and is capable of handling both binary and multi-class classification problems [32].

To clearly explain this method, one can consider a space with two sets of features separated by a linear hyperplane. The distance between the nearest points (*support vectors*) of these sets and the hyperplane defines the margin γ (Equation (3)). Maximizing this margin results in an optimal hyperplane. The hyperplane in an n-dimensional feature space is described by Equation (2) [33], where ω is the weight vector, x is the feature vector, and b is the bias term.

$$\omega \cdot x - b = 0 \tag{2}$$

$$\gamma = \frac{2}{\|\omega\|} \tag{3}$$

The margin is maximized by minimizing $\|\omega\|$.

As previously outlined, there are three categories (managerial, commercial, and technical), meaning a multi-class classification problem must be solved. In vector space, the groups of vectors that correspond to these categories may not be linearly separable. To address this, the Radial Basis Function (RBF) kernel was employed, which is particularly suited to non-linearly separable data. In practice, techniques such as One-vs-Rest or One-vs-One are often used. In One-vs-Rest, the model first distinguishes K_A from K_B and K_C, and so on. In or One-vs-One, the model differentiates K_A from K_B initially.

The RBF kernel, described by the kernel function in Equation (4) (where x_i and x_j are feature vectors, and r is a parameter that defines the influence range of a single training sample) [34], maps the data into higher dimensions to make the classes linearly separable, utilizing the Kernel Trick.

$$K(x_i, x_j) = \exp\left(-r\|x_i - x_j\|^2\right) \tag{4}$$

This method is complex, particularly in the context of non-linear separability and multi-class complexity. Therefore, this article focuses only on its characteristic components.

An SVM classifies new data based on support vectors, with the decision function represented by Equation (5) [35], where NS refers to the number of support vectors, α_i are the weights assigned to the support vectors, y_i are class labels, x_i are support vectors, and b is the bias term.

$$f(x) = \sum_{i=1}^{NS} \alpha_i y_i K(x_i, x) + b \tag{5}$$

The objective function for an SVM with an RBF kernel aims to maximize the difference between the sum of the Lagrange multipliers α_i and α_j and half the weighted sum of the dot products of the training vectors x_i and x_j, which allows the optimal hyperplane that maximizes the margin between the classes to be found, as shown in Equation (6) [36].

$$max \sum_{i=1}^{N} \alpha_i - \frac{1}{2} \sum_{i=1}^{N} \sum_{j=1}^{N} \alpha_i \alpha_j y_i y_j K(x_i, x_j), \text{ subject to} \sum_{i=1}^{N} \alpha_i y_i = 0,\ \alpha_i \geq 0 \tag{6}$$

where α_i and α_j are Lagrange multipliers, and y_i, y_j are class labels.

In practice, data labelling is conducted to enable the model to learn how to differentiate categories. Specific files are labelled as K_A, K_B, or K_C, corresponding to managerial, commercial, or technical aspects, respectively. As a result, when new EIR files are uploaded, the model can automatically classify paragraphs into the appropriate categories.

The performance of the trained model was evaluated using metrics such as accuracy, precision, recall, and F1 score. The scikit-learn library was utilized to calculate these parameters [37]. According to Equation (7), *accuracy* is the ratio of correct predictions to the total number of examples, $\hat{y}_i$ is the predicted value of the i-th sample, y_i is the actual value, $n_{samples}$ is the total number of samples, and $1(\hat{y}_i = y_i)$ is the indicator function, representing the number of correct predictions in the dataset.

$$accuracy(y, \hat{y}) = \frac{1}{n_{samples}} \sum_{i=0}^{n_{samples}-1} 1(\hat{y}_i = y_i) \quad (7)$$

Precision P (Equation (8)) measures the proportion of true positives T_p relative to the sum of true positives and false positives F_p. It indicates the percentage of instances that the model classified as positive which were actually positive. *Recall R* (Equation (9) also compares the number of true positives but relative to the sum of true positives and false negatives F_n, measuring the model's ability to capture all relevant positive examples.

$$\mathrm{P} = \frac{T_p}{T_p + F_p} \quad (8)$$

$$\mathrm{R} = \frac{T_p}{T_p + F_n} \quad (9)$$

For example, in a dataset of 50 paragraphs, where 12 belong to category K_A, if the model predicts that 10 paragraphs fall under K_A and 8 of these are correctly classified (i.e., T_p = 8), while 2 are incorrectly classified as K_A (i.e., F_p = 2), and 4 paragraphs that belong to K_A are not identified by the model (i.e., F_n = 4), then the precision P = 8/10 = 0.8 (80%) and the recall R = 8/12 = 0.67R = 8/12 = 0.67 (67%).

The *F1 score* is the harmonic mean of precision and recall, providing a balanced measure between the two, particularly when there is a disparity between positive and false predictions, according to Equation (10).

$$\mathrm{F1} = \frac{2 * T_p}{2 * T_p + F_p + F_n} \quad (10)$$

An analysis of text classification was performed for selected sentences to validate the model's effectiveness.

2.5. Step (7): Proposal for EIR Paragraph Generation and Appropriate Categorization

Although the text classification process could be considered complete at this point, it is the foundation for the subsequent step. This next phase involves the specification of the client's preferences through the use of a chatbot, which would assign appropriate sentences or paragraphs of a proposed EIR. This phase will be refined in the future as part of the development of an automatic, personalized EIR generation methodology, where a Generative Pre-trained Transformer (GPT), based on the Transformer architecture, will also be employed.

In this study, however, the performance of the existing model in text generation was preliminarily tested. Example phrases were provided, and the model was tasked with finding similar phrases by calculating the cosine similarity between the vector of the input phrase and the paragraph vectors in the database.

The *cosine similarity k(x,y)* (Equation (11)) between two vectors x and y is calculated as the ratio of the dot product of vectors x and y, with n-dimensions, to the product of the norms of x and y ($\|x\|$, $\|y\|$) [37]. In practice, it computes the cosine of the angle between these vectors in a multidimensional space. A result close to 1 indicates high similarity, while a result close to -1 signifies that the vectors are oriented in opposite directions.

$$k(x,y) = \frac{x \cdot y\prime}{\|x\|\|y\|}, \text{ where } x \cdot y\prime = \sum_{i=1}^{n} x_i \cdot y_i, \ \|x\| = \sqrt{\sum_{i=1}^{n} x_i^2}, \|y\| = \sqrt{\sum_{i=1}^{n} y_i^2} \quad (11)$$

If the model does not find a suitable match, it generates a new sentence containing the given phrase. These generated sentences are then categorized into the appropriate K_A, K_B, or K_C (managerial, commercial, or technical aspects).

This step lays the groundwork for future research on the automatic generation of EIRs, enabling more personalized and precise output aligned with stakeholder preferences.

3. Results

The dataset used to validate the model comprised 8494 paragraphs categorized under K_A (managerial), 5130 under K_B (commercial), and 944 under K_C (technical). The training and validation data were split in the commonly used 80:20 ratio [38]. The varying quantity of data across categories may have influenced the results. The model was expected to be more accurate in classifying K_A due to the larger volume of training data available for this category.

Figure 4 illustrates the vector for the token "detail" generated through text vectorization. Feature vectors were set to 100 dimensions, a crucial aspect for capturing semantic relationships between words. This dimensionality is key for conducting the semantic analysis required for this task, which reflects the complexity of the methods employed.

Token for "detail"				
	[$-7.35301822 \times 10^{-2}$	$2.26312488 \times 10^{-1}$	$6.59308136 \times 10^{-2}$	$-5.14289737 \times 10^{-3}$
	$9.67332255 \times 10^{-3}$	$-7.20760524 \times 10^{-2}$	$-1.10829920 \times 10^{-2}$	$1.58994019 \times 10^{-1}$
	$-4.89512235 \times 10^{-2}$	$-6.75689243 \times 10^{-3}$	$6.88483045 \times 10^{-2}$	$-1.50726750 \times 10^{-1}$
	$6.53875023 \times 10^{-2}$	$5.47512136 \times 10^{-2}$	$5.73545843 \times 10^{-2}$	$-1.18242744 \times 10^{-2}$
	$2.01056972 \times 10^{-1}$	$1.86026707 \times 10^{-1}$	$-5.27383387 \times 10^{-2}$	$-2.20020656 \times 10^{-2}$
	$9.07328427 \times 10^{-2}$	$6.79189526 \times 10^{-3}$	$-5.12664951 \times 10^{-2}$	$-2.52552748 \times 10^{-1}$
	$2.20263973 \times 10^{-2}$	$-8.84323344 \times 10^{-2}$	$-1.08524621 \times 10^{-3}$	$-5.74391223 \times 10^{-2}$
	$-1.61668509 \times 10^{-1}$	$-1.34903222 \times 10^{-1}$	$9.54405516 \times 10^{-2}$	$1.43698547 \times 10^{-2}$
	$-2.04434115 \times 10^{-2}$	$-5.02389967 \times 10^{-2}$	$-5.41473329 \times 10^{-2}$	$3.24597135 \times 10^{-2}$
	$2.73496546 \times 10^{-2}$	$-3.76884378 \times 10^{-2}$	$1.53292075 \times 10^{-1}$	$-2.57400453 \times 10^{-1}$
	$1.17873505 \times 10^{-1}$	$6.13079630 \times 10^{-2}$	$-1.00701474 \times 10^{-1}$	$-1.42019138 \times 10^{-1}$
	$5.61577827 \times 10^{-2}$	$-8.88353661 \times 10^{-2}$	$3.86460987 \times 10^{-3}$	$2.62816977 \times 10^{-2}$
	$4.47269678 \times 10^{-2}$	$5.52207492 \times 10^{-2}$	$-1.84518911 \times 10^{-4}$	$-1.17468409 \times 10^{-1}$
	$-5.12927137 \times 10^{-2}$	$1.60360094 \times 10^{-2}$	$1.55836165 \times 10^{-1}$	$1.07496241 \times 10^{-2}$
	$1.03818923 \times 10^{-1}$	$-5.22193797 \times 10^{-2}$	$-1.11454939 \times 10^{-2}$	$-1.32430941 \times 10^{-2}$
	$-1.79913528 \times 10^{-2}$	$5.61168343 \times 10^{-2}$	$2.60032147 \times 10^{-1}$	$-7.01491013 \times 10^{-2}$
	$-1.26845524 \times 10^{-1}$	$2.05462247 \times 10^{-1}$	$-7.60471150 \times 10^{-2}$	$3.74383740 \times 10^{-2}$
	$-2.99654901 \times 10^{-1}$	$1.09292649 \times 10^{-1}$	$-2.23372858 \times 10^{-2}$	$1.03836328 \times 10^{-1}$
	$2.15664014 \times 10^{-1}$	$2.84033138 \times 10^{-2}$	$7.57873580 \times 10^{-2}$	$-2.20980458 \times 10^{-2}$
	$-5.43334223 \times 10^{-2}$	$-2.74950489 \times 10^{-2}$	$-3.19890827 \times 10^{-1}$	$5.89186065 \times 10^{-2}$
	$-9.21700075 \times 10^{-2}$	$-1.05074167 \times 10^{-1}$	$1.68249488 \times 10^{-3}$	$1.04820758 \times 10^{-1}$
	$-6.24499172 \times 10^{-2}$	$-1.86762381 \times 10^{-2}$	$2.91833598 \times 10^{-2}$	$-5.47248544 \times 10^{-3}$
	$-5.12562282 \times 10^{-2}$	$9.93735064 \times 10^{-3}$	$8.50876123 \times 10^{-2}$	$-5.97740635 \times 10^{-2}$
	$2.58028973 \times 10^{-2}$	$-2.03926042 \times 10^{-2}$	$2.73750186 \times 10^{-1}$	$1.54974297 \times 10^{-1}$
	$6.13427088 \times 10^{-2}$	$-5.41452393 \times 10^{-2}$	$9.14488211 \times 10^{-2}$	$-3.80940177 \times 10^{-2}$]

Figure 4. Token for "detail" from text vectorization.

Figure 5 presents the evaluation metrics obtained. The overall accuracy, which represents the percentage of correctly classified paragraphs, was 86%. The precision, indicating the proportion of correct predictions among all predicted positives, averaged 91%, ranging from 85% for K_A to fully correct predictions for K_C. Recall ranged from 22% to 96%, with an overall F1 score averaging 0.7, more precisely 0.9 for managerial predictions, 0.81 for commercial, and 0.36 for technical aspects.

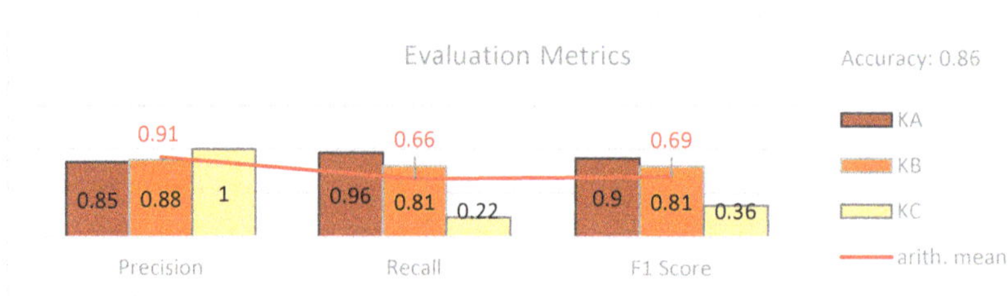

Figure 5. Evaluation metrics chart.

For selected sentences such as "The contractor shall compile in the BEP, for approval, the file formats that will be delivered to the client at each stage of the project," "Each model should be exported to the open Industry Foundation Classes (IFC) format," and "The final

BEP shall include information related to 'Project Roles & Responsibilities'," the model successfully classified them into the correct categories: K_A (managerial), K_B (commercial), and K_C (technical), as outlined by the ISO 19650 standard. The predicted classifications are shown in Table 1.

Table 1. Model-predicted categories for provided sentences.

No.	Sentence	Predicted Category
1	The contractor shall compile in the BEP, for approval, the file formats that will be delivered to the client at each stage of the project.	K_A
2	Each model should be exported to the open Industry Foundation Classes (IFC) format.	K_A
3	The final BEP shall include information related to 'Project Roles & Responsibilities'.	K_A
4	All issues should be promptly reported to the client in the common data environment.	K_A

where K_A refers to managerial aspects, K_B to commercial aspects, and K_C to technical aspects.

Subsequently, the model was tasked with proposing sentences related to the provided phrases: "energy analysis," "simulation of work sequences," and "architectural model designed in Revit." It searched the available database for similar sentences using cosine similarity, and if none were sufficiently similar, it generated new ones. The results, including predicted categories, are displayed in Table 2.

Table 2. Model-predicted sentences and categories for provided phrases.

No.	Phrase	Predicted Category	Most Similar Paragraph or New Paragraph	Similarity
1	Energy analysis	K_B	Energy analysis.	0.95
2	Simulation of work sequences	K_A	Project controls: The model will be capable of being utilized for identifying temporary as well as permanent works and provide a critical path analysis of site activities to prevent "trade clashes" and to provide the most efficient way of arranging working areas for operatives.	0.92
3	Architectural model designed in Revit	K_A	Developing constituent parts of the information model in connection with specific tasks.	0.91

where K_A refers to managerial aspects, K_B to commercial aspects, and K_C to technical aspects.

4. Discussion

The objective of this study was to propose an automatic text classification of Exchange Information Requirements (EIRs) according to ISO 19650 categories, using Support Vector Machines (SVMs) combined with text vectorization via Word2Vec. These methods proved sufficient to achieve this goal. In addition, the authors aimed to validate the model's potential by examining its predictions. While the results may vary with a larger dataset, the findings from the current, smaller dataset (with 14,568 paragraphs identified) are promising. As shown in Figure 5, the model achieved an overall accuracy of 86%, indicating a significant percentage of correctly classified paragraphs. The model was particularly effective in identifying managerial aspects, where it achieved an F1 score of 0.9, which is the harmonic mean between precision and recall. However, the 100% precision for the technical category must be considered in conjunction with its 22% recall. This can indicate that the model infrequently predicts this category, but when it does, the predictions are correct.

Understanding the reasons behind these results requires an examination of the complexity of the data and the method's limitations. The managerial category, which produced the best results, is typically the largest among K_A, K_B, and K_C. It is often written in language that is more easily interpreted by the Word2Vec algorithm and contain fewer tables (apart from the critical "roles and responsibilities" matrix) and less technical terminology than the other two categories. Both the commercial and technical categories may contain overlapping information, such as file formats, which can pertain both to data delivery schedules (K_B) and model requirements (K_C). Furthermore, the technical category, which yielded the poorest results in the evaluation metrics, is usually the shortest, contains specialized terminology, and often consists of tables. These tables, which present the levels of detail for geometric and alphanumeric information for various model components, are less conducive to classification through cosine similarity.

To improve performance, future iterations of the model could focus on balancing the dataset by increasing the pool of data from the technical category. Additionally, key terms or phrases associated with each category could be explicitly highlighted during training. For example, phrases such as "roles and responsibilities" could be tied to K_A, "BIM objectives" to K_B, and "file formats" to K_C.

When analyzing the model's predictions for the provided example sentences, as presented in Table 1, in sentence 1, the phrase "file formats" was used to test if the model would classify it under K_C (technical), yet it correctly classified it under K_A, recognizing the managerial context. Sentence 2, which involved file formats, should have been classified as technical, but the model assigned it to K_A, indicating an area for improvement. The third and fourth sentences were correctly classified under the category K_A. The authors conclude that three out of four sentences were acceptably classified.

Han et al. examined the application of Word2Vec and, while highlighting its efficiency in capturing semantic information, also pointed out a limitation: its inability to account for word order in sentences [24]. They observed that Word2Vec performs well with short sentences but struggles with more complex ones. In future studies, where the model will be trained on larger datasets, conducting a sensitivity analysis would be valuable to assess the impact of changes in input parameters on the model's performance, including sentence length and complexity, particularly with regard to technical terminology. It would also be beneficial to examine the effects of other parameters on classification, such as dataset size, the adjustment or the r parameter in SVM (which controls the margin between classes), the size of labelled data within each category, and the quality of the labelled data.

Table 2 presents the suggested similar paragraphs for the phrases provided. In the first case, the model identified "energy analysis" as a BIM use, which referred to specific requirements outlining how BIM models should be utilized for project management. These typically consist of a list of BIM uses, such as energy analysis, clash detection, and acoustic analysis, along with brief definitions of each. This section of the EIR pertains to commercial aspects; therefore, the phrase was correctly classified. However, the slight deviation in cosine similarity (0.95), despite an identical match, could be due to other paragraphs with similar semantics, affecting the similarity score. Sentence 2 returned an acceptable match regarding work sequences, though it did not fully capture the intended meaning of "simulation," indicating an area for refinement. Sentence 3 exhibited the model's limitation, as the proposed paragraph did not align with the query's intent, underscoring the importance of domain-specific training data.

This study acknowledges these limitations and identifies areas for improvement in future iterations, particularly in the domain of text generation. Future research will incorporate the use of GPT for more advanced text generation, which is expected to address some of the shortcomings encountered in this study.

5. Conclusions

This study investigates the application of Word2Vec for text vectorisation, Support Vector Machines (SVMs) for text classification, and the use of Word2Vec with cosine similarity for text generation. The findings demonstrate how Artificial Intelligence (AI) tools and methods can enhance information management in construction projects, specifically through the use of advanced methodologies such as Building Information Modelling (BIM) and Digital Twins.

The proposed model, which supports the creation of Exchange Information Requirements (EIR), particularly by categorizing text into managerial, commercial, and technical aspects according to the ISO 19650 standard, shows considerable potential. Even with a relatively small training dataset, the classification achieved an average F1 score—a harmonic mean of precision and recall—of 0.7, which is a promising result. Although this represents a limited sample, the initial tests indicate that the model is capable of classification, making it a valuable stepping stone for further research. This study highlights the effectiveness of Word2Vec for text vectorization and SVM with an RBF kernel for classification, which demonstrates the potential of these methods. However, the authors recognize the need to expand this exploration by integrating more advanced AI models, such as BERT, which could provide enhanced semantic understanding and overall classification accuracy. Additionally, it would be worthwhile in future research to investigate the impact of varying input data on the model's performance.

This study contributes both theoretically and practically to the field of information management in construction. Theoretically, it extends the application of AI in automating Exchange Information Requirements drafting for BIM/Digital Twin projects. Practically, this implementation could facilitate the adoption of these methodologies in the AECO industry, addressing existing barriers such as limited knowledge, the shortage of specialists, and alignment with contemporary trends, such as the European Union's digital transition.

Despite its contributions, this study is not without limitations. The relatively small dataset used for training constrains the generalizability of the results, and the current model requires further refinement to effectively handle larger and more diverse datasets. Additionally, while cosine similarity calculated using Word2Vec indicates semantic relationships between vectorized phrases, the generated sentences are not yet sufficiently accurate to be used directly in EIR documents. This points to the need for improvements in the text generation process, where more sophisticated models, such as GPT, could be employed to refine the generation of EIR fragments and improve overall text quality.

Looking forward, future research should focus on several key areas. First, the use of larger, more comprehensive datasets will be essential for improving the robustness and generalizability of the model. Secondly, exploring more advanced AI models, such as BERT and GPT, could provide deeper semantic understanding and enhance both the classification and generation of text. Our own future work will focus on integrating these advanced models into the automated EIR creation process. Furthermore, incorporating new EIRs into the database will be crucial for ensuring that the AI tools can adapt to changing trends in BIM/Digital Twin approaches.

In summary, while this study presents a robust method, there remains significant potential for future research to build on its findings. The implementation of such tools could significantly enhance the efficiency of complex construction processes involving advanced methodologies like BIM.

Author Contributions: Conceptualization, E.M.-K.; Methodology, E.M.-K.; Software, E.M.-K.; Validation, E.M.-K.; Formal Analysis, E.M.-K.; Investigation, E.M.-K.; Resources, E.M.-K.; Data Curation, E.M.-K.; Writing—Original Draft Preparation, E.M.-K.; Writing—Review and Editing, K.Z.; Visualization, E.M.-K.; Supervision, K.Z.; Project Administration, E.M.-K. All authors have read and agreed to the published version of the manuscript.

Funding: This research received no external funding.

Data Availability Statement: The original contributions presented in this study are included in the article; further inquiries can be directed to the corresponding author.

Conflicts of Interest: The authors declare no conflicts of interest.

References

1. Renda, A.; Schwaag Serger, S.; Tataj, D.; Morlet, A.; Isaksson, D.; Martins, F.; Mir Roca, M.; Hidalgo, C.; Huang, A.; Dixson-Declève, S.; et al. *Industry 5.0, a Transformative Vision for Europe—Governing Systemic Transformations towards a Sustainable Industry*; European Commission, Directorate-General for Research and Innovation: Brussels, Belgium, 2021. [CrossRef]
2. Mitera-Kiełbasa, E.; Zima, K. BIM Policy Trends in Europe: Insights from a Multi-Stage Analysis. *Appl. Sci.* **2024**, *14*, 4363. [CrossRef]

3. *ISO 19650-1:2018*; Organization and Digitization of Information about Buildings and Civil Engineering Works, Including BIM—Information Management Using Building Information Modelling—Part 1: Concepts and Principles. International Organization for Standardization: Geneva, Switzerland, 2018.
4. Moreno, C.; Olbina, S.; Issa, R.R. BIM Use by Architecture, Engineering, and Construction (AEC) Industry in Educational Facility Projects. *Adv. Civ. Eng.* **2019**, *2019*, 1392684. [CrossRef]
5. Suganya, R.; Buhari, S.M.; Rajaram, S. Different Applications and Importance of Digital Twin. In *Digital Twin Technology*; Wiley: Hoboken, NJ, USA, 2022; pp. 189–203. [CrossRef]
6. Opoku, D.G.J.; Perera, S.; Osei-Kyei, R.; Rashidi, M.; Famakinwa, T.; Bamdad, K. Drivers for Digital Twin Adoption in the Construction Industry: A Systematic Literature Review. *Buildings* **2022**, *12*, 113. [CrossRef]
7. Borkowski, A.S. Evolution of BIM: Epistemology, genesis and division into periods. *J. Inf. Technol. Constr.* **2023**, *28*, 646–661. [CrossRef]
8. Soon Ern, P.A.; Ooi, Y.Y.; Al-Ashmori, Y.Y. Comparative Study on the Perspective towards the Benefits and Hindrances of Implementing Building Information Modelling (BIM). *J. Sustain. Constr. Eng. Technol.* **2020**, *11*, 194–205.
9. Biswas, H.K.; Sim, T.Y.; Lau, S.L. Impact of Building Information Modelling and Advanced Technologies in the AEC Industry: A Contemporary Review and Future Directions. *J. Build. Eng.* **2024**, *82*, 108165. [CrossRef]
10. *ISO 29481-1:2016*; Building Information Models—Information Delivery Manual—Part 1: Methodology and Format. International Standard Organization: Geneva, Switzerland, 2016.
11. Gohel, P.; Dabral, R.; Lad, V.H.; Patel, K.A.; Patel, D.A. A comprehensive review on application of artificial intelligence in construction management using a science mapping approach. In *Artificial Intelligence Applications for Sustainable Construction*; Elsevier: Amsterdam, The Netherlands, 2024; pp. 285–300. [CrossRef]
12. Rangasamy, V.; Yang, J.B. The convergence of BIM, AI and IoT: Reshaping the future of prefabricated construction. *J. Build. Eng.* **2024**, *84*, 108606. [CrossRef]
13. Ma, S.; Wang, X.; Wang, X.; Liu, H.; Zhang, R. A Framework for Diagnosing Urban Rail Train Turn-Back Faults Based on Rules and Algorithms. *Appl. Sci.* **2021**, *11*, 3347. [CrossRef]
14. Zheng, Z.; Lu, X.Z.; Chen, K.Y.; Zhou, Y.C.; Lin, J.R. Pretrained domain-specific language model for natural language processing tasks in the AEC domain. *Comput. Ind.* **2022**, *142*, 103733. [CrossRef]
15. Dolhopolov, S.; Honcharenko, T.; Terentyev, O.; Savenko, V.; Rosynskyi, A.; Bodnar, N.; Alzidi, E. Multi-Stage Classification of Construction Site Modeling Objects Using Artificial Intelligence Based on BIM Technology. In *Proceedings of the 2024 35th Conference of Open Innovations Association (FRUCT), Tampere, Finland, 24–26 April 2024*; IEEE: Piscataway, NJ, USA, 2024; pp. 179–185. [CrossRef]
16. Bartal, A.; Jagodnik, K.M.; Chan, S.J.; Dekel, S. AI and narrative embeddings detect PTSD following childbirth via birth stories. *Sci. Rep.* **2024**, *14*, 8336. [CrossRef]
17. Huang, A.H.; Wang, H.; Yang, Y. A Large Language Model for Extracting Information from Financial Text. *Contemp. Account. Res.* **2023**, *40*, 806–841. [CrossRef]
18. Piazzi, M.A.; Feng, H.; Kassem, M. An investigation of concepts for the specification of graphical exchange information requirements in building information modelling. *J. Inf. Technol. Constr.* **2022**, *27*, 662–684. [CrossRef]
19. Goonetillake, J.F.; Renb, G.; Lia, H.; Yaob, J. A prototype tool to embed digital exchange information requirements in construction projects. In Proceedings of the 30th EG-ICE: International Conference on Intelligent Computing in Engineering, London, UK, 4–7 July 2023; pp. 1–10.
20. Tomczak, A.; v Berlo, L.; Krijnen, T.; Borrmann, A.; Bolpagni, M. A review of methods to specify information requirements in digital construction projects. *IOP Conf. Ser. Earth Environ. Sci.* **2022**, *1101*, 092024. [CrossRef]
21. Luhn, H.P. The Automatic Creation of Literature Abstracts. *IBM J. Res. Dev.* **1958**, *2*, 159–165. [CrossRef]
22. Shakil, H.; Farooq, A.; Kalita, J. Abstractive Text Summarization: State of the Art, Challenges, and Improvements. *Neurocomputing* **2024**, *603*, 128255. [CrossRef]
23. Iqbal, T.; Qureshi, S. The survey: Text generation models in deep learning. *J. King Saud. Univ. Comput. Inf. Sci.* **2022**, *34*, 2515–2528. [CrossRef]
24. Han, M.; Zhang, X.; Yuan, X.; Jiang, J.; Yun, W.; Gao, C. A survey on the techniques, applications, and performance of short text semantic similarity. *Concurr. Comput.* **2021**, *33*, e5971. [CrossRef]
25. Wang, Z.; Dou, J.; Zhang, Y. Unsupervised Sentence Textual Similarity with Compositional Phrase Semantics. In Proceedings of the 29th International Conference on Computational Linguistics, Gyeongju, Republic of Korea, 12–17 October 2022; pp. 4976–4995.
26. Mikolov, T.; Chen, K.; Corrado, G.; Dean, J. Efficient Estimation of Word Representations in Vector Space. In Proceedings of the International Conference on Learning Representations, Scottsdale, AZ, USA, 2–4 May 2013. [CrossRef]
27. Mikolov, T.; Chen, K.; Corrado, G.; Dean, J. Word2vec Code. 2024. Available online: https://code.google.com/archive/p/word2vec/ (accessed on 24 May 2024).
28. Mikolov, T.; Sutskever, I.; Chen, K.; Corrado, G.; Dean, J. Distributed Representations of Words and Phrases and their Compositionality. *Adv. Neural Inf. Process. Syst.* **2013**. [CrossRef]
29. Zima, K.; Mitera-Kiełbasa, E. Employer's information requirements: A case study implementation of BIM on the example of selected construction projects in Poland. *Appl. Sci.* **2021**, *11*, 10587. [CrossRef]

30. Polish Association of Construction Engineers and Technicians; Polish Association of Construction Employers; Association of Polish Architects. BIM Standard PL. 2020. Available online: https://globalbim.org/info-collection/bim-standard-pl/ (accessed on 20 October 2024).
31. Zima, K.; Mitera-Kiełbasa, E. Level of Information Need for BIM Models: Australia, New Zealand and ISO 19650. *Civ. Environ. Eng. Rep.* **2022**, *32*, 1–3. [CrossRef]
32. Poole, D.L.; Mackworth, A.K. *Artificial Intelligence*; Cambridge University Press: Cambridge, UK, 2023. [CrossRef]
33. Vapnik, V.N. *The Nature of Statistical Learning Theory*; Springer: New York, NY, USA, 2000. [CrossRef]
34. Schölkopf, B.; Smola, A.J. *Learning with Kernels*; The MIT Press: Cambridge, MA, USA, 2018. [CrossRef]
35. Burges, C.J.C. A Tutorial on Support Vector Machines for Pattern Recognition. *Data Min. Knowl. Discov.* **1998**, *2*, 121–167. [CrossRef]
36. Cortes, C.; Vapnik, V. Support-vector networks. *Mach. Learn.* **1995**, *20*, 273–297. [CrossRef]
37. Scikit-Learn Developers (BSD License). Scikit-Learn. Machine Learning in Python. 2024. Available online: https://scikit-learn.org/ (accessed on 13 September 2024).
38. Hobson, L.; Howard, C.; Hapke, H. *Przetwarzanie Języka Naturalnego w Akcji. Rozumienie, Analiza i Generowanie Tekstu w Pythonie Na Przykładzie Języka Angielskiego*, 1st ed.; Wydawnictwo Naukowe PWN: Warsaw, Poland, 2021.

infrastructures

MDPI

Article

Parametric Analysis as a Tool for Hypothesis Generation: A Case Study of the Federal Archive Building in New York City

Mike Christenson

School of Architecture, University of Minnesota, Minneapolis, MN 55455, USA; mike001@umn.edu;
Tel.: +1-612-624-7866

Abstract: This study investigates the epistemological potentials of *parametric analysis* for digitally modeling ordinary, existing buildings, addressing a gap in architectural research. While traditional digital modeling prioritizes geometric accuracy, it often limits the ability to generate new architectural insights, treating models as static representations rather than as tools for knowledge production. This research challenges the assumption that geometric accuracy is necessary for epistemological validity, proposing parametric analysis as a hypothesis-generating tool capable of uncovering latent spatial and morphological properties that conventional methods overlook. Using Suárez's *inferential conception of scientific representation* as a theoretical framework, this research employs a comparative case study methodology, contrasting direct and parametric digital models of the Federal Archive Building in New York City, analyzing their respective contributions to architectural knowledge. Existing documentation of the Federal Archive Building provides the primary data. The findings reveal that parametric modeling can enable the discovery of latent design properties by facilitating the systematic exploration of geometric variations while maintaining other logics, specifically by demonstrating how certain architectural features accommodate site irregularities while preserving visual coherence. This research advances theoretical discourse by repositioning parametric models from descriptive artifacts to instruments of architectural reasoning, challenging conventional associations between representational accuracy and epistemological validity. Practical applications are suggested in heritage documentation, comparative architectural analysis, and educational contexts where the interpretive exploration of buildings can generate new insights beyond what geometrically accurate models alone can provide.

Keywords: digital modeling; parametric modeling; architectural epistemology; building documentation; architectural representation; morphological analysis

1. Introduction

This article develops theories in *architectural epistemology*, which is the study of the ways in which knowledge about architecture is produced, organized, and disseminated [1]. The production of architectural knowledge has historically relied on representational artifacts such as drawings and models to document, analyze, and interpret the existing built environment [2–4]. Traditional approaches to digitally modeling existing buildings tend to prioritize geometric accuracy and isomorphic representation, often treating digital models as static surrogates for the buildings they aim to represent [5–7]. However, a singular emphasis on accuracy risks overlooking the epistemological potential of digital models as tools for critical inquiry [8,9]. *Parametric analysis* offers an alternative approach that extends beyond conventional representation, enabling hypothesis generation through systematic variation in model parameters [10]. The current study investigates how parametric modeling can function as a knowledge-generating tool, in effect relieving digital models of the need to be descriptive artifacts and enabling them to become instruments of architectural reasoning and critique.

The significance of parametric analysis in this approach lies in its ability to expose the latent spatial and morphological properties of buildings by encoding variable relationships rather than fixed geometries. Unlike *direct modeling*, which constructs a digital replica with fixed dimensions, *parametric modeling* introduces associative rules that allow for iterative transformations [11]. In a direct modeling approach, digital tools are used to construct models that focus on geometry, rather than on material or environmental parameters [12]. Models that are built using a direct-modeling method lack parametrically defined relationships. In particular, semantic relationships within direct models depend on subject-matter knowledge and layer organization [10]. By contrast, a parametric modeling approach involves explicit associative rules between related parameters [13]. This approach aligns with broader discussions in architectural epistemology, where digital tools are increasingly recognized not only for their capacity to aid in documentation and visualization but also for their potential to reveal underlying architectural logics [14,15]. By examining the role of digital models in architectural knowledge production, this study contributes to ongoing debates about the knowledge status of representation and the methodological implications of modeling [16,17].

Scientific representation is theorized in different ways, each offering distinct insights into how models can function as knowledge-generating artifacts. For example, Peirce emphasizes a three-part relationship between sign, object, and interpretant, framing representation as a system of referential symbols [18]. Goodman foregrounds the selective and often conventional nature of representation, suggesting that models do not merely depict but also encode conceptual priorities [19,20]. Gouvea and Passmore argue that models derive meaning from interpretive frameworks that go beyond their representational relationship to phenomena [21]. Suárez suggests an *inferential conception of scientific representation*, in which a model's epistemic value lies in its capacity to generate inferences rather than in its strict geometric correspondence to a referent [22]. The current study applies Suárez's framework to architectural modeling, arguing that parametric analysis enhances a model's inferential capacity by enabling structured experimentation with variable model parameters. Ultimately, Suárez's framework is selected here because it explicitly frames representation as *inferential* rather than *isomorphic*, thus supporting the study's focus on hypothesis generation.

To examine the epistemological implications of parametric analysis, this article focuses on a case study of the Federal Archive Building in New York City [23]. The building is a typical nineteenth-century warehouse–office hybrid in a dense urban location. Its design reflects broader historical patterns in its North American context. Its multi-story, grid-based structure is typical of many contemporary buildings with similar functions. From the point of view of digital modeling, the Federal Archive Building presents a unique challenge, which is that despite its overall symmetrical and rectangular appearance, it is constructed on an irregular polygon base. This irregularity is immediately apparent in a map or plan view, and less so in ground-level perspective views (Figure 1). By comparing a direct model of the building with a parametric model that encodes variable conditions, this study explores how digital modeling techniques may influence the interpretation of architectural form.

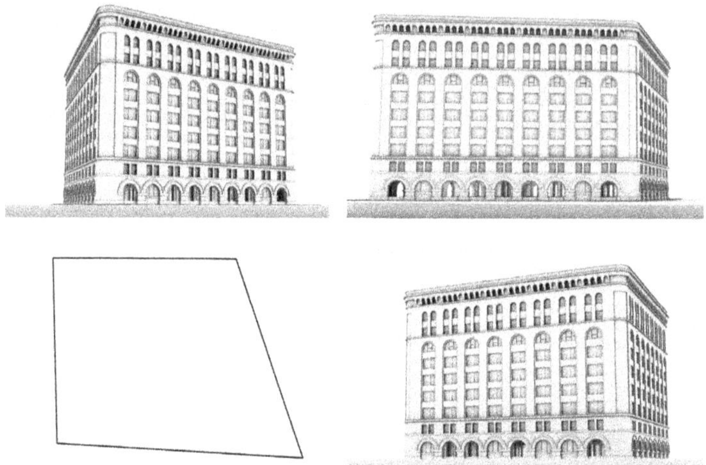

Figure 1. An outline plan view and three perspective views of a digital model representing the Federal Archive Building. The outline plan view visually demonstrates that the building is clearly constructed on a non-rectangular base polygon. However, this feature of the building is not equally obvious in the perspective views.

This study aims to examine the hypothesis-generating potential of parametric modeling for existing buildings. The specific objective is to critically assess how parametric models, through their structured variability, contribute to architectural epistemology by enabling insights that are inaccessible through direct modeling alone. This objective is pursued through a comparative analysis of direct and parametric modeling approaches, highlighting their respective epistemic affordances and limitations. The article begins with a discussion of the relevant theoretical background. It then describes the methodology employed to develop a specific approach to digital modeling aimed at generating architecturally relevant hypotheses. This is followed by a comparative discussion of direct and parametric modeling methods and a conclusion suggesting applications and future work.

2. Literature Review

2.1. Inferential Representation: Hypotheses and Meaning-Making

Critically, Suárez's inferential conception of scientific representation does not assume conventional expectations of accuracy, whether in measurable dimensions, configurations, or materials. More broadly, the inferential conception does not require a model to be isomorphic to its target. Instead, for a model to be considered a representation, it is sufficient for it to enable specific inferences; its value as a representation is characterized by the accessibility and usefulness of those inferences. The truth-value of the inferences themselves is not at issue. While truth-value or accuracy in models is often prioritized due to disciplinary demands for measurable precision, such approaches risk conflating accuracy with epistemological validity. As an example of how a model may achieve validity, prioritizing hypothesis generation over measurable accuracy, consider the visualization reproduced in Figure 2.

Figure 2. Maps of a freeway within a dense urban context, connecting endpoints (**A**) and (**B**).

Rows (1–6) illustrate a series of maps showing the route of an urban freeway connecting endpoints (A) and (B). Row (1) proposes an initial visual hypothesis, namely, that the freeway may be represented as a simple straight-line route connecting the two endpoints. Each successive row in the figure introduces distortions to this idealized route, progressively approximating the spatial constraints and conditions that are measurably present in the city, until row (6), which constitutes a detailed and relatively accurate map of the freeway in its urban context. Critically, the intermediate maps in rows (2) through (5) function as something more than static representations of the freeway's geographic reality. In fact, the maps, considered collectively, enable inferences about urban-scale forces that might influence the freeway's trajectory. These forces—whether topographical, infrastructural, social, or economic—may be inferred from the visible distortions observed in each successive intermediate map.

The point of this example is to illustrate that the sequence of maps does not rely on a conventional assumption that more detailed maps are inherently more useful in developing hypotheses. Instead, aligning with Suárez's inferential conception, the value of the maps resides in their collective ability to support inferences about how urban forces may shape the freeway's path. In this way, geographic accuracy becomes secondary to the maps' inferential capacity to suggest meaningful relationships between the route and its broader urban context.

2.2. *Critiques of the Model-Building Duality*

The relevance of this approach is further strengthened by a consideration of the *model-building duality* [9]. Briefly, this is the expectation that a model must be similar or isomorphic to its target to enable valid and meaningful inferences. This expectation is conventionally positioned in terms of geometric accuracy [24] and also in semantic consistency. Expectations for accuracy and consistency vary across disciplines [5,25]. Geometric accuracy or correspondence between model measurements and building measurements becomes critical for conservation and restoration projects [26–31].

Maintaining semantic consistency between model and referent may inform the practical uses of digital models in disciplines such as facilities management [32–35]. Semantic consistency may be enforced by frameworks such as the Industry Foundation Classes (IFCs), which aim to support neutral, objective, and accurate communication and interoperability [36]. Nevertheless, there are challenges in ensuring interoperability, especially concerning models of existing buildings. Unique architectural features may be difficult to accommodate within generalized data models [37–39]. Although the model-building duality can be seen as essential in specific use cases, it also may limit how models can meaningfully represent contextual and latent aspects of existing buildings, meaning those aspects which are not directly registered in measured or measurable geometry. A strictly assumed model-building duality risks an overemphasis on isomorphism. Relying on this idea causes models to become "surrogate[s] or replacement[s] for an original" [7], possibly limiting engagement with a target's latent qualities, such as alternative histories and spatial ambiguities.

In summary, an insistence on isomorphism, or, in general, on measurable accuracy, while obviously necessary for certain use cases, may inadvertently risk obscuring latent design properties and contextual relationships that can become accessible through more interpretive methodologies.

3. Methodology

The method described here aims to create and compare two digital models representing the Federal Archive Building: a *direct model* and a *parametric model* [11]. The comparative approach employed in this study aligns with established research demonstrating that parametric modeling extends beyond conventional documentation to facilitate hypothesis generation through structured variability [10,11,14,15,40]. By encoding relationships rather than fixed geometries, parametric models enable the systematic exploration of design logics and analytical alternatives, revealing spatial and morphological insights not accessible through direct modeling alone. The method also incidentally suggests the importance of *paradata,* which refers to the documentation of interpretive decisions, assumptions, and iterative processes [41]. In this way, the methodology is not limited to describing technical steps but is also a critical reflection upon how those steps shape and are shaped by a modelmaker's understanding of the evolving task. Robust paradata support interpretive accountability in the interest of maintaining scholarly transparency about modelmaker assumptions and model limitations.

The methodology includes three key components: first, data collection and direct modeling; second, variability determinations and parametric modeling; and finally, the visual comparison of artifacts derived from the modeling processes. These three components are visually summarized and explained in detail in each of the following three sections of the article.

3.1. Data Collection and Direct Modeling

The first methodological component was to construct a direct model of the subject building on the basis of collected data (Figure 3). Published drawings of the subject building were collected, inconsistencies (if any) were identified, and additional inferred and supplementary resources were brought to bear. Approximation techniques were used as necessary to complete the model.

In the mid-1970s, the Federal Archive Building was carefully documented by faculty at Columbia University as part of an adaptive reuse proposal [23]. In this way, this otherwise ordinary building was recorded and visually represented in a form suitable for developing the methodology described in this article. To coordinate the data, the published orthographic drawings, including plans and sections, were manually traced using Rhino (Figure 4).

The orthographic drawings make it clear that the building is structured on concrete columns and beams in its lower levels and steel in its upper levels. Thus, the structural grid assumes one of two basic configurations on each floor. Each of these configurations was modeled, and instances of each were located vertically relative to each other within the modeling environment, corresponding to the appropriate individual floor levels.

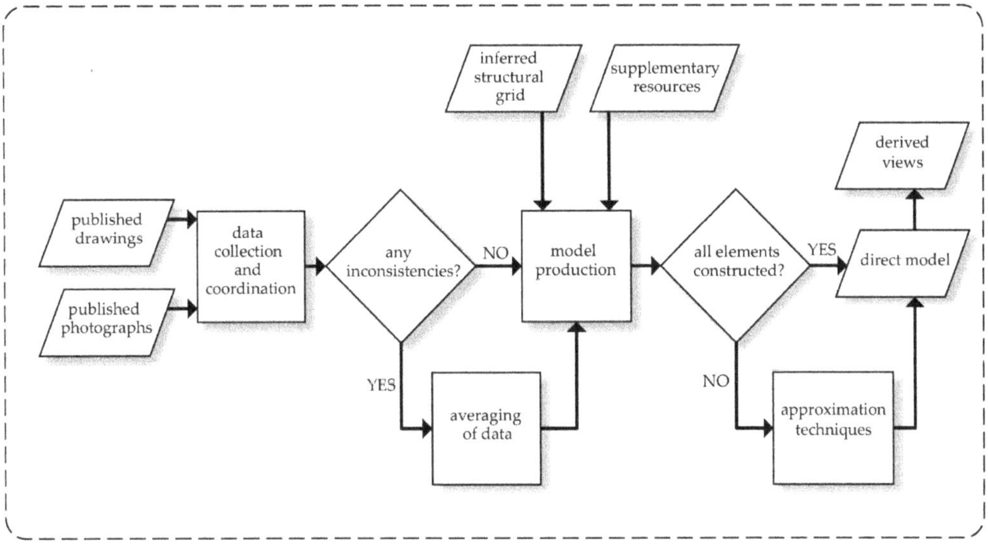

Figure 3. Process diagram for data collection and direct modeling.

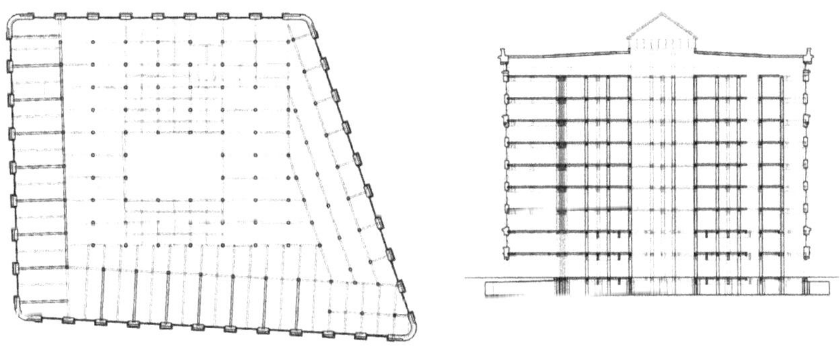

Figure 4. Plan and section representing Federal Archive Building, redrawn from [23].

Simultaneously, the building's exterior wall was modeled. The orthographic drawings were relied upon as a base, while volumetric approximations were derived from the inspection of extant photographs. In cases where published dimensions were inconsistent or irreconcilable, an averaging approach was used to construct a reasonable geometric approximation. To the extent possible, the model was visually validated during its construction with respect to the published documentation and photographs.

In general, the building's exterior wall consists of a solid masonry enclosure, thickest at the base and decreasing in thickness at the upper floor levels. The masonry enclosure is punctuated by a regularly repeating vertical bay of windows, organized vertically. Due to the building's non-rectangular base, this motif is repeated a different number of times on each of the building's four elevations. The exterior surface of the enclosure is generally flat, except at locations where it is bounded by horizontal cornices (e.g., at the top and at the base). These cornices were modeled as horizontal extrusions along the surfaces of the four walls. Where necessary, supplementary resources (e.g., contemporary photographs) were consulted to aid in decision-making. In any specific situations where the collected data were insufficient for modeling, approximation techniques were used to achieve the best conformance with the published documentation and supplementary resources.

As an example of the modeling technique, the four corners of the building's exterior enclosure were constructed as partial vertical cylinders (Figure 5, left). This design feature of the original building aesthetically absorbs the non-orthogonal walls into a rounded intersection at each corner (Figure 5, right). This architectural solution would later develop additional significance in the parametric analysis phase of the work.

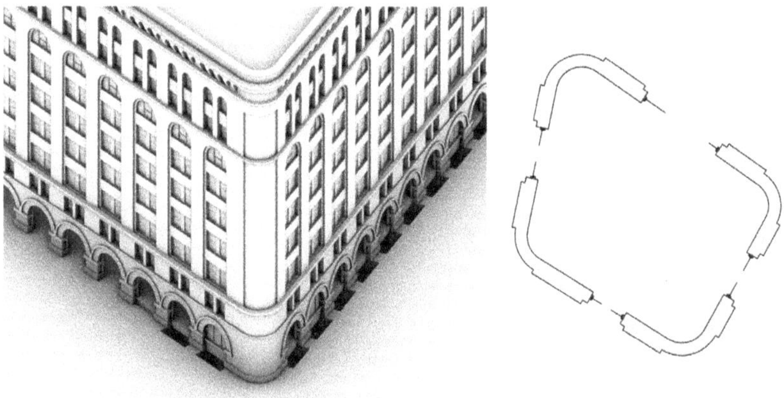

Figure 5. The rounded corners of the Federal Archive Building, as represented in an isometric view (**left**) and in plan details (**right**) of the direct model.

Although the published report does not include elevation drawings, the production of a detailed and geometrically accurate model of the building made it possible to produce elevations as derived artifacts (Figure 6).

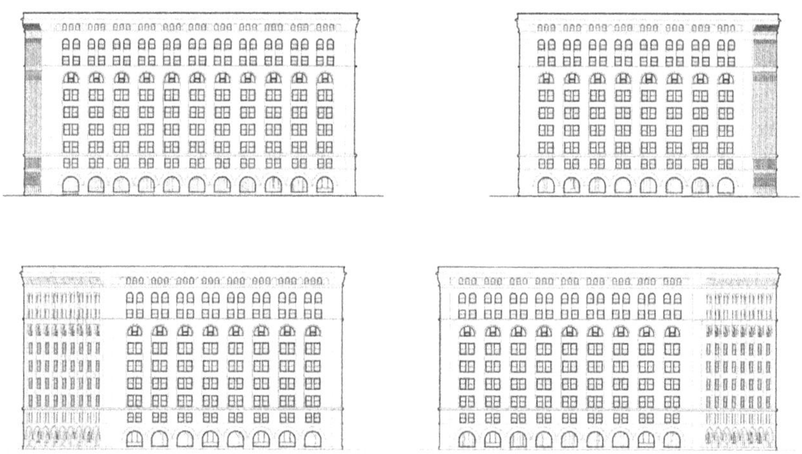

Figure 6. Elevation drawings, projected from the direct model representing the Federal Archive Building.

3.2. Variability Determinations and Parametric Modeling

In the next component, a parametric model was constructed to introduce variability into the representation of the building's base geometry while maintaining a consistent facade logic (Figure 7).

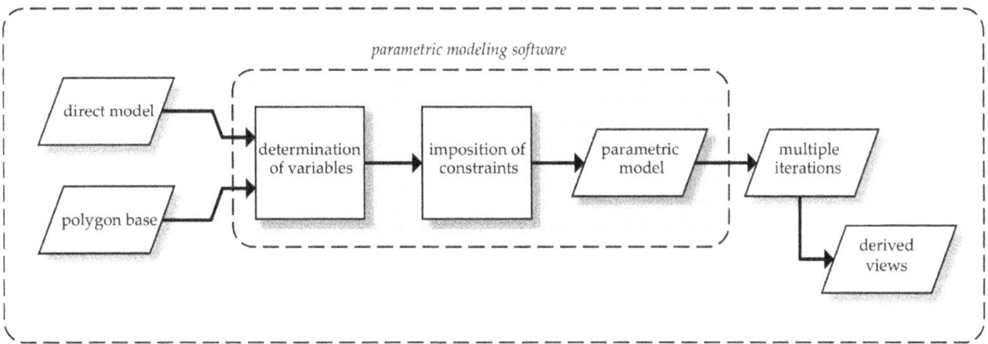

Figure 7. The process diagram for variability determinations and parametric modeling. The use of parametric modeling software to construct a single parametric model is highlighted.

Considering the direct model suggested that the Federal Archive Building could be formally understood as a distorted cube, or more precisely a vertically extruded polygon, with a uniform facade expression consisting of vertically configured, repetitive window bays (Figure 8). The decision to model the building as a vertically extruded polygon with variable base geometry had an immediate epistemological significance. As a modelmaking decision, it foreshadowed the kinds of parametric explorations that would be possible, as is discussed in a later section of this article.

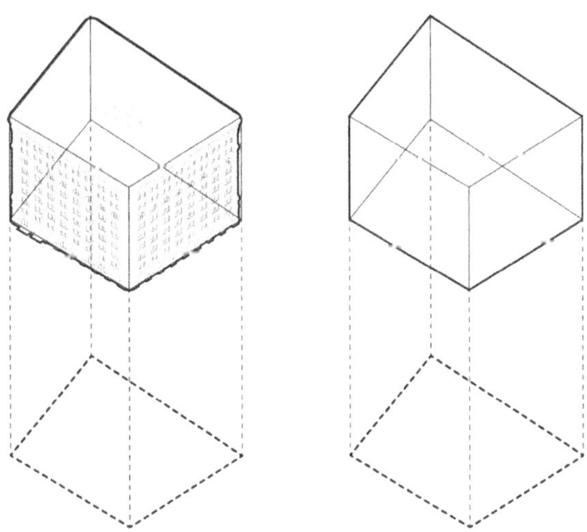

Figure 8. The direct model (**left**) and the initial iteration of the parametric model (**right**) representing the Federal Archive Building, understood as a vertically extruded polygon.

Following the decision to construct the model on the basis of a variable polygon, Grasshopper (a visual programming environment within Rhino) was introduced into the project workflow. Grasshopper was selected due to its established use in parametric workflows, allowing for the real-time manipulation of both the building's base geometry and the articulation of its facade. In essence, Grasshopper enables the assignment of semantic relationships among modeled geometry. This necessarily involves the subjective determination of variables and the imposition of modeling constraints.

To digitally model the building using a parametric method, a Grasshopper definition was prepared, i.e., an algorithm coded within Grasshopper, to generate a four-sided base polygon and extrude it into a vertical prism (Figure 9).

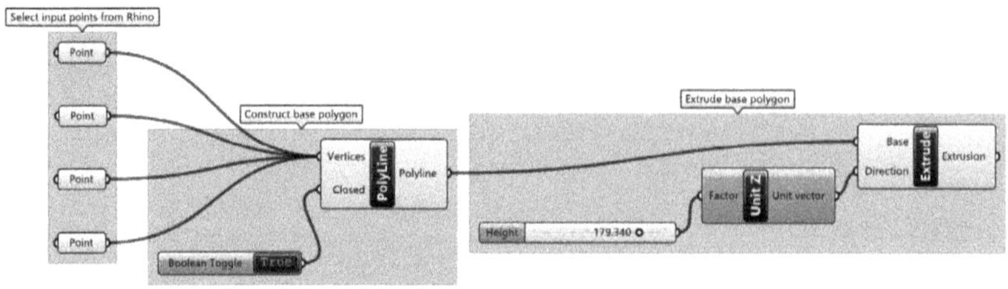

Figure 9. Grasshopper definition for variable base polygon. Rhino points selected (**left**), base polygon connects input points, and base polygon is extruded vertically (**right**).

Briefly, the Grasshopper definition synthesized point inputs to create a closed polyline (i.e., an irregular four-sided polygon) which was vertically extruded to form a boundary-represented solid in three dimensions. This parametric approach to modeling the building's mass allowed each one of the base polygon's four corners to be located on a common plane independently of each other; the points could be moved individually within Rhino. The corner locations of the base polygon—and hence the lengths of the polygon's sides—were thus effectively variable. (Alternatively, the points could be constructed in Grasshopper as responsive to x, y coordinate inputs.) Figure 10 illustrates three plan views of possible iterations of the variable base polygon, each resulting from a unique set of arbitrarily located corner points. Each of the plan views is visually paired with an isometric view of the model generated from the base polygon through vertical extrusion.

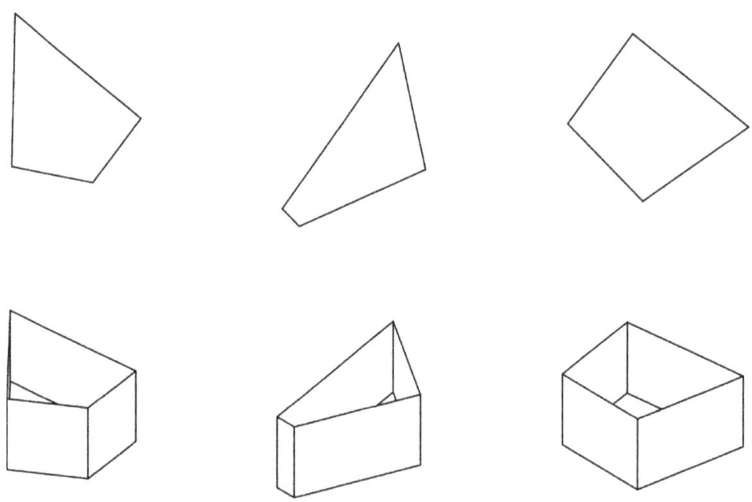

Figure 10. Variable base polygons coupled with 3D views.

Next, a single vertical bay of windows was drawn using standard Rhino tools (Figure 11a), using the direct model as a reference. This bay of windows was mapped, as a unit, repeatedly to each of the four vertical surfaces of the extruded cube according to the following method. This approach maintained a consistent distribution of windows, independently of any parametric changes in facade length (Figure 11).

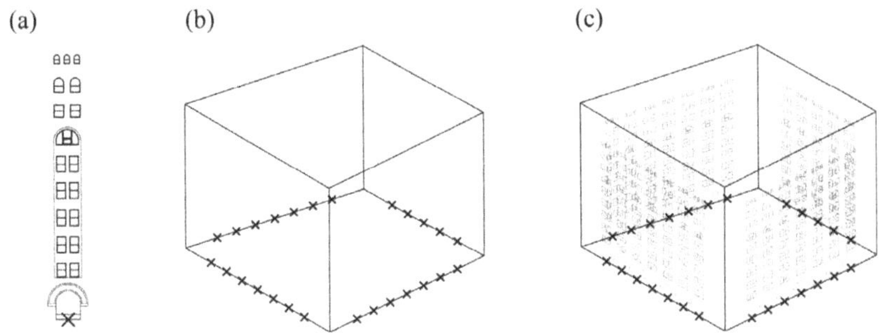

Figure 11. The vertical bay of windows mapped to the model. The figure illustrates (**a**) a typical bay of windows in elevation view; (**b**) an outline view of the extruded base polygon, with the base sides divided "by length"; and (**c**) a wireframe view of the extruded base polygon with the typical window bay "mapped" appropriately to each division point on the base polygon, resulting in a wireframe 3D model of the building with windows on all four facades.

More precisely, each of the base polygon's sides was divided by length using a fixed dimension from the existing building (22′-9″). Points were placed regularly along each of the polygon's four sides to mark these divisions (Figure 11b). A copy of the vertical bay was placed, vertically centered on each division point, and rotated to align with the orientation of the side on which it was placed. The resulting model was a variable wireframe simulation representing the building's form as variable (Figure 11c).

In the final component of the method, the models, iterations, and derived views were visually compared, with the aim of drawing out meaningful insights (Figure 12). If, in this component, no meaningful insights emerged, the previous component (variability determinations and parametric modeling) was repeated.

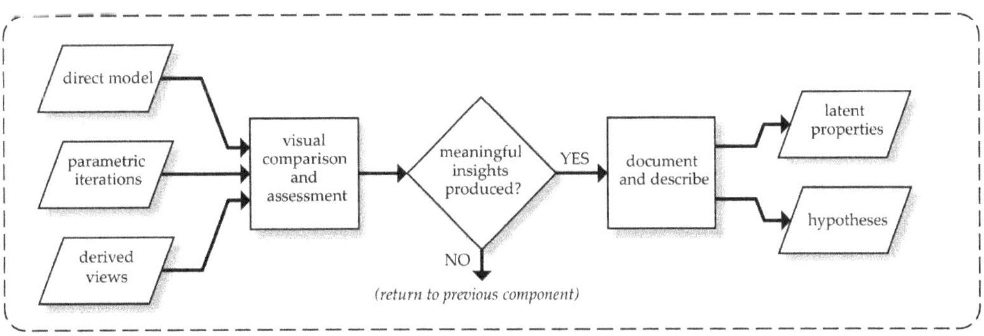

Figure 12. Process diagram for comparative analysis and interpretation.

Using this method, it was possible to generate multiple modeled iterations representing the Federal Archive Building. As the locations of the corner points were changed in real time, the model updated its configuration, mapping window bays appropriately to the modeled facades. In particular, the model may be configured on a square base (Figure 13, left), it may be rectangularly distorted (Figure 13, center), or the base may be distorted into a general quadrilateral (Figure 13, right).

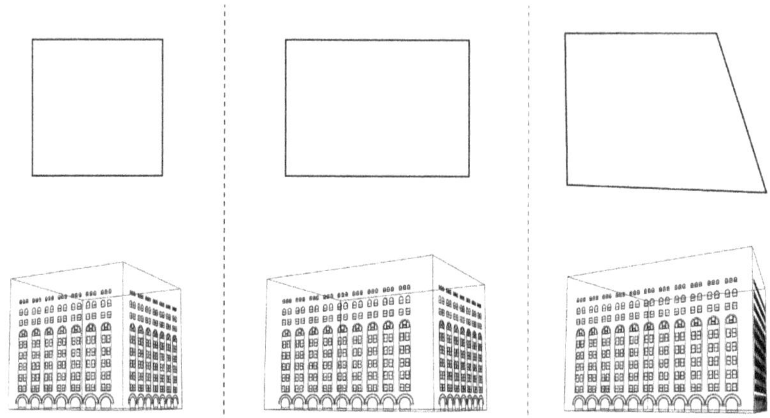

Figure 13. Plan views (**top**) and perspective views (**bottom**) of the parametric model representing the Federal Archive Building.

Despite the possible variability in the base polygon, the perspective views of the model remained generally consistent. To state this differently, perspective views of different models are not in any obvious way distinguishable from perspective views of a single model viewed from different locations. The different iterations, considered as a set, thus had the effect of visually problematizing relationships between the shape of the base and the proportions of the elevations. This visual problematization was specifically enabled by juxtaposing plan and perspective views.

The assessment of the iterations was ultimately based on the ability of each representation to facilitate inferences about the building's architectural logic. Rather than measuring effectiveness through quantitative metrics, the study evaluated parametric modeling's knowledge contributions in terms of its capacity to prompt new ways of seeing and understanding the building. The parametric model's iterative quality made explicit certain latent design properties, which were inferred through the comparative visualization of derived artifacts (specifically, plan and elevation views).

In summary, the methodology establishes a framework for using parametric modeling as a tool for architectural inquiry. The direct model was understood to be geometrically accurate with respect to archival documentation. The parametric model was variable in its dimensions and proportions, and hence not geometrically accurate in a strict sense. The following section will examine how the parametric model may prompt initial insights into the building's latent design properties and suggest how these insights may support inferences leading to hypothesis generation.

4. Discussion

Both of the models representing the Federal Archive Building—the direct model and the parametric model—possess *representational force* and *inferential capacity* according to the definitions provided by Suárez [22]. Representational force, in the case of both models, resides essentially in the claim that the models "point to" the Federal Archive Building, although a claim of representational force could also be established on the basis of the visual resemblance between images of either model and images of the building. The question of *inferential capacity* triggers an important distinction between the two models. First, the direct model enables inferences to be drawn on the basis of its isomorphism to the building, i.e., assuming that the model and the building reliably correspond to each other in a one-to-one manner, whole to whole and part to part. Using the direct model, having been built on the basis of documentation presumed to be accurate, measurements may be reasonably inferred from the model beyond those that were recorded on site (e.g., by measuring scale dimensions from the model). Similarly, the model enables the visualization of spatial configurations that are not directly visible on site (e.g., by cutting a section through the model). In short, the direct model's inferential capacity enables precise spatial measurements and validation against ground-truth and archival records. By contrast, the parametric model enables a different kind of inference to be drawn. The parametric model necessarily avoids strict isomorphism, replacing the one-to-one correspondence of parts with a correspondence that is topological and conceptual rather than strictly geometrical. Stated differently, the parametric model's inferential capacity is grounded not in geometric accuracy but in its ability to dynamically generate variations that remain conceptually linked to the building. In this way, the model supports the exploration of perceptual effects, adaptive design logics, and site constraints. Thus, the comparative question is as follows: what unique epistemological contributions does the parametric model offer that the direct model does not?

4.1. Key Hypotheses and Insights

A primary hypothesis emerged through the parametric model's unique ability to generate multiple variations of the building's base polygon while maintaining a facade configuration broadly consistent with the existing building's design. This suggested, at a minimum, that the building could have been constructed on various non-rectangular sites while remaining consistent with an implied set of facade-generating rules. When viewed in plan, these variations appear markedly different from each other. However, when these same variations are examined through simulated street-level perspectives, the perceptual differences between iterations are more nuanced. While the views can be distinguished from one another if carefully compared, they are generally consistent in their overall appearance. Precisely, the claim is that it is easier to visually distinguish variations in plan than to distinguish them in perspective.

This interplay between distinctly different plan geometries and their visually similar perspective appearances, as reflected in the model, suggests a latent design property of the building, namely, its capacity to maintain visual consistency despite allowing for significant variations in its base geometry. In turn, this insight suggests a historical hypothesis: the building's designers may have recognized, perhaps implicitly, that they could accommodate the irregular site constraints without compromising the building's perceived visual qualities as they would appear to a street-level pedestrian. In a sense, then, the parametric model makes this hypothesized design logic explicit, providing a kind of post hoc validation of a speculated decision-making process.

The critical point is that this insight emerged not from static documentation or from the construction of the geometrically accurate direct model, but from the parametric model's dynamic capacity to generate and visualize multiple iterations through multiple means—specifically, through plans and perspectives. The parametric model's systematic mapping of window bays onto the variable-length facades further reinforces the emerging hypothesis. As the base polygon changed, the Grasshopper definition automatically adjusted the distribution of window bays, maintaining consistent horizontal spacing despite variations in facade length—a feature observed in the existing building's four elevations. This systematic adaptation suggested a design logic that could accommodate site irregularities while preserving architectural coherence within facades—a strategy that was likely implicit in the original design process, but which became explicit through parametric analysis.

Taken together, these insights suggest a reconsideration of the archival documentation on which both models were based. Rather than simply confirming the building's irregular footprint as a response to pre-existing site constraints, parametric analysis suggests how this irregularity could have been sophisticatedly managed through the designers' careful attention to facade variability and street-level perception. The parametric model's inferential capacity makes it possible to generate a nuanced understanding of possible historical design strategies, suggesting how the building's architects could have worked to resolve tensions between non-negotiable site constraints and a desire for facade regularity. Indeed, it is in this final realization that the direct model's detailed, geometrically accurate representation takes on new significance. What initially appeared as an aesthetic choice in the direct modeling phase—that is, the cylindrical expression of the building's four exterior corners—is suggested by parametric analysis as a kind of metaphorical expression of "hinges" that allow the building's facades to meet at variable angles while maintaining visual coherence. The cylindrical corners may be said to represent a physical manifestation of the building's hypothesized geometric flexibility. In brief, this insight emerged from the interplay *between* direct and parametric modeling approaches: the careful documentation of physical features in the direct model suggested possibilities that were then explored and, in a way, validated through parametric experimentation.

4.2. Epistemic Contributions of Parametric Analysis

Generally, the epistemic contributions of parametric analysis can be organized into three major themes, each representing a distinct way in which the methodology advances architectural knowledge production. The first theme concerns knowledge production in general: parametric analysis triggers a fundamental shift in how architectural knowledge about a building can be produced. This shift follows from important distinctions between direct models and parametric models. Where direct models enable accuracy in documentation and prediction, parametric models enable a researcher to transition away from a reliance on static documentation into a mode of dynamic inquiry. This shift emphasizes processes of model construction and exploration over products, i.e., finished, objectively accurate models. By enabling the real-time manipulation of morphological relationships, parametric analysis may reveal design logics that are otherwise latent in static documentation.

Second, parametric modeling has unique characteristics as a knowledge-generating activity, deriving from its capacity for dynamic iteration. While maintaining topological connections to the target building, parametric models can generate multiple variations, exploring the building's underlying conceptual frameworks. This iterative capacity allows researchers to investigate how different geometric configurations might affect a building's visual expression, even as core design relationships are preserved. The ability to systematically generate multiple iterations for analysis distinguishes this approach from traditional documentation-based methods. As an example, parametric analysis offers potential for transferring insights between buildings across different contexts. Figure 14 visually juxtaposes a geometrically accurate elevation of the Federal Archive Building (a) with a parametrically modeled elevation (b), a hypothesized extended version of that elevation (c), and a similarly detailed but geometrically accurate elevation of the now-demolished Marshall Field Wholesale Store in Chicago (d).

The comparison in Figure 14 suggests how parametric analysis might reveal shared design strategies across seemingly different buildings. Such an approach could develop into a systematic method for comparative architectural analysis.

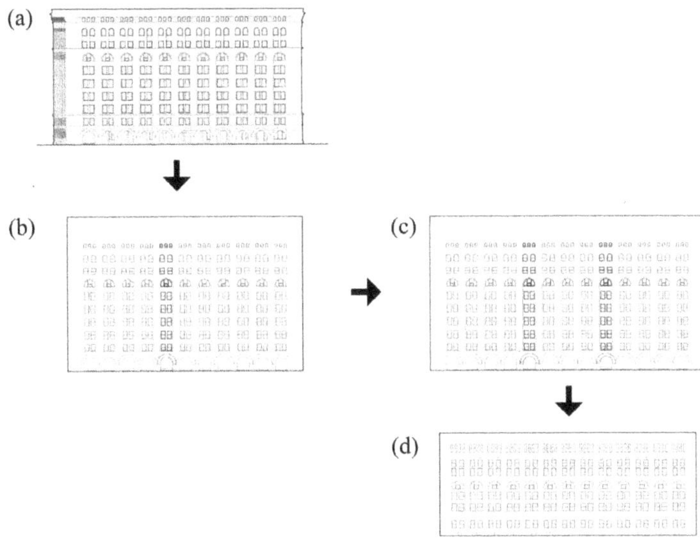

Figure 14. A comparison of a geometrically accurate elevation of the Federal Archive Building (**a**) with a parametrically modeled elevation (**b**), a hypothesized extended version of that elevation (**c**), and a similarly detailed but geometrically accurate elevation of the now-demolished Marshall Field Wholesale Store in Chicago (**d**).

Third, parametric models prompt new understandings concerning the role of traditional documentation (e.g., floor plans, elevations, and sections) in modelmaking. Rather than viewing limited historical documentation as a constraint on building complete and accurate models, parametric analysis transforms this limitation into an opportunity for interpretive exploration. Models become tools for generating new ways of perceiving and hypothesizing about latent design logics, even in the absence of complete historical records. This approach aligns with Suárez's inferential conception, as the parametric model's representational force remains present while its inferential capacity operates in fundamentally different ways than in direct modeling.

5. Conclusions

As the case study of the Federal Archive Building makes clear, this study establishes a methodological framework for analyzing ordinary buildings through parametric analysis, offering new ways to understand latent design logics. More broadly, it positions parametric digital models as hypothesis generators, demonstrating how dynamic, changeable models enable unique inferences about existing buildings. Finally, in a theoretical context, the study provides empirical support for Suárez's inferential conception of scientific representation within the context of architectural modelmaking.

Meaningful hypotheses concerning the existing building were brought to light using the parametric model, despite the model itself not being geometrically accurate in a strict sense. These hypotheses involved the building's latent design logics, its perceptual regularity, and insights into the decision-making processes of its designers. In this way, the approach described here is consistent with the aims of the *London Charter*, to the extent that it exemplifies principles of interpretive richness and experimental methodologies [42]. In fact, the approach described here suggests a way of meeting these principles in ways that static, geometrically accurate models do not.

5.1. Limitations and Challenges

While parametric analysis offers a productive framework for hypothesis generation, its epistemological assumptions introduce specific challenges. These challenges may be broadly categorized into three areas: first, the role of geometric accuracy and its relevance to architectural knowledge production; second, the risks of overinterpretation, where inferences exceed empirical support; third, the method's tendency to privilege formal and parametrically definable relationships over material, historical, or cultural dimensions of the built environment. Addressing these concerns requires both methodological transparency and a critical awareness of the limits of parametric modeling.

A central premise of parametric analysis is that architectural knowledge can be advanced by digitally modeling existing buildings in ways that deprioritize isomorphism. While this approach aligns with Suárez's inferential conception, it also challenges assumptions about knowledge production and use that are characteristic of domains such as conservation, facilities management, and structural analysis. In these contexts, geometrically accurate digital models are not only useful but essential. In contrast, parametric analysis functions heuristically: its value lies in its capacity to generate and test hypotheses rather than to inform the production of defensible representations of an objectively measurable physical reality. This requires that the scope and potentials of parametric analysis are clarified relative to traditional model-based analytic approaches. While parametric models may be capable of revealing or suggesting underlying logics, they should not be mistaken for tools of precise documentation. They are correctly positioned as complementary to, rather than ignorant of, direct models that rely on accurate measurement. Hybrid approaches integrating parametric and direct modeling may help to ensure that inferences and hypotheses remain appropriately contextualized, even if they are not empirically grounded in the strictest sense.

Because parametric models of existing buildings invite hypothesis generation, they also introduce the risk of speculative overreach. A modelmaker may reasonably infer relationships, patterns, or design intentions through parametric analysis, despite those inferences lacking explicit support in the form of historical or physical evidence. While this issue is not unique to parametric modeling—indeed, all architectural representations are shaped by their makers' assumptions—it becomes problematic as part of an approach aimed at revealing latent design logics. Thus, the question is how to distinguish plausible hypotheses from unwarranted speculation. In this light, robust paradata may be a mitigation strategy. Indeed, the *London Charter*, in addressing digital representations of existing buildings, highlights the importance of "manifold interpretative interventions", "multiple possible stories", and "highly speculative experiments ... evaluated on their own terms" [42]. The transparency provided by paradata is particularly important within comparative studies that test parametric inferences against archival documentation, site analysis, or historical accounts—any of which could help validate or challenge proposed hypotheses.

A third challenge concerns the potential for parametric analysis to privilege geometric and formal properties over material, historical, and cultural dimensions of buildings. For example, the systematic mapping of window bays in the case study highlights a formal design logic, but it does not inherently reflect upon construction techniques, material decay, or cultural significance. This challenge once again highlights a need to develop hybrid methodologies integrating parametric modeling with more traditional forms of material and historical research.

Ultimately, these challenges do not undermine the validity of parametric analysis, but rather highlight a need for its integration in context and a form of methodological reflexivity. That is simply to say that the epistemic contributions of parametric modeling should be balanced against an awareness of its limits.

5.2. Future Directions

This study opens several promising directions for future research. Future work should explore the application of parametric analysis to different building types, particularly those with irregular or historically contingent geometries. The work should be positioned to integrate with existing historical analysis methods, and could be explored as a potential field for methodological standardization. Its practical applications could extend to heritage documentation and interpretation, with potential impacts on adaptive reuse planning, urban analysis, and typological studies. For example, researchers in the field of heritage conservation could integrate parametric analysis to test hypothetical reconstructions of incomplete or ambiguous architectural and archeological records.

The current study prioritizes geometric relationships over material properties. Future research should explore ways to integrate parametric modeling with material analysis. This integration could include studies of aging and decay, with implications for conservation and adaptive reuse. By incorporating data on material change over time, parametric models could deepen the available insights into building lifecycles. In a related way, validation of parametric models against ground-truth data can be argued as essential to hypothesis credibility. To test this argument, future studies should systematically investigate validation approaches for parametric representations with respect to remote-sensing data, historical documents, site records, and other primary sources. This integration could potentially inform methodological standards for the use of parametric modeling in architectural research.

The cultural and symbolic aspects of architecture remain seriously underexplored in the parametric modeling literature. Future research should consider how parametric modeling can engage with historical narratives, spatial practices, and socio-political influences. Parametric workflows informed by culturally cognizant parameters could enable researchers to newly examine how buildings encode and reflect social meanings and broader historical contexts. Similarly, the integration of parametric analysis with ethnographic, historical, or archeological methods could strengthen the method's validity with respect to architectural knowledge. Combining parametric modeling of the existing built environment with established qualitative and interpretive research approaches could enable more nuanced and contextually grounded analysis.

Finally, each of the areas mentioned here also offer pedagogical opportunities, whether in addressing architectural history and theory, digital methodologies, or other related topics in both building science and the digital humanities.

Funding: This research received no external funding.

Data Availability Statement: The original contributions presented in this study are included in the article. Further inquiries can be directed to the corresponding author.

Conflicts of Interest: The author declares no conflicts of interest.

References

1. Christenson, M. *Theories and Practices of Architectural Representation*; Routledge: London, UK, 2019.
2. Blau, E.; Kaufman, E.; Evans, R. *Centre Canadien D'Architecture. Architecture and Its Image: Four Centuries of Architectural Representation: Works from the Collection of the Canadian Centre for Architecture*; Centre Canadien d'Architecture/Canadian Centre for Architecture: Montreal, QC, Canada, 1989.
3. Piotrowski, A. On the Practices of Representing and Knowing Architecture. In *The Discipline of Architecture*; Piotrowski, A., Robinson, J., Eds.; University of Minnesota Press: Minneapolis, MN, USA, 2001.

4. Lucas, R. Drawing Parallels: Knowledge Production in Axonometric, Isometric, and Oblique Drawings. *Arch. Des.* **2015**, *85*, 102–107.
5. Volk, R.; Stengel, J.; Schultmann, F. Building Information Models (BIM) for Existing Buildings: Literature Review and Future Needs. *Autom. Constr.* **2014**, *38*, 109–127. [CrossRef]
6. Tang, P.; Huber, D.; Akinci, B.; Lipman, R.; Lytle, A. Automatic Reconstruction of As-Built Building Information Models from Laser-Scanned Point Clouds: A Review of Related Techniques. *Autom. Constr.* **2010**, *19*, 829–843. [CrossRef]
7. Reilly, P. Towards a Virtual Archaeology. In *Computer Applications and Quantitative Methods in Archaeology 1990*; BAR Publishing: Oxford, UK, 1991; pp. 133–139.
8. Christenson, M. An Emergent Ontology for Digitally Modeling Existing Buildings: Examining Kyoto's Nishiki Market. *Archnet-IJAR Int. J. Archit. Res.* **2024**, *18*, 1–15. [CrossRef]
9. Christenson, M. Problematizing the Model-Building Duality: Examining the New Sacristy at S. Lorenzo, Florence, Italy. *Front. Archit. Res.* **2023**, *12*, 651–663. [CrossRef]
10. Christenson, M. Parametric Analysis of Existing-Building Geometry: Generative, Dissective, and Morphological Approaches. *Int. J. Archit. Comput.* **2024**, *22*, 14780771241310209. [CrossRef]
11. Tommasi, C.; Achille, C.; Fassi, F. From Point Cloud to BIM: A Modelling Challenge in the Cultural Heritage Field. *Int. Arch. Photogramm. Remote Sens. Spat. Inf. Sci.* **2016**, *41*, 429–436. [CrossRef]
12. Jabi, W. *Parametric Design for Architecture*; Laurence King: London, UK, 2013; pp. 196–197.
13. Sacks, R.; Eastman, C.M.; Lee, G. Parametric 3D Modeling in Building Construction with Examples from Precast Concrete. *Autom. Constr.* **2004**, *13*, 291–312. [CrossRef]
14. Burry, J.; Burry, M. Sharing Hidden Power: Communicating Latency in Digital Models. In *Proceedings, 24th International Conference on Education in Computer Aided Architectural Design in Europe*; Bourdakis, V., Charitos, D., Eds.; University of Thessaly School of Engineering: Volos, Greece, 2006; pp. 786–793. [CrossRef]
15. Barrios Hernandez, C.R. Thinking Parametric Design: Introducing Parametric Gaudi. *Des. Stud.* **2006**, *27*, 309–324. [CrossRef]
16. Drucker, J. *Graphesis: Visual Forms of Knowledge Production*; Harvard University Press: Cambridge, MA, USA, 2014.
17. Suárez, M. *Inference and Representation: A Study in Modeling Science*; University of Chicago Press: Chicago, IL, USA, 2020.
18. Peirce, C.S.; Hartshorne, C.; Weiss, P. *Collected Papers of Charles Sanders Peirce*; Belknap Press of Harvard University Press: Cambridge, MA, USA, 1965; Volume 2.
19. Goodman, N. *Languages of Art: An Approach to a Theory of Symbols*, 2nd ed.; Hackett: Indianapolis, IN, USA, 1976.
20. Elgin, C.Z. *Between the Absolute and the Arbitrary*; Cornell University Press: Ithaca, NY, USA, 1997.
21. Gouvea, J.; Passmore, C. 'Models of' versus 'Models for': Toward an Agent-Based Conception of Modeling in the Science Classroom. *Sci. Educ.* **2017**, *26*, 49–63. [CrossRef]
22. Suárez, M. An Inferential Conception of Scientific Representation. *Philos. Sci.* **2004**, *71*, 767–779. [CrossRef]
23. Herdeg, K.; Eckstut, S.; Bell, H.K.; Byard, P.; Young, J. *Working Paper 1: The New York Federal Archive Building*; Columbia University: New York, NY, USA, 1976.
24. Johnson, D.S. Testing Geometric Authenticity: Standards, Methods, and Criteria for Evaluating the Accuracy and Completeness of Archaeometric Computer Reconstructions. *Vis. Resour.* **2009**, *25*, 333–344. [CrossRef]
25. Radanovic, M.; Khoshelham, K.; Fraser, C. Geometric Accuracy and Semantic Richness in Heritage BIM: A Review. *Digit. Appl. Archaeol. Cult. Herit.* **2020**, *19*, e00166. [CrossRef]
26. Pepe, M.; Costantino, D.; Restuccia Garofalo, A. An Efficient Pipeline to Obtain 3D Model for HBIM and Structural Analysis Purposes from 3D Point Clouds. *Appl. Sci.* **2020**, *10*, 1235. [CrossRef]
27. Bagnolo, V.; Argiolas, R.; Cuccu, A. HBIM for Archaeological Sites: From SFM-Based Survey to Algorithmic Modeling. *Int. Arch. Photogramm. Remote Sens. Spat. Inf. Sci.* **2019**, *42*, 57–63. [CrossRef]
28. Dore, C.; Murphy, M. Integration of Historic Building Information Modeling (HBIM) and 3D GIS for Recording and Managing Cultural Heritage Sites. In Proceedings of the 2012 18th International Conference on Virtual Systems and Multimedia, Milan, Italy, 2–5 September 2012; pp. 369–376. [CrossRef]
29. Abualdenien, J.; Borrmann, A. Levels of Detail, Development, Definition, and Information Need: A Critical Literature Review. *J. Inf. Technol. Constr.* **2022**, *27*, 363–392. [CrossRef]
30. Bolognesi, C.; Garagnani, S. From a Point Cloud Survey to a Mass 3D Modeling: Renaissance HBIM in Poggio a Caiano. *Int. Arch. Photogramm. Remote Sens. Spat. Inf. Sci.* **2018**, *42*, 117–123. [CrossRef]
31. Murphy, M.; McGovern, E.; Pavia, S. Historic Building Information Modelling: Adding Intelligence to Laser and Image-Based Surveys of European Classical Architecture. *ISPRS J. Photogramm. Remote Sens.* **2013**, *76*, 89–102. [CrossRef]
32. Panah, R.S.; Kioumarsi, M. Application of Building Information Modelling (BIM) in the Health Monitoring and Maintenance Process: A Systematic Review. *Sensors* **2021**, *21*, 837. [CrossRef]
33. Wong, J.K.W.; Ge, J.; He, S.X. Digitisation in Facilities Management: A Literature Review and Future Research Directions. *Autom. Constr.* **2018**, *92*, 312–326. [CrossRef]

34. Carbonari, G.; Stravoravdis, S.; Gausden, C. Building Information Model Implementation for Existing Buildings for Facilities Management: A Framework and Two Case Studies. In *Building Information Modelling (BIM) in Design, Construction and Operations*; Mahdjoubi, L., Brebbia, C.A., Laing, R., Eds.; WIT Press: Southampton, UK, 2015; pp. 395–406. [CrossRef]
35. Kassem, M.; Kelly, G.; Dawood, N.; Serginson, M.; Lockley, S. BIM in Facilities Management Applications: A Case Study of a Large University Complex. *Built Environ. Proj. Asset Manag.* **2015**, *5*, 261–277. [CrossRef]
36. Froese, T.; Grobler, F.; Ritzenthaler, J.; Yu, K.; Akinci, B.; Akbas, R.; Koo, B.; Barron, A.; Kunz, J.C. Industry Foundation Classes for Project Management: A Trial Implementation. *Electron. J. Inf. Technol. Constr.* **1999**, *4*, 17–36.
37. Fai, S.; Sydor, M. Building Information Modelling and the Documentation of Architectural Heritage: Between the 'Typical' and the 'Specific'. In Proceedings of the 2013 Digital Heritage International Congress (DigitalHeritage), Marseille, France, 28 October–1 November 2013; Volume 1, pp. 731–734. [CrossRef]
38. Fai, S.; Filippi, M.; Paliaga, S. Parametric Modelling (BIM) for the Documentation of Vernacular Construction Methods: A BIM Model for the Commissariat Building, Ottawa, Canada. *ISPRS Ann. Photogramm. Remote Sens. Spat. Inf. Sci.* **2013**, *II-5*, 115–120. [CrossRef]
39. Penttilä, H.; Rajala, M.; Freese, S. Building Information Modelling of Modern Historic Buildings. In Proceedings of the Predicting the Future [25th eCAADe Conference Proceedings, Frankfurt, Germany, 26–29 September 2007; pp. 26–29. [CrossRef]
40. Jabi, W.; Potamianos, I. Parameterizing the Geometry and Visualizing the Lighting Method of Byzantine Church Domes. In *Digital Heritage: Progress in Cultural Heritage: Documentation, Preservation, and Protection: 6th International Conference, EuroMed 2016, Nicosia, Cyprus, 31 October–5 November 2016, Proceedings, Part I*; Springer International Publishing: Cham, Switzerland, 2016; pp. 171–183.
41. Bentkowska-Kafel, A.; Denard, H. Processual Scholia: The Importance of Paradata in Heritage Visualization. In *Paradata and Transparency in Virtual Heritage*; Bentkowska-Kafel, A., Denard, H., Baker, D., Eds.; Ashgate: Farnham, UK, 2016; pp. 320–335.
42. London Charter for the Computer-Based Visualization of Cultural Heritage. 2009. Available online: https://londoncharter.org/introduction.html (accessed on 14 March 2025).

infrastructures

Article

Intuitive and Experiential Approaches to Enhance Conceptual Design in Architecture Using Building Information Modeling and Virtual Reality

Balamaheshwaran Renganathan [1], Radhakrishnan Shanthi Priya [1,*], Geetha Ramesh Kumar [1], Jayanthi Thiruvengadam [2] and Ramalingam Senthil [3]

[1] School of Architecture and Interior Design (SAID), SRM Institute of Science and Technology, Kattankulathur, Chennai 603203, India; balamahr@srmist.edu.in (B.R.); geethak@srmist.edu.in (G.R.K.)
[2] Department of Biomedical Engineering, College of Engineering and Technology, SRM Institute of Science and Technology, Kattankulathur, Chennai 603203, India; jayantht@srmist.edu.in
[3] Department of Mechanical Engineering, SRM Institute of Science and Technology, Kattankulathur, Chennai 603203, India; senthilr@srmist.edu.in
* Correspondence: shanthir1@srmist.edu.in

Abstract: The conceptual design phase in architecture requires both intuitive and iterative approaches, which traditional Building Information Modeling (BIM) workflows fail to support properly. BIM provides data-driven decision-making and project coordination but does not offer affective or experiential feedback capabilities. BIM and Virtual Reality (VR) integration offers a promising solution to improve user-focused spatial assessments during initial design phases. The research follows three distinct phases, including a Systematic Literature Review to identify BIM-based conceptual workflow limitations, semi-structured interviews with architects to understand practical challenges and expectations, and the development of a BIM-based framework combining immersive VR for affective and visuospatial evaluation. A testing phase of the proposed framework occurred in the pilot study. The current BIM workflows show significant deficiencies in their ability to support creative flexibility, user engagement, and experiential validation. The BIM-VR framework implemented in the pilot study showed improvements in spatial cognition, emotional engagement, and iterative design decision-making during the conceptual design phase. Early-stage architectural design evaluation becomes more effective through VR integration into BIM workflows because it provides real-time immersive user feedback. The proposed framework helps develop BIM tools that are more intuitive for humans while advancing user-informed design practices in the architecture, engineering, and construction industries.

Keywords: building information modeling; sustainable design; well-being; pre-occupancy evaluation; affective experience; conceptual design; early-stage design; virtual reality; immersive technology

1. Introduction

The conceptual stage of architectural design is a fundamental phase in which many significant design choices are defined, directing the project's future development [1,2]. The design process at this stage requires designers to use an iterative problem-solving approach that combines spatial, functional, aesthetic, and environmental elements. Architectural design thinking represents a complex mental process that uses explicit reasoning and tacit knowledge while depending on intuition, experience, and iterative exploration [3].

The architectural design process requires architects to use diverse representational tools, including sketches alongside physical models and digital simulations, as a medium for ideation, problem-solving, and experimentation. The architectural design process often follows a non-linear, dynamic path that experts classify as a 'wicked problem', since each solution introduces new variables affecting the entire system [4,5]. Designers work across a complex problem domain where various contextual elements affect their choices, and the solution space demands constant improvement of alternative solutions. The evaluation process occurs through multiple rounds based on specified criteria, including functional requirements, aesthetic considerations, environmental factors, and socio-cultural aspects. Design evaluation frameworks structure the process by which the evolving design is checked against project targets and disciplinary standards [6,7].

As a result of this complexity, the AEC industry faces exceptional challenges when evaluating conceptual designs, because it needs methods that unite creativity with performance-based decision systems. The architecture, engineering, and construction (AEC) industry faces exceptional challenges when evaluating conceptual designs because it needs complete assessment methods that unite creativity with performance-based decision systems. The Design industry has long sought robust frameworks to facilitate informed design choices during the early stages, ensuring that spatial quality, sustainability, and user experience are effectively integrated [8–11]. However, traditional evaluation methodologies often lack consistency, making establishing data-driven insights that support iterative refinements challenging. The implementation of computational technology has both optimized design workflows and allowed architects to create novel architectural expressions, which were previously impossible [12,13]. Current digital tools like BIM have significantly transformed the architectural design process, yet their application during the early, conceptual stages remains limited. Traditional BIM workflows, which are documentation-heavy, struggle to accommodate conceptual design's exploratory and dynamic nature. Additionally, these tools often fail to integrate immersive, user-centered feedback that could enhance the evaluation of spatial and functional designs during these stages. This research fills this critical gap by proposing a novel integration of Virtual Reality (VR) and neurofeedback with BIM workflows to provide real-time, experiential design evaluation that includes user interaction and spatial perception. This hybrid approach will allow for a more intuitive and human-centered evaluation of designs, moving beyond the traditional, static assessment methods that dominate current practices.

1.1. History of Digital Design Tools

The current state of the art in digital design has exposed limitations in traditional CAD-based workflows when projects become data-rich and involve many stakeholders. Building Information Modeling (BIM) emerged as a response to these limitations by delivering advanced digital representation through data-driven intelligent processes that transformed architectural workflows [14]. BIM has transformed architectural practice by allowing users to digitally depict the built environment's physical attributes and functional characteristics [15,16]. As an integrated knowledge resource, BIM enables users to make data-driven decisions throughout every phase of the building lifecycle starting with design and ending with decommissioning [17]. BIM offers a unified digital ecosystem that allows different stakeholders, including architects, engineers, contractors, and facility managers to operate in a shared virtual environment jointly [18]. BIM platforms improve project coordination through its clash detection capabilities, performance analysis features, and real-time simulations to support informed decision-making [19]. Generative design and artificial intelligence (AI) technologies have accelerated BIM's development by enabling automatic design optimization alongside AI-based anomaly detection and digital twin real-time integration [20].

The implementation of BIM during initial design stages faces restricted adoption, even though it holds tremendous potential for transformation [21,22]. The detailed documentation-based structure of traditional BIM workflows struggles to accommodate the conceptual design's dynamic and exploratory nature. Many architects find sketch-based tools more suitable for their needs than BIM platforms because they believe the latter restricts their creative process during the initial stages of ideation [23–25]. The slow learning curve and technological complexities along with implementation resistance from small firms and the computational overhead, have limited BIM adoption [26,27]. Open-source BIM platforms together with WebBIM solutions, are entering the market, which makes BIM more adaptable [28,29]. BIM will only achieve its full acceptance when it penetrates the AEC workflows from the conceptual phase onwards.

1.2. Conceptual BIM and Standards

The governance and management of BIM and digital assets are increasingly standardized through frameworks such as ISO 19650 [30–32], supported by national and industry-specific guidelines including the National BIM Standard (NBIMS) [33,34], UK BIM Framework [35], and initiatives by BuildingSMART [36,37]. These frameworks facilitate interoperability, structured data management, and collaboration within the AEC industry. Recent literature underscores BIM's synergistic integration with architectural workflows, highlighting its role in improving multidisciplinary communication, reducing errors, and enhancing design efficiency. For instance, integration of BIM with the RIBA Plan of Work significantly streamlines design coordination and onsite communication, while BIM-based frameworks enriched with stakeholder collaboration, automated conceptual modeling, and Post Occupancy Evaluation (POE) data promote informed early-stage decisions and reduce project rework [23]. Despite these advancements, the application of BIM in conceptual stages remains limited due to its static, documentation-centric nature [38]. The obstacles to using BIM in the conceptual design stage must be solved to ensure early design integration and maximize architectural innovation and project efficiency.

1.3. Neuroarchitecture, BIM and VR

Neurophenomenological studies focus on understanding human spatial perception by measuring the brain's response to spatial environments [39]. These studies use advanced technologies, such as eye-tracking and motion tracking, to capture real-time physiological feedback that reflects a user's interaction with a space [40]. This research aims to incorporate this neurofeedback into BIM workflows to provide richer insights into how spaces are perceived cognitively. In traditional BIM workflows, spatial evaluation is limited to visual and functional analysis, but integrating neurofeedback makes the assessment more dynamic and user-centered. This approach will enhance architectural design by providing deeper insights into how users experience spaces, beyond what traditional tools can capture.

One promising approach to address these early-stage shortcomings is integrating Neuroarchitecture methods with BIM and Virtual Reality (VR) [41]. BIM+VR integration can enhance visuospatial design evaluation by combining BIM's data-driven decision-making with immersive experiential feedback [41,42]. Current BIM platforms provide streamlined workflows by enabling real-time collaboration [43,44], clash detection [45,46], and performance analysis [47,48], yet lack a means to understand intuitive spatial assessment capabilities. The limitation prevents BIM from adequately supporting conceptual design evaluation. VR addresses this gap by allowing architects and other stakeholders to experience proposed designs in an interactive space, where they can instantly adjust spatial arrangements, lighting, and materials to test design ideas [49]. The combination of BIM and VR systems enables non-specialist users to participate in the design process through active feedback on design variations at the beginning of projects [50,51]. Advanced VR technologies with eye-tracking and motion tracking capabilities enable methodologies like neurophenomenological studies that provide richer understanding of human spatial perception and user experience [40,52]. However, no BIM workflow incorporates such neurofeedback data into the conceptual stage of the design process—a gap this study aims to fill. Despite computational demands and software interoperability challenges, BIM-VR and neuroarchitecture integration represent a transformative shift in architectural workflows, bridging analytical precision with human-centered spatial evaluation for more informed and effective design outcomes.

1.4. Research Rationale

Current BIM workflows demonstrate essential limitations during conceptual design evaluation as previously explained. The current BIM workflows lack precise mechanisms for modeling interactions and do not provide experiential or affective feedback and fail to establish standardized methods for assessing spatial perception in virtual environments [53,54]. The current evaluation challenges demand an adaptable human-focused real-time assessment method to improve BIM-based conceptual design processes. To address this gap, the central research question guiding this study is: How can BIM workflows be optimized to enable affective and visuospatial evaluation during early-stage architectural design through immersive and user-centric feedback mechanisms?

In response to this question, the primary purpose of this research is to develop an optimized BIM framework to evaluate affective and visuospatial design aspects in early architectural processes for AEC professionals. The research has three specific objectives to reach its goal:

1. To systematically review current BIM workflows for early-stage design evaluation (via an SLR) and identify limitations and research gaps.
2. To conduct semi-structured interviews with practicing architects, gathering professional perspectives on current conceptual BIM design challenges and opportunities.
3. To develop a BIM-based framework for visuospatial conceptual design evaluation incorporating immersive, experiential feedback.

The structure followed in this research is shown Figure 1. To achieve the key research question and fulfill these objectives, a Systematic Literature Review (SLR) was conducted following the PRISMA framework [55,56], which highlighted a significant gap in current BIM workflows during the conceptual design phase, particularly in evaluating affective and visuospatial qualities. To verify and enrich these findings, semi-structured interviews [57–60] were conducted with practicing architects—specifically those in lead design roles responsible for conceptual ideation. This method was chosen for its methodological robustness, offering flexibility for participants to express nuanced perspectives and structure for researchers to probe targeted areas of concern. Insights from the SLR and expert interviews informed the development of a BIM-based evaluation framework, which was subsequently tested through a pilot study to assess its effectiveness in improving design decision-making during early-stage architectural processes.

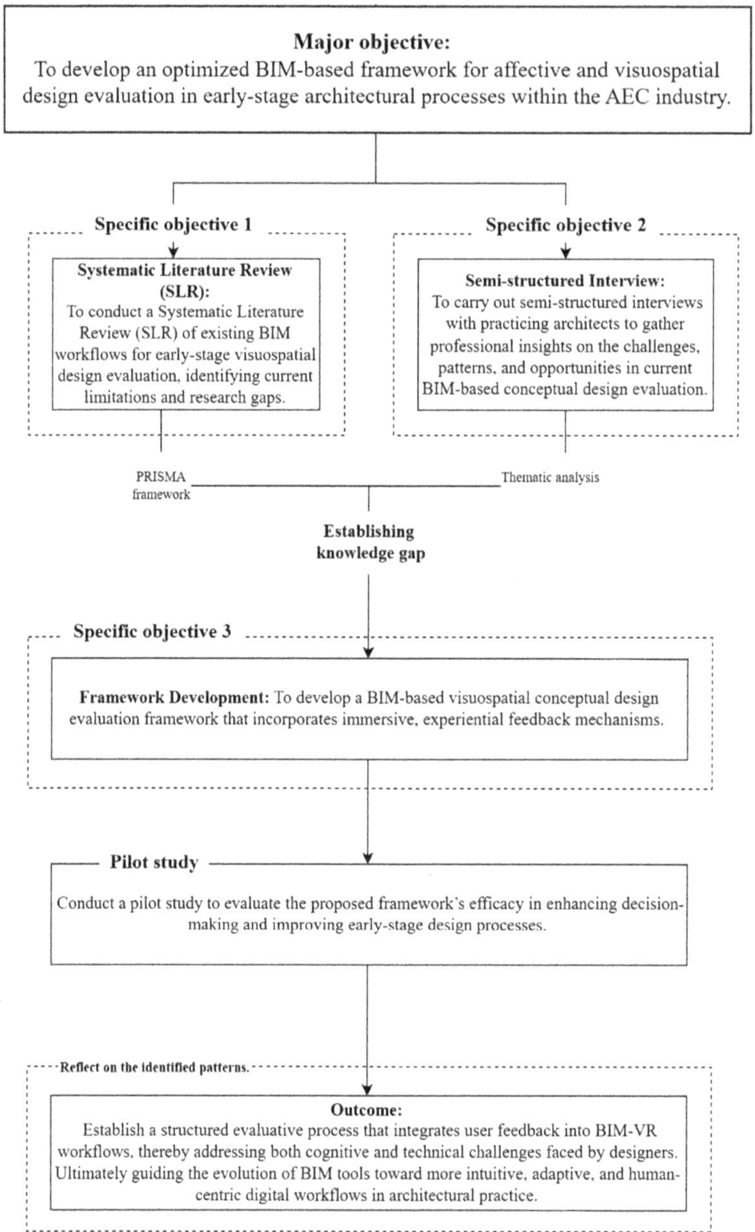

Figure 1. Research structure adopted in this study. Source-Authors.

This research is novel in connecting architecture and neuroscience through technology. Unlike previous BIM studies, we integrate real-time user neurofeedback and immersive VR into the conceptual design workflow—a first-of-its-kind approach to informing design decisions with human experiential data. The research aims to create a standardized evaluation method incorporating user feedback within BIM-VR workflows to solve designer challenges related to cognition and technology. The proposed approach will direct BIM tool development toward creating more user-friendly and adaptable digital workflows for architectural domains.

2. Materials and Methods

This study adopts the Design Science Research (DSR) [61,62] paradigm, which is well-suited for developing and evaluating iterative artifacts, products, and processes. This research follows a pragmatic approach, ensuring the development of a BIM-based framework that is both theoretically grounded and empirically validated. The study employs a structured build-and-evaluate methodology to create a validated BIM-based framework for affective and visuospatial evaluation in the early stages of architectural design. This approach ensures continuous refinement and assessment, ultimately leading to a framework that enhances BIM workflows for design professionals.

2.1. Systematic Literature Review

The SLR presented in this study was conducted following the PRISMA guidelines [55,56], using Scopus as the primary database because of its comprehensive indexing of high-impact architecture, engineering, and construction (AEC) research. The review period was set from 2005 to 2025, which was strategically chosen to include the evolution of BIM since its widespread adoption in the mid-2000s. A comprehensive keyword strategy was developed to capture a wide spectrum of relevant literature, including terms related to BIM, conceptual design evaluation, and early-stage architectural workflows. The final search string applied was ("BIM" OR "Building Information Modelling" OR "Building Information Modeling") AND ("Conceptual Phase" OR "Preliminary Phase" OR "Early Stage") AND ("Architectural Design" OR "Design Development"), resulting in an initial retrieval of 281 articles from Scopus.

The subsequent structured screening process involved multiple refinement stages. The title and keyword screening reduced the initial dataset from 281 to 273 articles. Further abstract screening narrowed the dataset down to 80 potentially relevant studies. In the subsequent phase, an in-depth, full-text assessment was conducted to critically evaluate methodological rigor, relevance to BIM-based design evaluation, and contributions specific to early-stage architectural workflows. Studies predominantly focused on construction management, energy analysis, lifecycle assessment, and non-architectural applications or those lacking a peer-review process were excluded from the review. Finally, 49 articles were finalized for inclusion.

The SLR focused on content analysis rather than META data, intentionally narrowing the scope to identify BIM workflows that support affective conceptual design evaluation, particularly through AI and VR integration. This targeted approach allowed for identifying relevant workflows and methodologies, aligning directly with the study's research objectives, rather than conducting a broad meta-analysis. This carefully selected body of literature provided essential insights into existing BIM workflow methodologies, highlighting prevalent challenges and identifying opportunities for workflow optimization. In particular, we found recurring themes of limited creative freedom in early BIM use, insufficient user feedback mechanisms, and poor integration of human factors, confirming that current approaches do not fully support conceptual design needs. The synthesized findings offer a comprehensive overview of current practices in BIM-based visuospatial design evaluation, underscoring the critical need for enhanced workflows to improve efficiency, accuracy, and informed decision-making during the conceptual phases of architectural design. Figure 2 shows the structure of the SLR study as per the PRISMA framework.

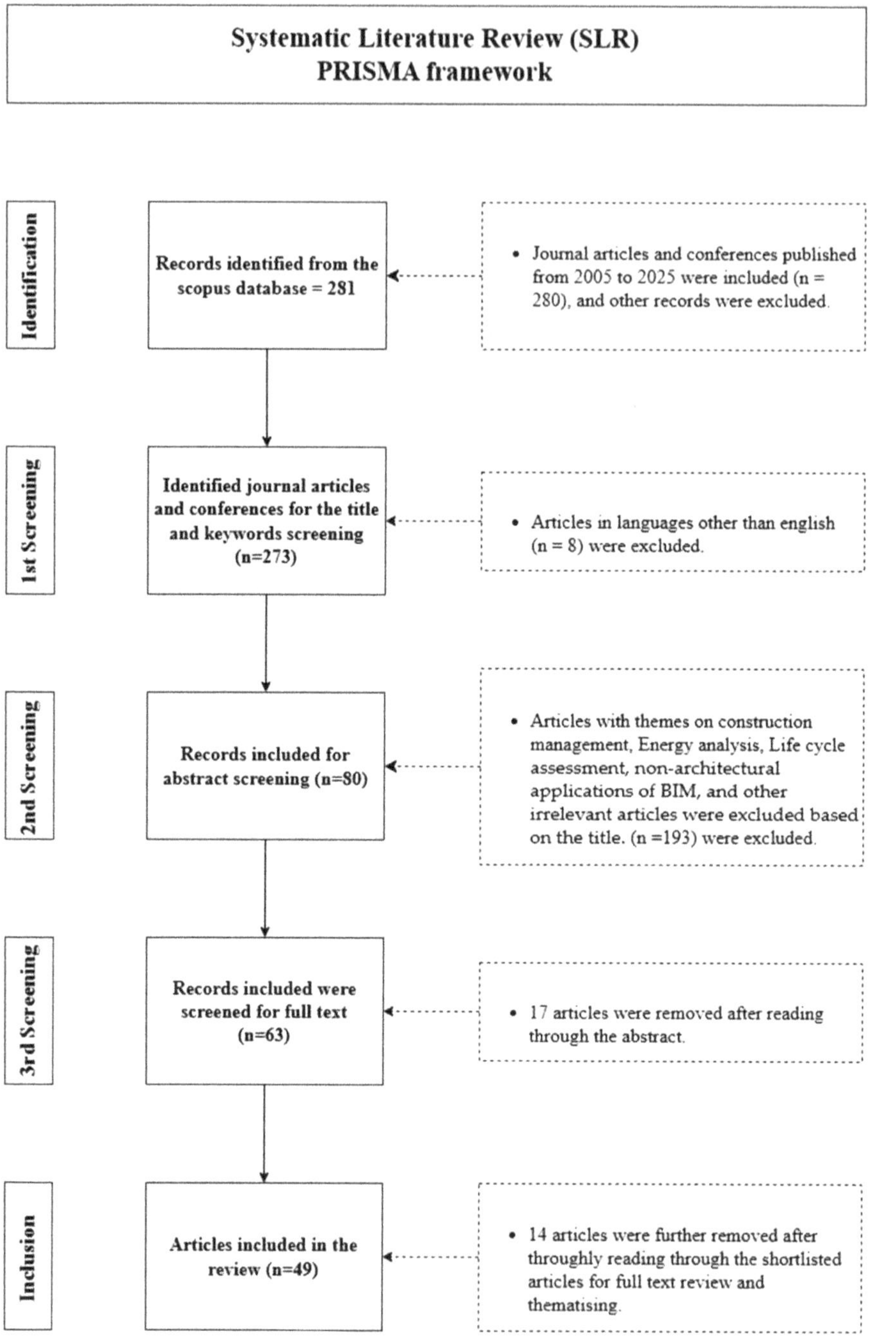

Figure 2. SLR of the study as per the PRISMA framework. Source: Authors.

2.2. Semi-Structured Interviews

The semi-structured interviews of practicing architects were conducted to investigate the practical application of BIM in conceptual design evaluation and to verify the concurrence of the knowledge gap identified from the SLR. The main goal was to reveal the most frequent difficulties, practical approaches, and possible ways to improve the workflow at the initial architectural design stage. The adoption of a qualitative approach was selected because the implementation of BIM in practice occurs within specific contexts, which cannot be captured through quantitative methods [63]. A semi-structured interview [59,64] was determined as the most suitable methodology for this study. Kvale's (1996) Seven Stages of Interviewing was chosen as the main framework for guiding the semi-structured interview [60]. The seven stages provided an extensive yet adaptable structure that perfectly matched the research goals of individual insights with broader workflow patterns.

Multiple analytical frameworks were assessed to determine which one would be most suitable. Grounded Theory [65] was examined for its potential to develop new theory from data, yet its inductive approach and theory-building focus did not fit the study's goal of workflow assessment. Gioia's methodology [66] provides systematic development of conceptually rigorous frameworks, yet focuses on organizational studies and abstract theorization, making it less relevant to BIM workflow evaluation in practical applications. Patton's [63] system for qualitative interviewing provides thorough classification of interview methods yet it does not contain a systematic analysis method. Spradley's [67] ethnographic approach proved unsuitable for this research because it emphasized cultural immersion instead of process evaluation. Multiple participants in workflow-oriented inquiries benefit from pattern recognition through the long interview approach developed by McCracken [68], yet this method is most effective for obtaining deeply personal narratives. Kvale's model provides a systematic yet flexible approach to qualitative data analysis that balances scientific standards with interpretive flexibility (Figure 3).

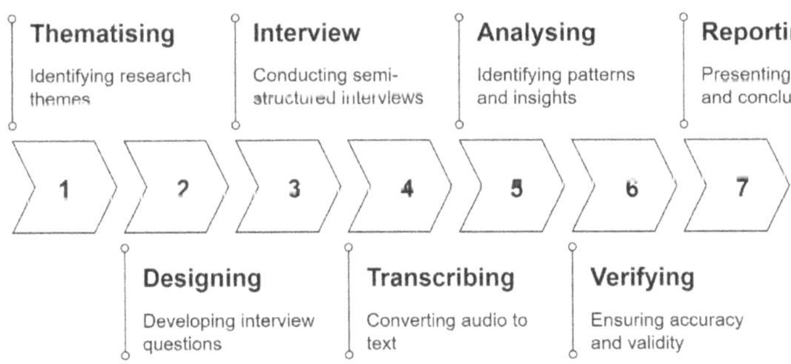

Figure 3. Kvale's (1996) [60] Seven stages of interviewing. Source: Authors.

In this study, 10 semi-structured interviews were conducted with participants (A1 to A10), each representing a diverse range of experience levels and expertise within the architectural field. The sample size of 10 was intentionally chosen, as it aligns with well-established practices in qualitative research, where a sample size of 8 to 12 participants is considered appropriate for exploratory or pilot studies, particularly when engaging expert professionals [69]. The primary objective of this study was to explore initial insights into the BIM framework, with a particular focus on integrating emerging technologies such as AI and VR into the architectural design process. Limiting the sample to 10 participants ensured that the data collected was rich and deep while avoiding unnecessary extension of the data collection process.

The participants were selected using purposive sampling [70], a non-random method that includes individuals with specific knowledge or experience relevant to the research. In this case, the interviewees were chosen based on their BIM background and active involvement in the early-stage design process. This purposive sampling strategy enabled the collection of targeted, expert insights, making the findings more relevant to the research objectives. All the industry experts are from Chennai, India, resulting in a homogeneous sample. The interviewees had an average experience of 9.6 years in the industry, with extensive expertise in BIM-based workflows, and they were active decision-makers in their respective firms' design processes. This wealth of experience ensured that the participants were well-positioned to provide informed and insightful responses about BIM workflows and integrating emerging technologies into conceptual design. A total of 11 questions (Table 1) were framed as a part of the interview process. Thematic saturation [71,72], a critical concept in qualitative research, was achieved when no new significant themes emerged after interviewing 8 participants. The 10th interview fully reached thematic saturation, as no new themes emerged, confirming that the data had sufficiently covered all relevant aspects of the research.

Table 1. Semi-structured interview questions.

QC	Questions
Q1	Describe your typical workflow using BIM during conceptual design.
Q2	How is BIM used differently in conceptual design compared to later stages?
Q3	Which BIM tools/software do you use for conceptual design, and why?
Q4	What are the biggest challenges you face when evaluating conceptual design options using BIM?
Q5	How do you handle or integrate feedback from clients/stakeholders during the conceptual design phase?
Q6	Do BIM workflows constrain your design creativity during conceptual design? If so, how do you overcome this?
Q7	Have you observed any best practices or strategies to improve BIM workflows for conceptual design?
Q8	How does your firm handle design iterations and decision-making using BIM?
Q9	What improvements or innovations could enhance BIM workflows for conceptual design evaluation?
Q10	Have you tried integrating AI, VR, or generative design into your BIM workflows? What has been your experience?
Q11	What would your ideal BIM-based workflow for conceptual design look like?

Note: QC—Question Code.

To ensure the reliability and validity of the findings, reflexivity [73] measures were incorporated throughout the research process. The first author regularly engaged in self-reflection to minimize any potential biases arising from their background in architecture and BIM. Peer debriefing sessions were also conducted to ensure further the accuracy and integrity of the coding and thematic analysis process. In addition, memos were used to document the researcher's reflections during the coding process, which helped maintain transparency and rigor. The complete list of codes that emerged from the interviews is presented in Supplementary Materials, providing a comprehensive overview of the themes and subthemes identified during the analysis. This methodologically rigorous approach ensured that the themes were grounded in the participants' responses and reflective of the central issues related to BIM workflows, hybrid design processes, and the integration of emerging technologies. By the end of the 10 interviews, the data had reached thematic saturation, providing a solid foundation for the conclusions drawn from the study. These conclusions informed the development of a proposed BIM framework, highlighting key factors influencing architectural design processes, including creativity, flexibility, and integrating AI and immersive technologies. The integration of Kvale's Seven-Stage Framework [60] and Braun and Clarke's [74] thematic analysis framework ensured a methodologically rigorous approach to the coding and analysis process, providing a solid foundation for the study's conclusions.

2.3. Framework Development

The SLR and semi-structured interviews produced essential insights that led to identifying new themes and keywords that guided the creation of a framework to optimize

BIM workflows in early-stage architectural design. The research methods exposed new important areas including WebBIM, Desktop BIM and BIM–VR integration, which would improve visuo-spatial evaluation in the AEC industry. The analysis of academic sources and professional opinions demonstrated the existing workflow problems and scattered implementation of BIM visualization tools in the conceptual phase. Advanced visualization technologies and platform-specific functionalities should be studied to determine their strategic potential for supporting more intuitive, flexible, and immersive design processes. The process followed for developing this framework has precedence from the literature [75–77]. The three-stage conceptual framework emerged to address these workflow gaps by streamlining processes and using visualization for spatial clarity before enhancing early-stage architectural and interior design decision-making.

2.4. Pilot Study Design

The proposed conceptual framework underwent feasibility assessment through a pilot study [78,79], which also tested the robustness of the experimental workflow. A three-dimensional immersive virtual reality (IVR) scenario was developed to replicate a real-world interior gallery environment, incorporating varied surface articulation as experimental stimuli. The main goal during this phase was to confirm environmental simulations and study how participants reacted emotionally and cognitively, particularly regarding their attention levels, engagement, relaxation states, and cognitive load. To achieve this, the study incorporated Electroencephalography (EEG) [80–82]—a non-invasive neurophysiological method that measures electrical activity in the brain through scalp electrodes. EEG captures brainwave patterns across multiple frequency bands, including Beta (attention and active thinking), Theta (relaxation), Gamma (cognitive load), and Alpha (emotional states). When integrated with VR environments, EEG enables real-time, moment-to-moment tracking of users' affective and cognitive responses to spatial stimuli.

This capability makes EEG particularly powerful for architectural design evaluation [83,84], as it provides quantitative, human-centered feedback on how users perceive and interact with spaces beyond what is visually apparent. By mapping brainwave data to spatial features, designers can assess whether a space induces cognitive clarity, mental fatigue, emotional comfort, or attentional engagement. The research design includes descriptions of experimental conditions, together with participant selection criteria, psychological measurement tools, and data analysis methods, all aimed at demonstrating how EEG-informed insights can guide iterative, evidence-based design decisions during the conceptual phase.

2.4.1. Interior Gallery Environment, VR HMD, and EEG Device

The virtual environment consisted of six gallery spaces measuring 3 × 3 × 3 m each. The spaces followed a linear pattern to maintain a constant visitor path, as shown in Figure 4a–c. The gallery being presented in grayscale clay model form reduced visual biases by eliminating color as an active variable in brain activity. The gallery artifacts followed Christopher Alexander's wholeness theory by showing different levels of visual complexity. Figure 4a–c shows the elevation, plan, and 3D sections, respectively.

The HP Reverb G2 Omnicept Edition head-mounted display (HMD) provided participants with environmental experience through its built-in sensing capabilities like muscle movement, gaze, pupil size, and pulse rate. The Emotiv 5 channel EEG device with five channels recorded real-time neural responses from brain activity while participants experienced the VR environment. The image in Figure 5a,b shows both the gallery spatial arrangement and participant navigation route.

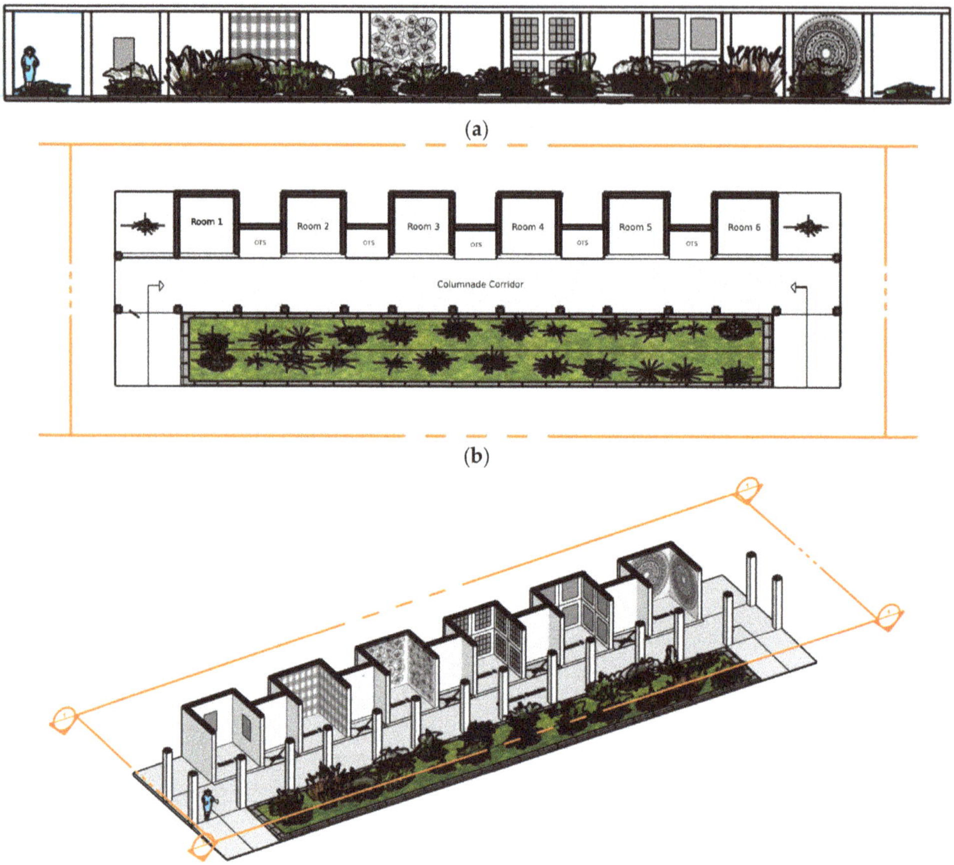

(**a**)

(**b**)

(**c**)

Figure 4. (**a**). Elevation of the experimental gallery model showing various patterns on the wall surfaces. Source-Authors. (**b**). Plan of the experimental gallery model. Source-Authors. (**c**). 3D section of the experimental gallery model. Source-Authors.

(**a**) (**b**)

Figure 5. (**a**). Emotiv 5 Channel EEG. (**b**). HP Reverb G2 Omnicept HMD. Source: Authors.

The participants moved through all gallery spaces one after another as EEG data acquisition occurred without interruption. The research team used an Emotiv 5-channel EEG headset to monitor brain wave data from participants who viewed six virtual galleries through a VR Head-Mounted Display (VR-HMD). The EEG recording system monitored five electrode locations Anterior F3, T7, Pz, T8, and Anterior F4, which follow the International 10–20 System to detect emotional and cognitive and sensory neural responses. The participants received comfortable seating while researchers calibrated the EEG device to obtain precise data. To ensure safety and preserve EEG data integrity, participants remained seated while exploring the VR scenes, avoiding movement that could cause collisions or introduce motion-related artifacts. Before the experimental trials, a 2–3 min neutral VR environment was provided to help participants acclimate to the interface and practice seated navigation with a joystick. This phase reduced anxiety, improved comfort, and stabilized physiological responses, minimizing EEG fluctuations from disorientation or unfamiliarity.

2.4.2. Experimental Protocol

The experimental procedures started after each participant gave their informed consent at the beginning of the study. The HMD fitting process was followed by a short training session, which led participants through a test scenario to learn VR navigation skills. The participants received their introduction to the experimental gallery setup after their acclimation period. Figure 6 shows the participant experience journey during the experiment. The experimental gallery spaces were presented to participants in an incomplete counterbalanced randomization sequence to reduce sequence and repetition effects. After the initial VR setup, a sample VR environment was shown to make the participants comfortable with the VR environment. Participants carried out a sustained attention task throughout their VR exposure period to sustain their mental focus. The six virtual environments appeared sequentially to each participant while EEG recordings continued without interruption. After the VR experience ended, researchers performed an unstructured interview to gather participants' thoughts about their experience and followed by their design suggestions. The survey used structured open-ended interview questions to obtain the participants' feedback and to validate the framework. Each participant took an average of around 15 min to complete the entire pilot study.

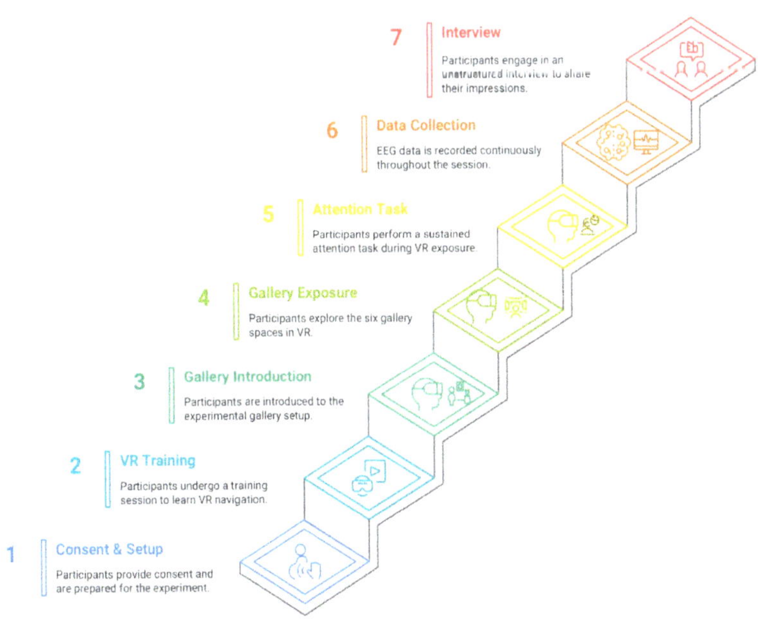

Figure 6. Participant experience journey during the experiment. Source: Authors.

2.4.3. Participants, Psychological Measures, and Data Analysis

A total of 10 participants (8 male, 2 females; average age = 21.45 years) were invited for the study. In exploratory VR-EEG studies, a sample size of 10 participants is both methodologically and practically justified [85]. Pilot experiments are designed to test feasibility, refine protocols, and observe initial trends rather than to generalize findings to larger populations [78]. This is a common and accepted practice in cognitive neuroscience [86] and HCI [87,88], particularly in high-tech setups involving complex equipment like EEG and VR headsets. Each participant in such studies provides a rich dataset across multiple conditions, making small-N designs efficient for within-subject comparisons [89,90]. Additionally, for example, the study by Sparke et al. [91] has explicitly stated that power calculations may be deferred in early-stage research, and samples of 10 are consistent with other proof-of-concept EEG trials. In usability and VR experience testing, sample sizes of 5–12 users are often sufficient to uncover dominant experiential patterns, technical issues, and early cognitive or affective responses to spatial design.

Moreover, architecture students are a suitable convenience sample for VR-based spatial cognition studies due to their domain familiarity and ability to interpret and engage with design stimuli meaningfully. Within-subjects designs, where each participant experiences all virtual conditions, increase statistical sensitivity and reduce inter-subject variability, enhancing the robustness of the results despite the small sample. Literature in neuroarchitecture and immersive design has shown that even with $N \approx 10$, significant EEG-based differences and behavioral trends can be detected, especially when using high-density EEG and repeated-measures designs [92]. The goal of such pilot studies is not hypothesis confirmation but demonstration of measurable effects—such as neural signatures correlating with architectural features or shifts in emotion-linked EEG bands. These outcomes validate the experimental approach and serve as a foundation for scaling the study.

Eligibility criteria included (1) age between 18 and 24 years, (2) Architecture/Design students currently enrollment in a university program, (3) Indian nationality to ensure cultural consistency, (4) corrected vision using contact lenses (for either myopia or hyperopia), and (5) abstention from any substances that could alter cognitive performance (e.g., caffeine) within 24 h before the session. All participants completed the study in full, and participants' consent was obtained through a format at the beginning of the experiment. Figure 7 shows the experimental setup with participants receiving basic training in the VR environment. The data collected from the device are exported to Excel sheets for further data analysis across all the participants. A one-factor repeated-measures ANOVA was performed on the EEG band-power values with room type as the within-subjects factor to examine whether EEG responses differed significantly across room types. To complement the inferential statistics, EEG band-power data were visualized across the six room conditions. A line plot was created to show trends in power distribution for each frequency band in the results Section 3. Based on the results, the proposed BIM-VR-EEG framework will be scaled for integration into design studios and firms, enabling real-time affective feedback during early-stage architectural decision-making through immersive, data-driven collaboration. After the data collection, the results were shared with the student participants, and their impressions and feedback were recorded through the following questions to understand the framework's efficacy. Q1: Can you describe your overall experience of exploring the six virtual rooms? Which room stood out to you the most and why? Q2: After your feedback was incorporated, did you notice any meaningful changes or improvements in the refined design version? Describe what stood out. To what extent did the re-fined design feel more aligned with your expectations or preferences compared to the initial version?

Figure 7. Participant undergoing basic VR training before the experiment. Source: Authors.

3. Results and Discussion

3.1. Reviewing Results from SLR

The research findings from the systematic literature review are presented in this section through five thematic areas which resulted from the analysis are shown in Figure 8: (1) BIM, conceptual design and design processes, (2) BIM-integrated generative and computational design, (3) Human-centered and behavioral studies with VR in BIM, (4) Interoperability, data exchange, and workflow integration, and (5) Spatial reasoning and case studies addressing early-stage design. The analysis explores each theme through methodological approaches and essential findings together with recognized knowledge gaps to establish a complete understanding of current research and future directions for BIM-based architectural design practices.

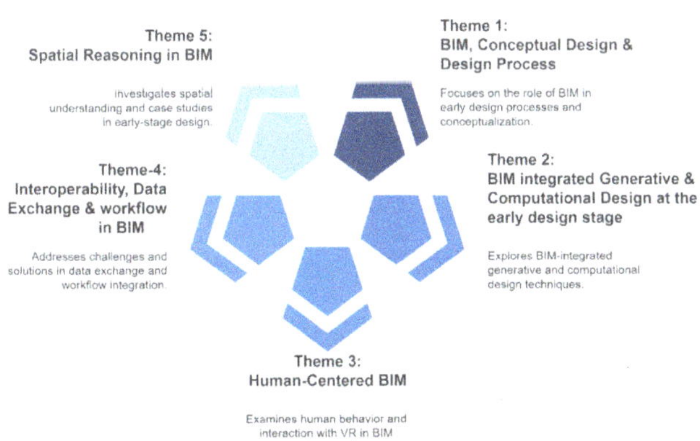

Figure 8. Themes identified through SLR: BIM workflows focusing on visuospatial design evaluation of early-stage design. Source: Authors.

3.1.1. Theme-1: Bim, Conceptual Design and Design Process

The conceptual design phase is essential for architectural design since it establishes spatial planning, functional requirements, and aesthetic considerations. Throughout history, this stage has relied on manual sketching, instinctive choices, and theoretical evaluation. BIM has revolutionized design processes through its computational capabilities, which improve decision-making efficiency and data-driven approaches. Implementing BIM for conceptual design faces ongoing debate because practitioners raise issues about its complicated nature, inflexibility, and unsuitability for initial architectural design phases. During conceptual design development, architectural projects receive their fundamental design elements, which include spatial relationships together with massing strategies and initial design intentions. Table 2 shows the relevant literature curated for theme-1.

Table 2. Key journal articles identified under theme-1.

Ref.	Methodology	Key Findings	Research Gaps
[93]	Conceptual Framework	Introduces the Polymorphic Diagram, a conceptual BIM environment facilitating spatial thinking.	Limited validation through practical architectural projects.
[94]	Ontology-Based Approach	Develops a lightweight BIM knowledge representation to support reasoning in early design.	Lack of real-world application and industry validation.
[95]	Pilot study: Remote user study approach, Qualitative study and Analysis.	Sketching is crucial in conceptual design. The study identifies common patterns in architectural sketching, aiding the development of sketch recognition systems.	Challenges in enabling computer recognition of varied sketching techniques. Further refinement is needed in automated CAAD integration.
[96]	Case-Based Reasoning (CBR)	Developing a CBR system for early-stage conceptual design, integrating precedent retrieval, parametric modeling, and ontology classification for energy performance.	Limited evaluation beyond commercial buildings, requiring further validation across diverse design typologies.
[97]	Case Study & Diagrammatic Reasoning	Explores constraint-based parametric diagrams in BIM, investigating their role in conceptual design by mediating between abstract reasoning and physical form.	Needs more empirical testing on large-scale architectural projects to validate effectiveness.
[98]	Qualitative Study	Explores the role of BIM in conceptual design, focusing on its effectiveness in assisting architects with decision-making related to functionality, aesthetics, and environmental relationships.	Further empirical studies required to validate findings beyond student case studies and into professional practice.
[99]	Case Study	Examines the role of BIM in early-stage architectural design, highlighting its effectiveness in performance analysis and decision-making rather than as a direct design creation tool.	Limited exploration of BIM's potential as a generative design tool rather than just a support system.
[100]	Decision Support System (DSS)	Introduces a DSS tool integrated with BIM for high-rise housing envelope design, enabling simulation of zoning regulations and real-time analysis.	Lack of real-time integration with evolving design parameters and user adaptability challenges.
[101]	Generative Modeling & Simulation	Proposes a BIM-integrated design dashboard for early-stage analysis, offering real-time structural, cost, and performance data visualization.	Challenges in implementing interactive, user-friendly interfaces for architects and real-time decision-making in evolving projects.
[102]	Case Study	Explores the integration of emerging digital analytical tools in the conceptual design phase.	Lack of a clear framework for integrating digital tools seamlessly in conceptual design.
[103]	Qualitative Case Study	Explores BIM as a design tool in the conceptual phase, analyzing student BIM-based design cases to identify knowledge and skill constraints.	Need for structured BIM training frameworks to improve conceptual design and performance analysis capabilities.
[104]	Literature Review & Secondary Data Analysis	Examines BIM's role in smart cities, its integration into the construction process, and multidisciplinary collaboration.	Unclear transition guidelines for ensuring BIM usability beyond conceptual design into construction phases.

Table 2. *Cont.*

Ref.	Methodology	Key Findings	Research Gaps
[105]	TRIZ & Function Modeling	Focuses on conceptual design automation in AEC projects, integrating modern construction software with idea generation techniques. Proposes a TRIZ-based computer-aided BIM approach for innovative conceptual solutions.	Needs further validation in diverse architectural contexts to assess practical usability.
[106]	Workshops & Industry Case Study	Examine BIM adoption challenges in the AEC/FM industry, identify key barriers, and propose a contingency-based BIM decision support framework.	Framework requires refinement through extensive industry applications to improve real-world implementation.

The literature reviewed under theme-1 reveals a consistent effort to extend BIM's capacity beyond its conventional role in documentation and coordination, toward supporting the intuitive and iterative processes of early-stage design. Several scholars advocate for embedding design cognition and reasoning within BIM workflows to align them more closely with how architects think and operate. For instance, Hamadah [93] explores diagrammatic reasoning as a means to communicate spatial thinking, while Kim and Grobler [94] develop ontological frameworks that support iterative conceptualization, acknowledging the non-linear nature of architectural ideation. Ziegler et al. [95] emphasize the transitional value of analog sketching in bridging traditional and computational workflows, suggesting that hybrid approaches can ease the conceptual use of digital tools. Additional efforts to address this conceptual gap are evident in reasoning-based approaches. Cavieres et al. [96] propose a case-based reasoning (CBR) system that links precedent retrieval to parametric modeling, enabling designers to access typological knowledge while maintaining semantic logic. Likewise, Al-Assaf and Clayton [97] argue that embedding parametric elements within BIM can significantly improve the efficiency and adaptability of early-stage design workflows. Collectively, these studies point to an emerging consensus: that making BIM cognitively compatible with design thinking requires new frameworks that prioritize flexibility, intuition, and knowledge reusability.

Despite these progressive strides, the literature also exposes deep-seated limitations in BIM's application to conceptual design. While some researchers praise BIM for enabling informed decision-making, others warn that its demand for detail and rigid data structures can hinder creative exploration. Ömer Halil Çavuşoğlu and Gülen Çağdaş [98] acknowledge BIM's capabilities in supporting environmental, functional, and structural evaluations at early stages. However, BIM's strength in analytical precision is often undermined by its inability to accommodate the exploratory and speculative nature of ideation. Studies by Çayaoğlu [99] and Lobos and Donath [100] affirm BIM's usefulness in early massing and feasibility analyses, yet Wu and Shin [101] argue that BIM still lacks robust decision-support tools for conceptual spatial reasoning. The most critical insight comes from Abdelhameed's series of studies. His 2016 work [102] emphasizes that early-stage design requires flexible analytical and generative approaches, which many BIM tools do not currently support. In his 2018 research [103], he introduces adaptive modeling workflows allowing iterative transitions from concept to documentation. His 2021 study [104], however, reiterates BIM's persistent parametric rigidity, which restricts open-ended design processes. These findings clearly illustrate the disconnect between the linear logic of BIM platforms and design ideation's non-linear, iterative nature—a gap that this research seeks to bridge.

Recent developments provide a glimpse of a more adaptable BIM future. Renev et al. [105] show how AI-driven automation can enhance conceptual productivity, while London et al. [106] integrate knowledge-based systems into BIM, allowing for greater design flexibility without compromising data reliability. These advancements suggest that BIM's next evolution lies in merging artificial intelligence, generative design, and user-informed customization, positioning the platform as a responsive partner in early-stage design rather than a rigid documentation tool. Implication for this research: The reviewed literature shows that while efforts are underway to enrich BIM with cognitive and generative capabilities, there remains a critical absence of user-centric experiential feedback within these workflows.

3.1.2. Theme-2: BIM Generative and Computational Design

The literature reviewed under theme-2 as shown in Table 3 highlights the growing convergence between Building Information Modeling (BIM) and generative design methodologies, illustrating a significant shift in how early-stage architectural and urban design workflows are conceptualized and executed. A key observation across multiple studies is that parametric and computational design approaches provide structured yet adaptable digital frameworks that enable iterative development and performance-driven exploration [107]. As noted by Park [108], the integration of BIM in early design stages offers a structured environment that supports improved decision-making. This is further supported by studies indicating that computational automation enhances design continuity while filling key support gaps in AEC operations [109].

Table 3. Key journal articles identified under theme-2.

Ref.	Methodology	Key Findings	Research Gaps
[107]	Parametric Design & BIM Integration	Explores parametric design for conceptual design decision-making, integrating Grasshopper, Revit, and 3D printing for digital fabrication.	Needs validation on large-scale projects beyond controlled case studies.
[108]	Experimental Design Studio	Examines early-stage BIM adoption in the design process through an experimental studio setup, identifying key challenges.	Limited real-world application beyond academic settings.
[109]	Survey & Conceptual Framework	Investigates the integration of BIM and generative design for conceptual design innovation, proposing a virtual generative workspace.	Lacks real-world implementation of the proposed generative BIM workspace.
[110]	Computational Design Study	Proposes a BIM-integrated generative design environment to improve conceptual design iteration and automation.	Requires empirical testing and software validation in real projects.
[111]	Generative BIM (G-BIM) Framework	Develops a G-BIM framework linking parametric tools with BIM applications for better information transition.	Limited exploration of industry-wide applicability and user adaptability.
[112]	Prototype Development & Focus Group Testing	Develops and tests a G-BIM framework to enhance process integration in construction projects at the conceptual design stage.	Further testing in real-life design contexts is needed to assess effectiveness.
[113]	BIM & Evolutionary Computing	Integrates genetic algorithms with BIM to optimize conceptual design, creating a Revit plug-in for interactive design evaluation.	Needs industry-scale validation to determine feasibility and scalability.
[114]	Generative & Visual Programming	Explores automation of urban conceptual design using generative programming and AI-based evaluation.	Further AI-driven enhancements are needed to reduce human oversight in decision-making.

This body of work emerges from a dual narrative: on one hand, computational design increases design productivity and control; on the other, it reveals persistent fragmentation in process continuity and platform interoperability. For instance, generative BIM (G-BIM) frameworks, as discussed by [110,111], aim to embed generative processes into BIM environments, creating virtual workspaces that enhance collaboration, improve design iteration, and facilitate seamless information exchange. These frameworks demonstrate that BIM, when embedded with generative logic, can evolve from a documentation tool into an intelligent design partner, enabling designers to explore various options early in the process.

Notably, prototype developments reviewed in the literature have been successful in mitigating process fragmentation, improving design cognition, and strengthening early-stage decision-making processes [112]. These efforts reflect the architectural discipline's broader ambition to make BIM a repository of information and a computationally intelligent system capable of driving design synthesis. As described by [113], the incorporation of evolutionary computing into BIM workflows allows for performance metric-based optimization. These adaptive workflows enable designers to iterate through increasingly refined solutions, bringing them closer to their conceptual and environmental goals through data-informed feedback.

Another promising area is the application of automated generative design in urban and residential planning contexts. According to [114], these approaches use visual and generative programming to automate zoning, environmental compliance, and space optimization. Such automation increases efficiency but retains the necessary human oversight, reinforcing the notion that computational intelligence can enhance—not replace—creative decision-making.

What this means for the current study is that while generative BIM frameworks show potential in bridging gaps between data-rich modeling and exploratory design, they still fall short in capturing the lived, emotional, and cognitive dimensions of spatial experience. The reviewed literature does not sufficiently address how these automated systems impact user perception, comfort, or experiential feedback in early-stage environments. This is the critical gap this research intends to fill.

3.1.3. Theme-3: Human-Centered BIM

The literature under theme-3 as shown in Table 4 reflects an emergent shift in BIM research toward human-centered and behaviorally responsive design approaches, primarily through the integration of Virtual Reality (VR) and real-time feedback systems. Several studies affirm that BIM can enhance spatial performance and user experience by incorporating immersive simulations, behavioral data, and affective evaluation tools during the conceptual design phase. For instance, body sensor networks (BSNs) and neurophenomenological techniques have been introduced to support pre-occupancy evaluation, enabling designers to assess cognitive and emotional reactions before construction begins [53]. However, despite these innovations, literature acknowledges a persistent resistance to adopting these methodologies and highlights the absence of dedicated tools to measure affective response within BIM workflows—marking a critical gap that limits the realization of experiential design evaluation.

This observation reveals that although conceptual BIM design has evolved to support user well-being, its practical implementation in mainstream design practices remains limited due to tool deficits and cultural inertia. VR-enhanced BIM is increasingly proposed to bridge this divide. For example, ref. [115] demonstrates how VR-enabled BIM models improve user interaction and iterative flexibility, reducing the need for later-stage modifications. Complementing this, ref. [116] proposes an intuitive parametric modeling methodology, enabling users to engage more meaningfully with design iterations across all stages. The use of automated frameworks, such as those described by [117], further shows how BIM plug-ins can facilitate Universal Design and Age-in-Place requirements, offering cost-effective pathways for inclusive, accessibility-driven outcomes.

Yet, the benefits of human-centered BIM are not limited to individual user experience—they also extend to stakeholder collaboration and predictive spatial planning. Tools like BIMFlexi [52] enable real-time VR collaboration between architects, engineers, and planners, transforming the traditional communication process. In specialized contexts such as healthcare design, VR-integrated BIM environments support pre-design evaluation (PDE), which, according to [118], allows architects to identify design errors and anticipate post-occupancy challenges—thereby enhancing facility performance through early user feedback.

Nonetheless, several challenges continue to inhibit the widespread adoption of these frameworks. As illustrated by [119], while cloud-based BIM-VR systems integrating Universal Design and Age-in-Place principles show promise, validation, and scalability remain unresolved. Similarly, studies such as [120] demonstrate that phenomenological approaches are being applied in BIM research to uncover socio-political and economic factors influencing adoption, especially in developing regions, where policy and infrastructure remain barriers. This suggests that technical advancement must be matched by institutional support.

Further literature calls for the development of multi-agent systems and enhanced user feedback models in immersive virtual environments (IVEs), noting their potential to improve behavioral simulations and spatial personalization [121]. A prime example is Safe BIM, a real-time human behavior interaction model developed to analyze user patterns during iterative design [122], showcasing how behavioral informatics can improve real-time design responsiveness.

For this research, the reviewed studies signal both the opportunity and necessity of advancing user-centric design evaluation within BIM environments. While immersive and sensor-based technologies are being explored, they remain under-validated regarding real-time cognitive and affective feedback integration.

Table 4. Key journal articles identified under theme-3.

Ref.	Methodology	Key Findings	Research Gaps
[53]	SLR & e-Delphi Survey	Examines BIM in conceptual design for user experience and well-being, highlighting gaps in pre-occupancy evaluation and advocating for neurophenomenology integration.	Resistance to new methodologies and lack of evidence-based design tools for affective responses in BIM.
[115]	BIM & VR Integration	Explores integrating BIM and Universal Design to enhance accessibility in conceptual design, using VR for immersive user interaction.	Lack of automation and predefined criteria to streamline Universal Design in BIM models.
[116]	Parametric Design & Experimental Study	Investigates intuitive modeling in BIM's parametric design for architectural workflows.	Limited empirical validation of proposed intuitive design workflows.
[117]	BIM & Universal Design Framework	Develops a BIM plug-in for automating Universal Design and Age-in-Place requirements at the conceptual stage.	Needs real-world validation in large-scale housing projects.
[123]	BIM & VR Collaboration Workflow	Introduces BIMFlexi, a multi-user VR environment for interdisciplinary communication in industrial building planning.	Limited assessment of BIMFlexiâ€™s long-term impact on design efficiency.

Table 4. *Cont.*

Ref.	Methodology	Key Findings	Research Gaps
[118]	Pre-Design Evaluation (PDE) & VR Simulation	Examines on how PDE using VR can anticipate design errors in healthcare facilities, reducing post-occupancy issues.	Needs broader validation beyond healthcare settings.
[119]	Cloud-Based BIM & VR Model	Develops a cloud-based model integrating BIM, UD, and VR to enhance Age-in-Place housing design.	Scalability concerns and need for further industry testing.
[120]	Phenomenological Study	Investigates macro-adoption of BIM in Uruguay, highlighting barriers and external influences on industry implementation.	Need for policy frameworks to support BIM adoption in developing economies.
[121]	VR & User Experience Study	Explores immersive virtual environments (IVEs) for user feedback in architectural design.	Further research is required on integrating IVEs with multi-agent systems for optimized design.
[122]	BIM & Human Behavior Simulation	Develops SafeBIM, a real-time simulation system in BIM for assessing user interactions with design iterations.	Limited real-world applications beyond controlled experiments.

3.1.4. Theme-4: Interoperability, Data Exchange, and Workflow in BIM

The reviewed literature under Theme-4 as shown in Table 5 identifies BIM interoperability and data exchange as critical enablers of collaborative design workflows in the AEC industry. One of the most consistently emphasized challenges in traditional architectural processes is the lack of seamless information sharing across disciplinary teams, which often leads to inefficiencies, miscommunication, and rework. BIM, by contrast, offers a shared digital environment that facilitates unbroken and coordinated workflows among architects, engineers, and consultants [124]. However, the transition between design authoring tools and BIM platforms continues to generate interoperability issues, resulting in data loss and workflow fragmentation—particularly in the early conceptual stages where fluidity and flexibility are paramount [125].

To address this, several frameworks and models have emerged to support interoperability, such as the BIM Interoperability Adoption Model (BIAM) [126], which provides organizations with tools to assess their compliance and improve collaborative integration. Additionally, studies show that incorporating AI-driven methodologies and digital platforms into BIM workflows has the potential to improve both information clarity and real-time decision-making. For example, web-based platforms that support adaptive detailing and live expert input have been shown to enhance early-stage design accuracy [127], while smart building technologies linked with BIM enable bidirectional interaction between users and systems, promoting energy efficiency and occupant-centric solutions [128].

The literature also indicates that the future of interoperability lies not merely in data transfer, but in intelligent, semantic data structuring that supports spatial reasoning and automated logic-based design decisions. Research demonstrates how semantic knowledge graphs and AI-enhanced BIM environments can reorganize topological data to improve analytical depth and reduce human error in complex design coordination [129]. What these studies collectively show is a movement from BIM as a static documentation system to one that is dynamic, context-aware, and collaborative. For instance, cloud-based and multi-user collaborative platforms that integrate tactile and haptic interfaces have improved team communication, especially in interdisciplinary and international projects [130]. Efforts to bridge the gap between high-tech BIM environments and low-tech conventional practices are also gaining traction, suggesting the need for scalable solutions adaptable to various organizational capacities [131]. In China's AEC sector, BIM-assisted workflows are already demonstrating measurable improvements in project delivery and design efficiency [132], further underscoring its global relevance.

Table 5. Key journal articles identified under theme-4.

Ref.	Methodology	Key Findings	Research Gaps
[124]	Process Model Comparison	Examines how BIM enhances project information flow, fostering collaborative interactions throughout the project lifecycle.	Further validation required on BIMs impact in large-scale, multi-disciplinary projects.
[125]	Empirical Study & Survey	Investigates challenges in transitioning from 3D digital design to BIM, focusing on data loss and workflow inefficiencies.	Limited generalizability beyond academic environments; industry validation needed.
[126]	Interoperability Framework & BIM Adoption Model	Introduces the BIM Interoperability Adoption Model (BIAM) to address technical, organizational, semantic, and legal interoperability challenges.	Needs further industry-wide adoption studies to assess real-world impact.
[127]	Web-Based Decision-Making Tool	Develops an online collaborative platform integrating adaptive detailing in BIM for early-stage decision-making.	Challenges in shifting traditional communication practices to digital environments.
[128]	Smart Building & User Interaction Study	Explores how smart buildings integrate user feedback through a mobile application to enhance occupant comfort and energy efficiency.	Limited real-world implementation outside academic and test environments.
[129]	AI & Semantic Building Fingerprints	Proposes a method to store topological building information using knowledge graphs for AI applications in design and construction.	Further research needed to standardize AI-driven spatial analysis in BIM workflows.
[130]	Cloud-Based BIM Collaboration Study	Examines cloud-based collaboration in early architectural design, integrating tactile interfaces with BIM workflows.	Requires further exploration of integration across different BIM platforms and team structures.
[131]	Interdisciplinary BIM Usability Study	Compares traditional analog tools with BIM in preliminary design to assess usability and efficiency in cooperative workflows.	Needs broader validation across different project types and scales.
[132]	BIM-Assisted Workflow Study	Analyzes BIM-assisted workflows in China's AEC industry, showing efficiency improvements over traditional methods.	Further studies required to evaluate long-term adoption and scalability in complex projects.
[133]	Multi-LOD BIM Model & Uncertainty Management	Develops a multi-LOD meta-model to improve early-stage design evaluation and uncertainty management in BIM.	Requires validation in large-scale projects and integration with other BIM performance assessment tools.

Nevertheless, despite technological advancements, the management of complex design information in early-stage BIM remains a significant hurdle. Research into multi-LOD meta-models has introduced structured ways to manage uncertainty and performance in conceptual phases, laying a foundation for data-informed decision-making [133]. Still, the challenge persists in creating adaptable, cross-platform systems that can seamlessly integrate the exploratory nature of early-stage design with the precision of BIM protocols.

For the current research, these findings reinforce the importance of integrating user-centric experiential evaluation into BIM workflows without disrupting interoperability. While data exchange and AI-enhanced systems have evolved considerably, there remains a gap in capturing how design decisions affect users' cognitive and affective experiences—particularly in immersive, conceptual environments.

3.1.5. Theme-5: Spatial Reasoning and Key Case Study Addressing Early-Stage Design

The literature reviewed under theme-5 as shown in Table 6 underscores the central role of spatial reasoning in shaping conceptual design decisions, particularly before transitioning into BIM environments. Spatial reasoning enables designers to intuitively and analytically assess spatial relationships, circulation, and hierarchy, which form the foundation for early-stage ideation. With the advancement of computational design tools, researchers have developed methods to formalize and optimize spatial reasoning processes, thereby supporting both intuitive exploration and data-informed design strategies.

One of the most significant contributions comes from non-manifold topology (NMT) modeling tools like Topologic, which combine geometry, semantics, and topological structure to preserve design intent over form fidelity, allowing for greater abstraction and flexibility in early spatial layout planning [134]. Similarly, integrating graph theory into BIM workflows has enabled quantitative circulation analysis, aiding designers in evaluating spatial efficiency and connectivity from a systems-thinking perspective [135]. These developments demonstrate that computational techniques can enhance both cognitive design processes and rational decision-making in the earliest stages of architectural development.

Further contributions push this idea toward democratization and simplification. For example, TopoBIM is introduced as a lightweight, topology-driven BIM interface that simplifies spatial modeling for non-experts and stakeholders, thereby reducing authoring complexity while preserving data consistency [136]. These efforts reveal a common goal: to empower all participants in the design process with intuitive, yet structured tools that facilitate collaboration and early evaluation.

Experimental design studios have also played a key role in testing integrated workflows, combining parametric design, digital fabrication, and interactive media to assess how computational methods perform across different stages of architectural programming [137]. In practice, BIM has shown relevance even beyond structural form-making; for instance, in interior decoration and public project management, where it supports more efficient and coordinated design delivery [138]. However, many of these academic developments still face scalability and usability challenges when transferred to real-world, high-stakes environments.

To bridge theory and practice, several case studies demonstrate the value of BIM-based approaches in specialized domains. In healthcare infrastructure, for example, scenario-based design within BIM environments is used to build flexible, future-proof spaces, highlighting the importance of adaptability in early-stage planning [139]. Another case focuses on envelope optimization for high-rise residential buildings, where a BIM-driven decision support system improves both design confidence and project acceleration during the conceptual phase [140]. These studies offer compelling evidence that BIM-integrated spatial reasoning methods can improve architectural outcomes when applied in domain-specific settings.

The implications of this research are twofold. First, while computational tools have expanded the capacity for spatial analysis in early-stage design, they still lack integration with experiential feedback mechanisms—particularly those that reflect users' cognitive and emotional responses. Second, despite the success of case-specific BIM applications, a universal and scalable framework that combines spatial reasoning, immersive interaction, and real-time human feedback remains absent.

Table 6. Key journal articles identified under theme-5.

Ref.	Methodology	Key Findings	Research Gaps
[134]	Software Development & Usability Study	Introduces Topologic, a non-manifold topology (NMT) modeling library to support hierarchical spatial representation in early architectural design.	Limited empirical validation of Topological impact on BIM workflows in real-world projects.
[135]	Graph Theory & BIM Integration	Explores the use of graph theory in BIM for circulation analysis, proposing a tool that abstracts building layouts into topology graphs for quantitative evaluation.	Needs broader validation across different building typologies beyond case studies.
[136]	Web-Based Topology & BIM Workflow	Proposes TopoBIM as a lightweight, topology-driven workflow for early design participation, reducing BIM authoring complexity.	Further research is needed on its usability and adoption by stakeholders in real-world projects.
[137]	Experimental Design Studio	Explores BIM integration across parametric design, digital fabrication, and interactive technology through an experimental studio.	Further validation needed in real-world architectural projects beyond academic settings.
[138]	BIM in Interior Decoration Design	Analyzes BIM's role in interior decoration for public buildings, highlighting improvements in design management and efficiency.	Challenges remain in adapting BIM workflows to diverse interior design project requirements.
[139]	Scenario-Based Design & BIM	Investigates BIMâEUR^{TM}s role in future-proof healthcare design, using scenario-based design to develop flexible infrastructure solutions.	Needs broader validation on balancing flexibility and standardization in healthcare facilities.
[140]	BIM-Based Decision Support System	Examines early-stage design for high-rise residential buildings, proposing a BIM-based tool for optimizing envelope designs.	Requires further testing on the scalability and adaptability of the decision support system.

3.1.6. Findings and Conclusion from SLR

The literature review reveals a sustained effort to evolve BIM workflows beyond traditional documentation and coordination functions, incorporating cognitive reasoning, generative algorithms, and performance-based evaluations. Despite these advancements, a significant gap persists in integrating user-centric experiential feedback—particularly in the early stages of architectural design, where ideation, perception, and spatial intuition play a crucial role.

While generative BIM frameworks show promise in enhancing design iteration and decision-making, they remain predominantly data-centric, often overlooking the emotional, perceptual, and cognitive dimensions of user experience. Few studies have investigated how these systems influence or respond to user comfort, spatial cognition, or affective engagement within immersive design environments. This lack of attention to the lived spatial experience creates a methodological blind spot in current BIM research and development.

Moreover, although immersive and sensor-based technologies—such as VR, EEG, and BSNs—are being explored in adjacent studies, their integration into mainstream BIM workflows is still at a nascent stage. These technologies remain underutilized and under-validated in capturing real-time affective and cognitive responses, particularly in the context of conceptual design evaluation.

The findings also emphasize that advancements in AI-enhanced interoperability, data exchange, and semantic structuring have improved workflow efficiency. However, parallel affective and experiential evaluation innovations have not matched these developments. Consequently, existing BIM tools lack the capacity to holistically assess how early design decisions affect human perception, comfort, and emotional resonance.

Two key implications emerge for the current study. First, while computational tools have substantially increased the analytical scope of spatial reasoning and layout optimization, they still fall short of incorporating experiential feedback mechanisms that reflect user cognition and emotion. Second, despite successful implementations of BIM in domain-specific case studies (e.g., healthcare, high-rise residential design), no scalable framework seamlessly integrates spatial reasoning, immersive interaction, and physiological user feedback.

3.2. Semi-Structured Interview

The analysis of interview data through Kvale's seven-stage framework [60] showed architects' BIM experiences during conceptual design. Braun and Clarke's [74] thematic analysis method were used for data analysis in this study to achieve analytical rigor and depth while being embedded in Kvale's interview inquiry framework. Braun and Clarke's method supplied detailed systematic procedures for coding and theme development and interpretation while Kvale's framework provided a structured design process for conducting interviews and reporting results. The combined analytical approach enabled researchers to perform transparent and reflexive analysis, which produced interpretive insights about the lived experiences of participating architects. The key stages of the analytical process followed are shown in the Figure 9. The analysis results in this section show the identified themes and their interpretive narratives.

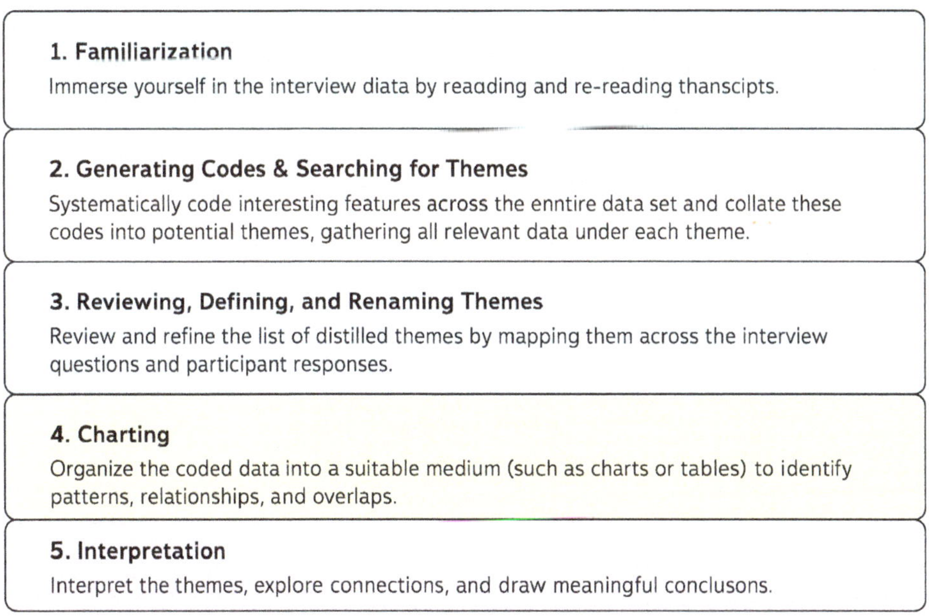

Figure 9. Visual summary of analyzing stage in semi-structured interview. Source: Authors.

3.2.1. Familiarization

The researchers started their familiarization process through multiple readings of interview transcripts to build complete understanding of participant stories. Through multiple readings researchers developed extensive knowledge about the content along with tone and context of each participant's account. The researchers recorded their initial observations about creative frustration and hybrid tool use, performance bottlenecks, and strategic workarounds that architects used to overcome BIM limitations during this phase. The researchers focused on emotional expressions through words that showed frustration and aspiration and innovation while studying detailed descriptions about workflows and collaborative routines and perceived tool deficiencies in BIM. The initial stage enabled researchers to recognize patterns while developing their ability to observe how architects handled creative freedom versus technological constraints during conceptual design.

3.2.2. Generating Codes and Searching for Themes

This stage systematically examines the interview data to identify meaningful patterns and recurring ideas. The process begins with coding, where significant words, phrases, or data segments are labeled based on their relevance to the research questions. These themes were derived through a systematic coding process using manual coding process. These codes act as analytical building blocks, capturing the essence of the participants' responses [141]. Figure 10 shows a glimpse of some of the codes generated in the preliminary coding. Figure 11 shows the six major themes that emerged from the text analyzed. Once the coding is complete, the next step involves organizing and grouping similar codes to search for broader patterns or emerging themes. This process helps simplify large volumes of qualitative data by clustering related insights, laying the groundwork for deeper analysis and interpretation.

Figure 10. Word cloud depicting some of the codes generated while analyzing the semi-structured interview. Source: Authors.

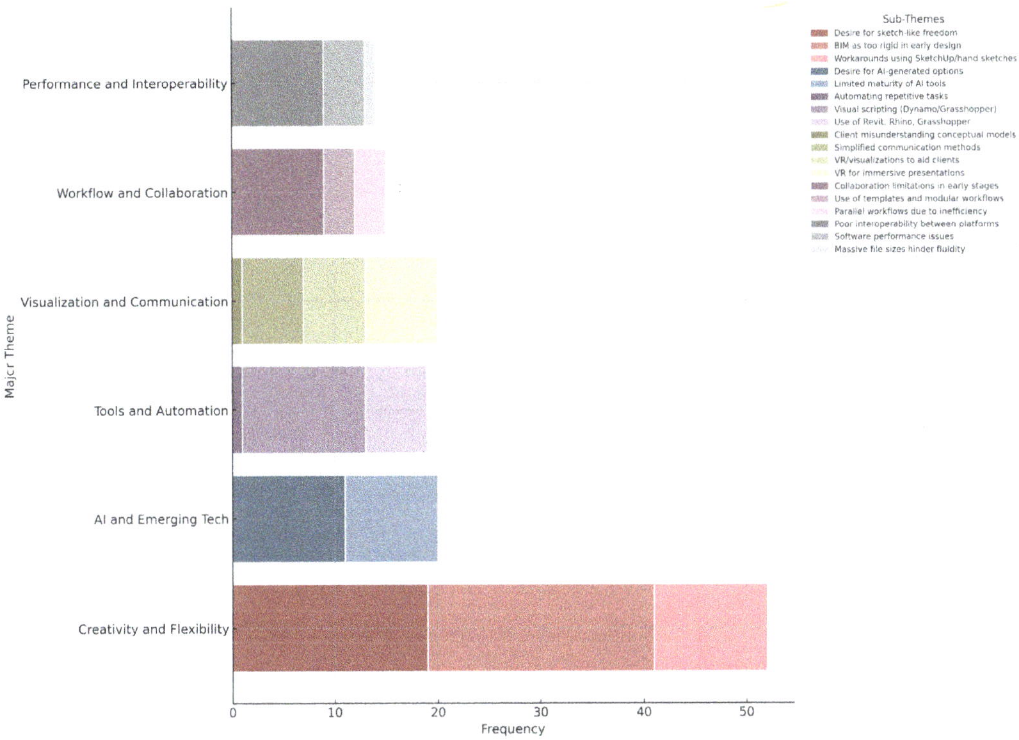

Figure 11. Major and Minor Categories generated by aggregating codes from the semi structured interview. Source Authors.

3.2.3. Reviewing, Defining, and Renaming Themes

The evaluation process of theme identification along with definition and renaming was conducted to guarantee that thematic categories effectively represented the fundamental data patterns and subtle details found in interview results. The initial analysis revealed six primary themes: Creative Constraints in BIM, Hybrid Workflow Preferences, Client & Stakeholder Communication Barriers, Workflow Strategies & Templates, AI, VR and Generative Design Integration, and Vision for Ideal Conceptual Design Workflow. The thematic categories underwent critical evaluation by reviewing coded excerpts to establish their alignment with the thematic categories. Participant A1 expressed the sentiment "Sometimes, I feel, uh, BIM, it, uh, limits my creativity" (Participant A1) under Creative Constraints in BIM, and Participant A5 stated "We rely heavily on parametric modeling to explore different scenarios quickly" (Participant A5) under Workflow Strategies & Templates. The review process received additional support from the comparative charts that were presented. The Comparative Chart of Themes Across Participants (Figure 12) displays the distribution of themes among the ten interviewees A1 through A10. Most participants discussed Creative Constraints in BIM and Ideal Conceptual Workflow Vision, thus showing their widespread occurrence, but AI, VR, and Generative Design Integration received attention only from particular participants who introduced new ideas. The Theme Mapping Across Interview Questions (Figure 13) showed how different themes appeared throughout the eleven interview questions. The validation process of theme relevance emerged from assessing each theme against the corresponding interview guide components.

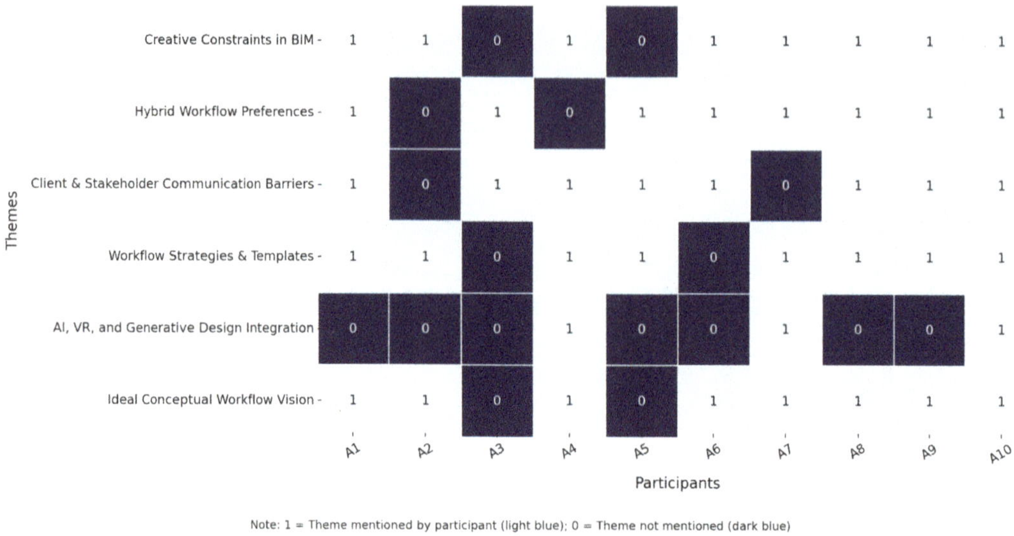

Figure 12. Comparative chart of themes across the participants. Source: Authors.

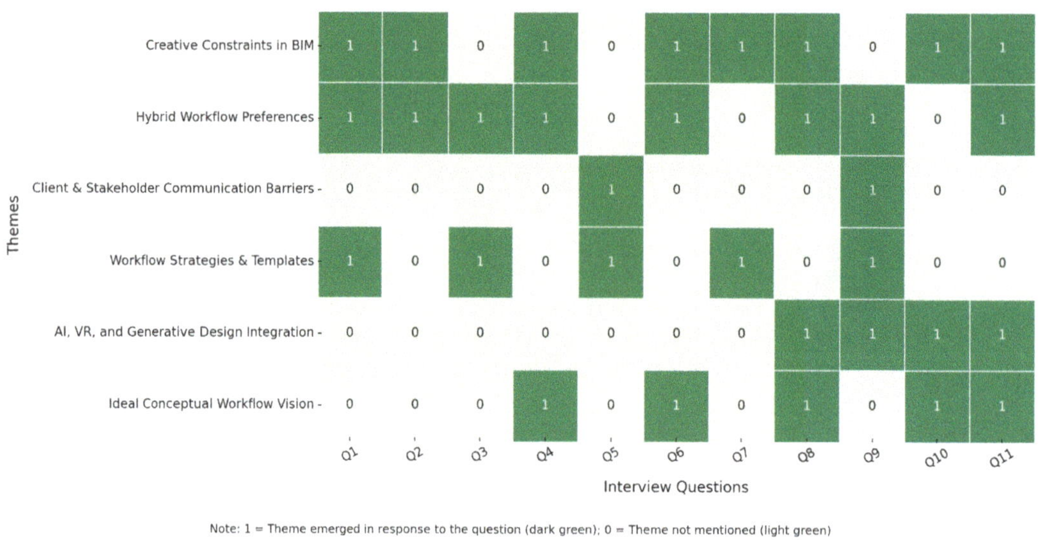

Figure 13. Comparative chart of themes across the questions. Source: Authors.

The repeated review process led to the combination or transformation of some sub-themes for better conceptual understanding and thematic cohesion. The analyzed thematic structure delivers a precise and simplified overview of participant perspectives about BIM implementation during conceptual design. The studied themes became the foundation for additional interpretation that led to the discussion of this research.

3.2.4. Charting the Themes and Sub-Themes

The process of charting the themes and sub-themes was carried out to visually organize and quantify the prominence of each thematic category that emerged from the interview data. This exercise facilitated a clearer understanding of the distribution and frequency of sub-themes across the six major themes identified in the analysis. Figure 14 provides a comprehensive visual summary of the how participants' individual accounts are linked to the major subject.

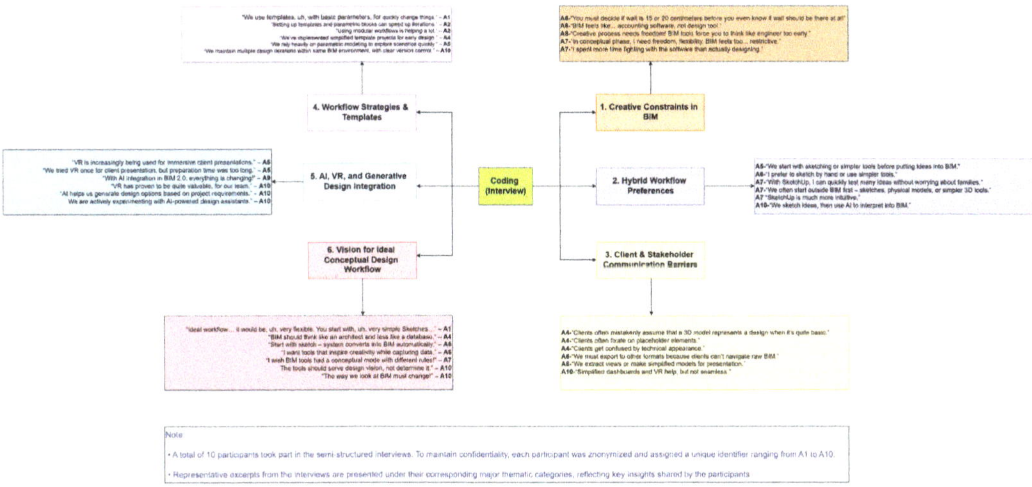

Figure 14. A sample of major categories mapped against the coded narrative from the semi-structured interview. Source: Authors.

For instance, in Figure 12, under the theme Creativity and Flexibility, sub-themes such as Desire for sketch-like freedom, BIM as too rigid in early design, and Workarounds using SketchUp/hand sketches were identified as significant concerns, with high mention frequencies. Similarly, sub-themes like Desire for AI-generated options and Limited maturity of AI tools were mapped under AI and Emerging Tech, indicating participants' interest in integrating emerging technologies in BIM workflows. Other prominent sub-themes such as Use of templates and modular workflows and Parallel workflows due to inefficiency were grouped under Workflow and Collaboration, while challenges like Client misunderstanding conceptual models and Simplified communication methods were categorized under Visualization and Communication. Additionally, interoperability and software performance issues were captured under Performance and Interoperability. Charting these sub-themes against their respective major themes provided a visual hierarchy of the participants' responses and highlighted the concentration of challenges and opportunities associated with using BIM during the conceptual design phase. This structured representation enabled a deeper analysis of patterns, overlaps, and unique insights shared by the interviewees, thereby strengthening the thematic framework for further interpretation.

3.2.5. Interpreting the Themes

The last stage of thematic analysis included explaining the uncovered themes and sub-themes to capture the meaning of the interview data. This was based on the thematic patterns that emerged during the charting process and participants' narratives. The analysis indicated that BIM is used in architectural practice; however, its usage during the conceptual design phase is often seen as limiting. This was because Creative Constraints in BIM and Hybrid Workflow Preferences showed that there is a need for architects to have more flexibility and freedom in the early stages of design, and, therefore, they would use hand sketches and other simple digital tools to enhance BIM workflows. Furthermore, the theme of Client & Stakeholder Communication Barriers revealed the discrepancy between the technical nature of BIM outputs and client understanding, thus pointing to the need for simplified and more immersive communication strategies. The recurring sub-themes under Workflow Strategies & Templates revealed that participants employed templates, modular workflows, and automation to solve operational inefficiencies. Also, the theme AI, VR, and Generative Design Integration captured the growing desire to integrate new technologies to enhance conceptual design capabilities. The theme Vision for Ideal Conceptual Workflow captured the participants' vision of the future BIM environment that is more intuitive, flexible, and creative and supports early-stage design thinking. The analysis of these themes provided a thorough understanding of the complex relationship between BIM tools, design processes, and user experiences, which formed the basis for the discussion and recommendations of the study.

3.2.6. Verifying

Following Kvale's framework, the stage of Verifying was aimed at ensuring the analytic rigor, credibility, and trustworthiness of the thematic analysis. To this end, triangulation was used by combining content analysis frequencies with thematic patterns from the interview transcripts. This methodological cross checking helped to provide a more robust validation of the emerging themes. The themes were systematically cross-checked by going back to the raw transcripts and the frequency of thematic occurrence to ensure that the analysis was still based on the participants' actual responses. Also, outlier perspectives were kept, ensuring diversity and avoiding thematization. For instance, while most of the participants complained about the inflexibility of Revit during the conceptual design phase, one participant (A9) provided a contrary view by pointing out the parametric possibilities of Revit when used with Dynamo, thus adding depth and complexity to the understanding of BIM tools in design ideation. Furthermore, cross participant validation was performed to identify patterns of convergence and divergence. Common frustrations such as the inflexibility of BIM and the technical limitations placed on creativity were identified in at least eight participants, and there was a clear consensus on using hybrid workflows such as SketchUp and hand sketching before using BIM. The divergent cases, like Participant A9's emphasis on AI driven workflows, were recognized as new and innovative practices but were noted as being outside the norm. This verification process increased the reliability of the findings and the analysis outcomes.

3.2.7. Reporting: Thematic Patterns and Knowledge Gap

The thematic analysis of the interview data revealed six dominant patterns reflecting the lived experiences of practicing architects concerning the use of Building Information Modeling (BIM) during the conceptual design phase. These patterns, distilled from participant narratives, illustrate the persistent tensions between creative ideation and the structural rigidity of BIM tools, alongside emerging practices that attempt to bridge this gap. Representative quotations are used here to foreground participants' voices and highlight the underlying knowledge deficit in existing BIM workflows.

Creative Constraints in BIM

A recurrent theme in the participants' narratives was the restrictive nature of BIM tools in facilitating creative exploration during the early design phase. Participants expressed frustration over the rigid parameters and precision requirements embedded within BIM environments, which were perceived as antithetical to the fluid and iterative nature of conceptual design. Participant A6 noted, "You must decide if wall is 15 or 20 cm before you even know if wall should be there at all!". This sentiment was echoed in the metaphorical comparison of BIM to "accounting software, not design tool" (A6), underscoring the tool's technical orientation over creative flexibility. Furthermore, participants highlighted how BIM tools prematurely force design decisions, with A6 remarking, "Creative process needs freedom! BIM tools force you to think like engineer too early". This rigidity was perceived as a barrier to design innovation, as captured by A7: "In conceptual phase, I need freedom, flexibility. BIM feels too... restrictive". The emotional toll of this constraint was evident in A7's reflection: "I spent more time fighting with the software than actually designing". These reflections reveal a significant gap between the current capabilities of BIM platforms and architects' cognitive and creative demands in the conceptual design stage. Existing BIM tools appear ill-equipped to support early design ideation's open-ended, iterative nature.

Hybrid Workflow Preferences

In response to these creative limitations, participants described adopting hybrid workflows that rely on non-BIM tools in the initial design stages. Sketching, physical models, and intuitive modeling platforms such as SketchUp were preferred to explore and iterate design ideas without the procedural constraints of BIM. As A5 explained, "We start with sketching or simpler tools before putting ideas into BIM", while A6 similarly affirmed, "I prefer to sketch by hand or use simpler tools". Participant A7 emphasized the intuitive advantage of alternative platforms: "With SketchUp, I can quickly test many ideas without worrying about families". Many participants indicated that their typical practice involves moving from analog or flexible digital tools into BIM only after core design ideas have been resolved: "We often start outside BIM first—sketches, physical models, or simpler 3D tools" (A7). A10's comment further exemplified this workflow fragmentation, "We sketch ideas, then use AI to interpret into BIM". The prevalence of such hybrid workflows underscores a knowledge and technology gap: current BIM platforms do not adequately support the iterative and exploratory needs of early-stage design, compelling practitioners to rely on disconnected workflows that complicate data continuity and collaboration.

Client & Stakeholder Communication Barriers

Participants also identified persistent communication barriers when using BIM outputs for client engagement and stakeholder presentations. According to participants, many clients misinterpret the technical representations generated in BIM models, mistaking schematic visuals for final design solutions. As A4 observed, "Clients often mistakenly assume that a 3D model represents a design when it's quite basic". This misunderstanding often leads clients to fixate on placeholder elements or minor visual details: "Clients often fixate on placeholder elements" (A4), and "Clients get confused by technical appearance" (A6). Participants routinely resorted to creating simplified visualizations, exporting to other formats, or using additional tools to bridge this comprehension gap: "We must export to other formats because clients can't navigate raw BIM" (A6), and "We extract views or make simplified models for presentation" (A8). However, these efforts were not always seamless, as A10 acknowledged: "Simplified dashboards and VR help, but not seamless".

These findings indicate a critical gap in how BIM models communicate design intent to non-technical stakeholders, necessitating additional effort, time, and tools to make design concepts accessible.

Workflow Strategies and Templates

Despite the limitations of BIM in conceptual design, participants shared strategies to streamline their workflows and maximize efficiency within these constraints. The use of templates, parametric blocks, and modular workflows emerged as common practices. Participant A1 noted, "We use templates, uh, with basic parameters, for quickly change things", while A2 highlighted that "Setting up templates and parametric blocks can speed up iterations". Modular approaches were seen as beneficial: "Using modular workflows is helping a lot" (A2), and some firms institutionalized this strategy by developing "simplified template projects for early design" (A4). Parametric modeling was also leveraged to facilitate rapid scenario testing: "We rely heavily on parametric modeling to explore scenarios quickly" (A5). Moreover, iterative version control within the BIM environment was reported: "We maintain multiple design iterations within same BIM environment, with clear version control" (A10). These strategies reveal how practitioners negotiate the limitations of BIM tools yet also expose a knowledge gap: current BIM systems require considerable customization and procedural workarounds to enable even a limited degree of flexibility.

AI, VR, and Generative Design Integration

Several participants discussed emerging technologies such as AI, VR, and generative design as potential enablers of more flexible and immersive design workflows. VR was noted for its increasing use in client presentations: "VR is increasingly being used for immersive client presentations" (A5), though preparation time remained a concern: "We tried VR once for client presentation, but preparation time was too long" (A5). A9 highlighted the integration of AI-driven design tools: "With AI integration in BIM 2.0, everything is changing!" Participant A10 elaborated on the practical application of these technologies: "VR has proven to be quite valuable, for our team" and "AI helps us generate design options based on project requirements". Experimental use of AI-powered design assistants was also mentioned: "We are actively experimenting with AI-powered design assistants" (A10). These narratives reflect an emerging but underexplored area in practice: the integration of advanced computational tools to overcome BIM's conceptual design limitations. However, the use of such technologies remains experimental and fragmented, pointing to a clear knowledge and implementation gap in mainstream architectural workflows.

Vision for an IDEAL Conceptual Design Workflow

Finally, participants articulated a vision for an ideal BIM workflow that prioritizes creative flexibility, intuitive usability, and seamless integration of conceptual and technical phases. Participant A1 envisioned, "Ideal workflow... it would be, uh, very flexible. You start with, uh, very simple sketches...". A common desire was for BIM tools to align more closely with the thought process of designers: "BIM should think like an architect and less like a database" (A4). Automation of transitions from sketch to BIM was suggested: "Start with sketch—system converts into BIM automatically" (A6), alongside calls for tools that "inspire creativity while capturing data" (A6). A7 articulated the need for differentiated modes within BIM: "I wish BIM tools had a conceptual mode with different rules!". Participants emphasized that digital tools should augment rather than dictate the design process: "The tools should serve design vision, not determine it" (A10), and underscored the need for a paradigm shift: "The way we look at BIM must change!" (A10). This shared vision illustrates a fundamental knowledge and technology gap in existing BIM ecosystems. The absence of integrated, flexible, and creativity-oriented workflows hinders the effective use of BIM during the early stages, necessitating further research and development to bridge this divide.

3.3. Findings from the Semi-Structured Interview

The findings from the semi-structured interviews reveal that while BIM is increasingly used in architectural design, it poses significant challenges during the conceptual design phase. The need for greater creative freedom and flexibility is clear, with many architects relying on hybrid workflows that combine BIM with other design tools. Additionally, the integration of AI, VR, and generative design tools into BIM workflows was identified as a promising direction for addressing these limitations. Finally, the communication barriers between BIM-generated outputs and client understanding indicate a need for improved visualization and communication tools in BIM platforms. These findings underscore the knowledge gaps in current BIM systems and the importance of developing a more integrated, user-friendly platform that supports both creative ideation and technical accuracy in the early stages of design. The insights gained through the interviews are foundational to developing the proposed BIM evaluation framework that integrates immersive technologies and human feedback.

3.4. Design and Development of Framework

The SLR conducted in this study revealed a recurring gap in early-stage BIM-based architectural design workflows—specifically, the absence of mechanisms for incorporating experiential, human-centered feedback into digital modeling environments. While advancements in automation, generative design, and simulation-based performance metrics have improved technical capabilities, few frameworks account for how users emotionally and cognitively experience space. This shortfall was further substantiated through semi-structured interviews with practicing architects, who consistently identified BIM's limitations in supporting creative ideation, emotional resonance, and intuitive spatial interpretation during the conceptual phase. Participants emphasized the pressing need for evidence-based and perceptually rich methods that extend beyond visual accuracy and functional criteria to evaluate spatial quality from a user's point of view.

In response, this study proposes a three-stage experiential framework as shown in Figure 15, that directly integrates findings from both the SLR and practitioner interviews. The framework bridges the gap between design ideation and performance-based refinement, aligning architectural workflows with human-centered design principles.

Stage-1: Experiential Design Ideation responds to the creative constraints in traditional BIM environments by integrating sketch-based modeling, parametric tools, and AI-driven generative design platforms. This stage promotes flexible and iterative exploration of spatial concepts, accommodating emerging WebBIM and openBIM workflows that support interoperability and early-stage design experimentation.

Stage-2: Immersive Evaluation introduces VR-integrated simulations and neurophysiological feedback [142,143] to assess spatial experiences in real-time. Drawing from both the literature and interview data, this stage addresses the need for user-centered evaluation methods. To operationalize this, the study employs Electroencephalography (EEG) [82]. EEG's high temporal resolution enables moment-to-moment tracking of affective and cognitive responses as users navigate virtual environments. This allows architects to observe how elements such as form, scale, materiality, and lighting influence user engagement—offering insights unattainable through traditional performance simulation tools [80,81,83,144–147].

Stage-3: Experience-Informed Evidence-Based Design (EBD) refinement ensures that insights gained through immersive evaluation are translated into actionable design updates. This stage uses software like Autodesk BIM 360—software is version 16.14.0.2573—and other real-time collaboration tools to support iterative modifications and integrated stakeholder collaboration, thus addressing persistent issues of workflow continuity and cross-platform interoperability identified in the SLR. By embedding EEG-enabled feedback loops into the BIM-VR pipeline, the framework introduces a quantifiable, evidence-based approach to experiential design evaluation. It enables the real-time capture of user-centered insights to inform conceptual decisions, ultimately supporting the development of architectural environments that are not only technically optimized but emotionally resonant and cognitively aligned. This multi-modal framework, thus, lays a foundation for more adaptive, intuitive, and immersive BIM workflows in early-stage design beyond visual accuracy and functional criteria to evaluate spatial quality from a user's point of view.

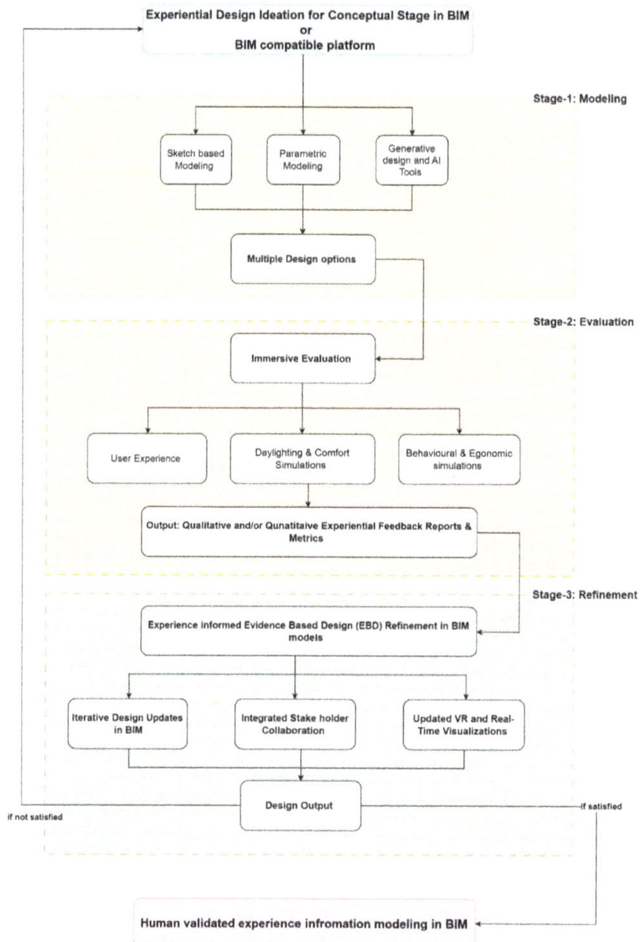

Figure 15. Experience evaluation framework for conceptual model in BIM. Source: Authors.

3.5. Pilot Study

A pilot study was carried out to operationalize the proposed framework for human-centric experiential evaluation in the conceptual design phase by testing the integration of EEG-based neural data with immersive VR environments and BIM workflows. The main purpose of the pilot study was to determine the possibility of gathering real-time user experience data during the early design evaluations and to find out if this kind of feedback can be helpful for iterative refinements within BIM platforms. This study directly corresponds to Stage-2 (Immersive Evaluation) of the framework, which highlights the need for both qualitative and quantitative metrics to assess the affective, cognitive, and perceptual responses of users when interacting with spatial design concepts.

The pilot study examined how portable neuroelectrical tools, such as the Emotiv 5-channel EEG headset, can be used during immersive VR walkthroughs of six conceptual room designs. This approach is in line with the framework's objective of supporting evidence-based refinements in Stage-3, by using human validated experiential data to inform design decisions in BIM. This pilot's findings are the first step towards integrating human experience more systematically into architectural workflows, closing the gap between creative ideation and experiential validation in real-time digital environments. The

EEG signals were acquired with the Emotiv 5-channel headset which has a sampling rate of 128 Hz and has five electrodes (Anterior F3, T7, Pz, T8 and Anterior F4). Signal acquisition was performed through EmotivPRO standard software, which gave raw time-series data for each of the five electrodes. Electrode impedance was checked before recording to ensure data quality, and only sessions with satisfactory contact quality were included in the analysis. Each participant's session involved six 1 min EEG recordings, one per virtual room. The raw EEG data were examined visually and filtered using a 0.5–30 Hz band pass filter to eliminate high frequency noise and low frequency drift. Emotiv's "EmotivPRO standard" EEG software measures brainwave activity as band power in microvolts squared (μV^2), indicating the strength of signals in each frequency band. These values are calculated using FFT and reflect the brain's average power in bands like Delta, Theta, Alpha, Beta, and Gamma. Table 7 provides an overview of EEG frequency bands, their typical power values, and associated mental states, which facilitates the interpretation of recorded EEG signals [148,149].

Table 7. Classification of EEG Frequency Bands with Typical Power Ranges and Associated Mental States.

Band	Frequency Range (Hz)	Typical Power Range (μV^2)	Associated Mental States
Delta	0.5–4 Hz	0.5–5 μV^2	Deep sleep, unconsciousness (not typically active during wakefulness)
Theta	4–8 Hz	2–10 μV^2	Relaxation, meditation, creativity, internal focus
Alpha	8–12 Hz	5–15 μV^2	Calm wakefulness, relaxed alertness, mental coordination
Low Beta	12–15 Hz	4–12 μV^2	Quiet focus, low engagement, active thinking
High Beta	15–30 Hz	6–20 μV^2	Alertness, stress, high engagement, active concentration
Gamma	30–100 Hz	3–12 μV^2	Higher cognitive functions, learning, memory, and perception

For the analysis, the EEG signals were processed and aggregated in a systematic manner. First, the data were segmented into 1 min epochs for each of the virtual rooms. Then, channel-wise preprocessing was performed, in which noise artifacts such as motion or muscle activity linking to the artifacts were detected and eliminated through thresholding techniques. The channel-wise data at the participant level were then averaged across all six virtual room conditions to obtain individual level metrics. Finally, these signal values, including band power or mean amplitude, were averaged across all participants for each electrode and room condition to obtain group-level insights. The following subsections present the detailed analysis of the collected data.

3.5.1. Brain Wave Distribution Across VR Galleries

The box plot in Figure 16 illustrates the distribution of EEG frequency bands recorded from participants across six VR galleries, each with distinct spatial patterns. The analysis revealed observable differences in neural activity across environments, suggesting varied cognitive and emotional responses. In the boxplot, the central box represents the interquartile range (IQR), encompassing the middle 50% of the data (Q1 to Q3), with the horizontal line denoting the median. Whiskers extend to data points within 1.5 × IQR from the quartiles, while values beyond this range are treated as statistical outliers. In this study, diamond-shaped markers represent EEG amplitude outliers across specific brainwave bands (Theta, Alpha, etc.) under six distinct VR room conditions, indicating instances of elevated or suppressed neural activity outside the normative range.

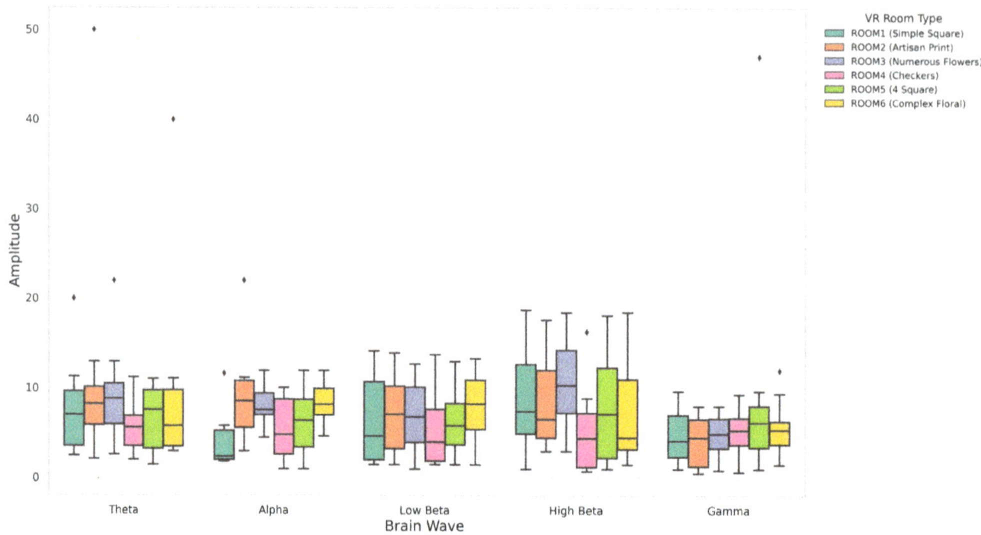

Figure 16. Distribution of Brain waves across VR Rooms.

In Room 1 (Simple Square), Theta and Alpha waves showed high variance. This observation suggests that while some participants experienced a relaxed state (as indicated by elevated Theta/Alpha), others may have found the environment monotonous or understimulating—leading to inconsistent engagement. Theta wave distributions included extreme outliers in Room 2 (Artisan Print), indicating that some participants may have reached intense relaxation or disengagement. Meanwhile, bursts of Beta activity in this room point to moments of cognitive engagement despite the overall relaxed state.

In Rooms 3 and 4, increased dispersion in Beta waves was observed, likely due to the visually complex patterns, which suggest heightened cognitive processing and attentional demand. Concurrently, lower Alpha and Theta levels imply that these rooms evoked weaker relaxation responses. Room 5 (4 Square) exhibited a more balanced distribution of Theta and Beta waves, suggesting a spatial experience supporting relaxation and active engagement. Finally, Room 6 (Complex Floral) displayed stable Beta and Gamma wave activity without extreme outliers, indicating sustained cognitive engagement without signs of mental overload or fatigue.

3.5.2. Average EEG Activity in Different VR Rooms

Table 8 displays the EEG band-power values recorded across six immersive virtual room conditions, highlighting distinct patterns of neural modulation across Theta, Alpha, Beta, and Gamma frequency bands. A one-factor repeated-measures ANOVA was conducted with room type as the within-subjects factor to assess whether these variations were statistically meaningful. The analysis revealed no statistically significant main effect of room type on EEG band power ($F = 1.635$, $p = 0.189$). However, the test yielded a moderate effect size (partial $\eta^2 = 0.254$), indicating that approximately 25.4% of the variance in neural responses could be explained by spatial condition—a non-trivial proportion, particularly within a small sample pilot context. While the lack of statistical significance precluded the execution of post hoc pairwise comparisons, as per standard analytical guidelines, the magnitude of the effect remains noteworthy. The results suggest an emerging trend of room-specific neural differentiation, which may reflect subtle but consistent modulations in spatial perception or cognitive-affective engagement across environments. Given the sample size constraints ($n = 10$), the study may have been underpowered to detect statistically significant effects, especially for mid-range differences.

Table 8. Average EEG Band Power (μV^2) Across Different VR Room Environments.

S. No	Rooms	Theta μV^2	Alpha μV^2	Low Beta μV^2	High Beta μV^2	Gamma μV^2
1	ROOM1 (Simple Square)	7.37	3,74	6.27	9.18	4.85
2	ROOM2 (Artisan Print)	11.79	8.13	7.14	8.14	3.78
3	ROOM3 (Numerous Flowers)	8.12	7.37	6.58	10.25	4.53
4	ROOM4 (Checkers)	5.34	5.07	5.33	4.71	4.54
5	ROOM5 (4 Square)	6.7	6.21	6.05	7.8	9.02
6	ROOM6 (Complex Floral)	9.127	8.05	8.27	7.13	5.57

Note: The EEG power values presented in the table represent absolute band power measured in microvolts squared (μV^2). Each value is the average across 10 participants for the corresponding rooms.

To interpret these patterns, EEG power band values were plotted for each room type (Figure 17). The observed trend shows that Room 2 (Artisan Print) elicited the highest Theta and Alpha power, frequencies commonly associated with relaxation and inward-focused attention, implying that participants may have experienced this space as calming and immersive. Rooms 3 (Numerous Flowers) and 5 (4 Square) showed elevated Beta activity, with Room 3 peaking in High Beta, which is typically linked to increased cognitive processing and attentional engagement, suggesting these spaces may have induced higher mental workload. Gamma activity, observed most prominently in Rooms 5 and 6 (Complex Floral), is associated with higher-order cognitive integration, possibly reflecting deeper perceptual or learning-related processing. In contrast, Room 4 (Checkers) consistently showed the lowest power across all bands, indicating it may have elicited a relatively neutral or under-stimulating experience. Lastly, the uniform elevation across all bands in Room 6 suggests a multisensory and cognitively rich environment, engaging both affective and cognitive neural systems. Together, these observations reinforce the notion that distinct spatial patterns in VR environments can influence EEG signatures in meaningful ways, even without statistical significance, supporting the value of this framework for immersive architectural design research.

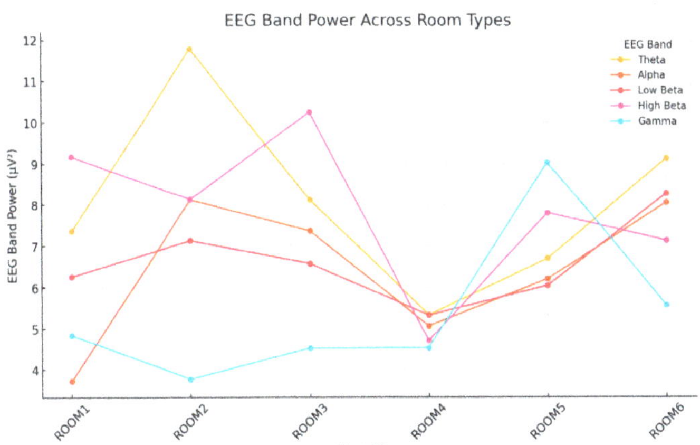

Figure 17. EEG band-power across six virtual room types, showing average μV^2 values for five frequency bands and highlighting spatial variations in neural responses.

3.5.3. Correlation Between Brain Waves

The heatmap in Figure 18 presents the Pearson correlation coefficients [150,151] between EEG brain wave bands—Theta, Alpha, Low Beta, High Beta, and Gamma—recorded while participants experienced various virtual reality gallery environments. The observed correlation values range from −1 to +1, where +1 indicates a perfect positive correlation, −1 a perfect negative correlation, and 0 no linear relationship. In the heatmap, red hues represent strong positive correlations, blue hues denote negative correlations, and lighter tones indicate weak or negligible associations. The most prominent finding was a moderate positive correlation between Theta and Alpha bands (r = 0.32), indicating co-activation during states characterized by calmness and internalized attention. This pattern is consistent with prior research associating Theta activity with meditative focus and Alpha with relaxed wakefulness [152–154]. Additionally, positive correlations between Low Beta and High Beta (r = 0.21) and between High Beta and Gamma (r = 0.29) were observed, suggesting that these fast-frequency bands are jointly activated during tasks requiring heightened attentional control, sensory integration, and executive processing [155–158]. In contrast, weak negative correlations were found between Theta and Beta bands (ranging from −0.17 to −0.18) and between Theta and Gamma (r = −0.19), as well as between Alpha and High Beta/Gamma (r = −0.10 and −0.064, respectively). These inverse relationships point to a functional trade-off between affective immersion and cognitive arousal: when individuals are deeply relaxed or emotionally engaged, brain activity associated with external task focus tends to diminish, and vice versa [152,159]. These findings are particularly relevant in the context of VR-based spatial immersion, where design parameters such as scale, enclosure, color, and ambient qualities can potentially bias the user toward one of these cognitive-emotional states [160,161]. Moreover, research on neuroadaptive environments has shown that EEG indicators like Theta, Alpha, and Beta can be dynamically tracked and modulated in real-time to optimize user experience and attentional balance in VR contexts [162]. Thus, the heatmap not only quantifies neural synchrony but also reveals a bifurcation in spatial experience—between spaces that invite reflection, stillness, and introspection (Theta–Alpha dominant), and those that stimulate action, awareness, and cognitive load (Beta–Gamma dominant). For architectural designers, these results underscore the importance of spatial decisions not merely as aesthetic or functional choices, but as tools to modulate mental states through measurable neurophysiological responses.

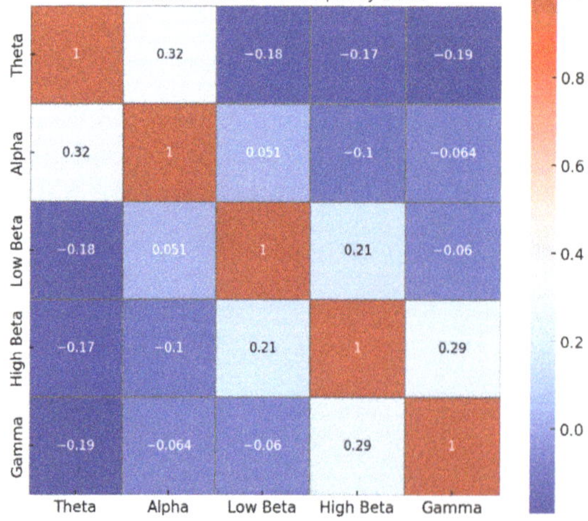

Figure 18. Correlation Matrix of Brain Wave Band Powers.

3.5.4. Comprehensive Analysis of the Experimental Study

The "Comprehensive Analysis of the Experimental Study" table (Table 9) presents a synthesized summary of observed EEG responses across six VR room types. It highlights each room's dominant neural activity patterns and interprets them into corresponding cognitive or emotional experiences and design implications. The Simple Square room was characterized by elevated High Beta activity, an EEG indicator typically associated with focused attention and mental engagement. This observation suggests that space may be well-suited for environments requiring concentration and cognitive control, such as workstations or study zones.

Table 9. Summary of Dominant Experiences, EEG Indicators, and Design Implications Across Different Room Types.

S.No	Room	Dominant Experience	EEG Indicators	Design Implication
1	Simple Square	Cognitive Engagement	High Beta dominance	Suitable for focused, attention-demanding spaces
2	Artisan Print	Deep Relaxation	High Theta, Low Gamma	Ideal for meditation rooms, calm environments
3	Numerous Flowers	Cognitive Stimulation	High Beta, Low Theta	Use in interactive galleries, learning spaces
4	Checkers	Neutral/Mild Relaxation	Low overall activity	Can be avoided; lacks stimulation or relaxation benefit
5	4 Square	Higher-Order Cognitive Processing	Gamma dominance	Suitable for problem-solving spaces, learning environments
6	Complex Floral	Relaxation with Positive Engagement	High Theta and Gamma	Ideal for wellness spaces that require subtle engagement

The Artisan Print room showed increased Theta and reduced Gamma activity, a pattern commonly linked to deep relaxation and reduced cognitive load. This implies that space could effectively support meditative or calming environments, making it ideal for wellness areas or rest zones. The Numerous Flowers room elicited high levels of High Beta and low Theta, indicating heightened cognitive stimulation and alertness. This combination suggests it would be most effective in interactive settings, such as educational galleries or creative learning environments.

In contrast, the Checkers room displayed minimal EEG activity across all bands, suggesting a neutral or mildly relaxing response, which lacks strong cognitive or emotional impact. Based on this observation, it may have limited design value and could be less effective for environments where targeted user experience is desired. The 4 Square room exhibited dominant Gamma activity, often associated with higher-order cognitive processing, suggesting its suitability for problem-solving tasks or educational applications.

Finally, the Complex Floral room demonstrated a combination of elevated Theta and Gamma waves, indicating a blend of relaxation and positive cognitive engagement. This neural signature implies the room could be particularly effective in wellness-focused spaces where subtle mental stimulation is beneficial—such as therapeutic environments or reflective public areas. Overall, these findings offer a bridge between neuroscientific data and design application, enabling evidence-based spatial planning that aligns emotional and cognitive responses with intended environmental functions.

3.6. Validation of Stage-2 and Stage-3 Through Feedback Interviews: Immersive Evaluation and Experience-Informed Refinement

The proposed framework requires validation of Stage-2 (Immersive Evaluation) and Stage-3 (Experience-Informed Refinement in BIM) effectiveness in decision-making. The participants experienced six virtual room environments while answering two open-ended questions to self-assess their immersive design experience and the effects of design refinements. The participants received anonymous treatment to maintain ethical standards and protect their privacy. The study required real-time experiential feedback refinement, so researchers used Revit's native Twinmotion plugin which integrated with Unreal Engine for immersive visualization and spatial interaction.

The two questions posed were as follows:

Q1: Can you describe your overall experience of exploring the six virtual rooms? Which room stood out to you the most and why?

Q2: After your feedback was incorporated, did you notice any meaningful changes or improvements in the refined design version? Describe what stood out. To what extent did the refined design feel more aligned with your expectations or preferences compared to the initial version?

The results from qualitative data analysis of the answers confirm the framework by showing how immersive simulations produce useful experiential feedback (Stage-2) and how this feedback combined with collaborative BIM-based refinement (Stage-3) produces more satisfactory and user-aligned design outputs. The results of Question 1 are presented in Tables 10 and 11, and those from Question 2 are summarized in Table 12.

Table 10. Results from Question 1: Participant Room Preferences and Subjective reasoning.

Participant	Standout Room	Subjective Description
P1	Room 3	Engaging and clear.
P2	Room 5	Cognitive clarity and balance.
P3	Room 2	Calming and repetitive.
P4	Room 4	Practical flow but lacked emotion.
P5	Room 6	Emotionally grounded.
P6	Room 5	Definitive arrangement.
P7	Room 1	Minimalist and less engaging.
P8	Room 3	Balanced, not too complex.
P9	Room 4	Familiar but emotionally flat.
P10	Room 6	Positive ambiance and sensory feedback.

Table 11. Results from Question 1: Participant room preferences, Corresponding EEG-Based Spatial Design Matches.

Room	EEG Pattern Observed	Suggested Design Use
Room 1	High Beta >> Cognitive Engagement	Focus/Attention-heavy spaces
Room 2	High Theta & Low Gamma >> Deep Relaxation	Calm zones or meditation areas
Room 3	High Beta & Low Theta >> Cognitive Stimulation	Interactive/learning spaces
Room 4	Low Overall Activity >> Neutral/Mild Relaxation	Least stimulating or neutral environments
Room 5	Gamma Dominance >> Higher-Order Cognition	Problem-solving or educational settings
Room 6	High Theta & Gamma >> Relaxation + Positive Engagement	Wellness or reflective environments

Table 12. Results from Question 2: Participant Feedback on Design Refinements and Resulting Spatial Alignment.

Participant	Changes Made as Refinement to the Existing Model	Design Alignment
P1	Enhanced spatial proportions.	Refined Room 3 felt more engaging and cognitively stimulating.
P2	Adjusted enclosure and openness.	Refined Room 5 better supported deeper cognitive tasks.
P3	Improved flow and calming transitions.	Room 2 became more immersive for relaxation purposes.
P4	Accentuate hierarchy and spatial rhythm.	Room 4 remained less stimulating; preferred more cognitive engagement.
P5	Better materiality and ambient effects.	Room 6 felt richer in multisensory cues, improving positive engagement.
P6	Better zoning and clarity in layout.	Room 5 supported clearer problem-solving and cognitive focus.
P7	Retained minimal abstraction.	Room 1 remained suitable for attention-heavy environments.
P8	Need for Stronger behavioral cues and transitions.	Room 3 fostered more interactive engagement, as desired.
P9	More performance feedback is integrated to reduce monotony.	Room 4 improved in function but still lacked emotional depth.
P10	Need to add texture to enhance sensory balance and user engagement.	Room 6 was closest to wellness-based experiential intent.

Table 10 reports participants' subjective responses to Question 1, highlighting which virtual room stood out most and why. The observed data show that participants expressed a wide range of preferences—some, like P1, P2, and P8, favored rooms (Room 3 and Room 5) that they described as offering "engagement", "clarity", or "balanced configuration". These responses suggest that these rooms were perceived as cognitively stimulating or supportive of focused thinking. Others, such as P5 and P10, identified emotional and atmospheric qualities, describing Room 6 as "emotionally grounded" or possessing "positive spatial feedback", pointing to a deeper affective resonance with the environment.

In contrast, participants like P4 and P9 mentioned Room 4 (Checkers) as familiar but emotionally lacking, reflecting a more neutral or underwhelming spatial experience. These observations imply that participants could discern and communicate nuanced differences in the experiential quality of each room. The thematic diversity in responses—from mental clarity to emotional grounding—demonstrates that immersive VR effectively evokes a range of cognitive and affective impressions. This supports the framework's validity in using VR environments to elicit meaningful user feedback in early-stage architectural design evaluations.

Table 11 presents the observed EEG-based classifications for the six virtual rooms, identifying their dominant frequency bands and associated cognitive-affective states. For instance, Room 2 (Artisan Print) showed elevated Theta and reduced Gamma activity—an EEG pattern commonly associated with deep relaxation and inward-focused attention, suggesting the room induced a calming, meditative experience. Similarly, Room 5 (4 Square) was characterized by Gamma dominance, which is often linked to higher-order cognitive functions such as problem-solving and integrative thinking, indicating its potential effectiveness in environments designed to support focused mental tasks.

In contrast, Room 4 (Checkers) exhibited uniformly low EEG activity across all frequency bands, implying the room elicited minimal cognitive stimulation or emotional engagement. This helps explain participants' more neutral or underwhelming subjective responses to this space. These physiological patterns objectively validate the user-experienced spatial qualities recorded in Table 11. The match between measurable brainwave activity and participant-reported experiences reinforces the reliability of the framework, demonstrating its ability to connect immersive spatial design with real-time biometric feedback.

By aligning neural data with experiential feedback, this EEG-based classification system supports the broader goal of developing evidence-based, user-centered design approaches in architecture. It highlights the value of physiological metrics as a rigorous method for evaluating affective and cognitive responses to space, paving the way for more responsive and personalized spatial environments.

Table 12 presents observed changes in design outcomes based on participant feedback during the validation phase. Specific modifications—such as spatial proportion enhancements (P1), enclosure adjustments (P2), and improved zoning (P6)—were implemented in Rooms 3 and 5. These changes led to clearer spatial organization and heightened cognitive stimulation, aligning these rooms more closely with their intended interactive and problem-solving functions. Similarly, participants seeking relaxation or wellness (P3, P5, P10) identified improvements in spatial flow, materiality, and sensory textures. These refinements enhanced Rooms 2 and 6, making them more immersive and emotionally resonant for meditative or restorative experiences.

In contrast, feedback from P7 and P8 emphasized maintaining minimalist abstraction or integrating stronger behavioral cues—reflecting diverse user expectations related to cognitive engagement and spatial legibility. Room 4, which received layout and feedback optimizations from P4 and P9, still failed to evoke a strong emotional connection, reaffirming earlier findings from immersive evaluation that it was the least affectively engaging environment.

These results indicate that the feedback-driven refinements were meaningful and effectively aligned design outputs with user expectations. The findings underscore the interpretive value of combining immersive VR evaluation with EEG-based tracking, which captures subtle affective and cognitive responses. As previous research supports, this dual-method approach reveals experiential patterns that might be overlooked in conventional design reviews. Consequently, EEG-informed experiential feedback—when integrated into BIM-based design workflows—can enhance the architectural design process by aligning spatial intent with end-user experience, advancing the field of human-centered, evidence-based design.

4. Discussion

This study explores how user-centered neurophysiological feedback can meaningfully enhance conceptual design evaluation within BIM workflows—an area historically dominated by technical and performance-based metrics. By employing EEG and immersive VR within a BIM-integrated framework, this research demonstrates a novel, yet feasible, approach for capturing real-time cognitive and affective responses to spatial stimuli. The Systematic Literature Review (SLR) and semi-structured interviews were instrumental in identifying gaps in current BIM workflows, particularly the underutilization of affective metrics and practitioners' desire for more intuitive, feedback-driven design tools. These findings informed the structure of the three-stage framework and ensured its alignment with both theoretical and professional design practices.

While the pilot study utilized a small participant pool ($n = 10$), the observed moderate effect size (partial $\eta^2 = 0.254$) aligns with similar exploratory EEG-VR studies in architecture. For example, Xia et al. [163] and Banaei et al. [164] reported statistically significant findings using 10–15 participants, but only under distinct stimulus contrasts. These precedents affirm that smaller architectural manipulations may not always yield statistically significant outcomes, but they can still produce meaningful neural trends worth further investigation. In this study, the absence of statistically significant differences in EEG band-power across six rooms ($p = 0.189$) was accompanied by notable perceptual differentiation. Approximately 25% of the variance in neural response could be attributed to room type—a result comparable to previous VR-EEG studies such as Jelic et al. [161], where moderate η^2 values were interpreted as indicators of early-stage differentiation. This study, therefore, adopts a similar interpretation: it views these EEG patterns as exploratory signals that, while not definitive, contribute to a growing body of knowledge on neuroarchitecture.

This interpretation gains further credibility from Taherysayah et al. [84], who argue that small-N EEG studies in neuroarchitecture should not be dismissed when effect sizes are moderate and supported by qualitative convergence. In this research, subjective reports aligned with EEG trends—cognitively stimulating spaces elicited higher Beta/Gamma activity, while emotionally grounding environments elevated Theta/Alpha rhythms—supporting the framework's underlying assumption that architectural stimuli can be meaningfully linked to measurable experiential feedback.

Furthermore, Stage-2 and Stage-3 of the framework—Immersive Evaluation and Experience-Informed Refinement—were validated qualitatively through post-experience feedback interviews. Participants acknowledged improvements in spatial clarity, zoning, and ambiance following iterative design modifications based on experiential input. This validation reinforces the practical feasibility of integrating biometric data with subjective assessment in early-stage BIM design workflows.

This study contributes to the expanding discourse on neurophysiological methods in architectural evaluation [143,164–166]. The integration of EEG [145,146] into a BIM-VR framework [41,123,167–169] not only confirms the feasibility of such a workflow but also emphasizes EEG's potential in capturing attention, relaxation, and cognitive load in a spatial context. The framework's methodological foundations are well-aligned with recognized design models such as the RIBA Plan of Work [23], the Double Diamond [170,171], and IDEO's Human-Centered Design approach [172]. This alignment underpins its versatility across both architecture and interdisciplinary design domains that prioritize empathy, iteration, and feedback.

The proposed framework can be selectively adopted depending on project size and complexity. For smaller projects, the emphasis can be placed on experience prototyping and rapid iteration to facilitate quick design decisions, with tools like WEBBIM being particularly useful for such scenarios. In contrast, larger projects benefit from the framework's capacity to support multi-stakeholder visualization and collaborative review, leveraging legacy software to manage more detailed processes. This framework is designed to handle both scenarios effectively, though it must be tested across various case-based scenarios to ensure its versatility and scalability.

Lastly, while the framework demonstrates clear adaptability to areas like Wellbeing [165,173,174], Universal Design [117], and Ambient Assisted Living [175], its long-term impact will depend on addressing social and cultural variability in spatial experience. Future iterations should account for individual differences in perception shaped by age, gender, culture, and design literacy. Additionally, incorporating multimodal sensing technologies such as eye-tracking, GSR, or heart rate variability will improve granularity, increase validity, and expand the scope of experiential design research.

5. Conclusions

This pilot study presents a novel, interdisciplinary framework that bridges a long-standing gap in early-stage architectural design: integrating real-time, user-centered experiential feedback into BIM workflows. The framework introduces three interconnected stages—experiential Design Ideation, Immersive Evaluation, and Experience-Informed Refinement—that directly embed affective and cognitive insights into the design process using immersive VR and EEG data.

Grounded in a Systematic Literature Review and architect interviews, the research identified recurring limitations in current BIM practices and strong practitioner interest in intuitive, immersive, and evaluative technologies. To operationalize this insight, a pilot study was conducted, in which participants experienced six VR-designed room typologies. While statistical significance was not achieved, the study observed clear neural trends correlated with subjective spatial impressions, underscoring the framework's potential for affective differentiation in architectural stimuli.

5.1. *Key Contributions*

This study presents a framework combining BIM, VR, and EEG to assess spatial experience in design. Empirical results show EEG's sensitivity to cognitive and emotional responses, while practical insights illustrate how experiential feedback can enhance early design through data-driven integration. The primary contributions are as follows:

- Proposed a new framework (methods) that unifies BIM, VR, and EEG to evaluate spatial experience during conceptual design;
- Demonstrated EEG's sensitivity to cognitive and emotional responses even in a small-N, proof-of-concept setting;
- Practical: illustrated how experiential feedback can be systematically embedded into early design workflows, offering a data-rich complement to intuitive ideation.

5.2. *Limitations*

This study, while offering valuable insights into integrating experiential feedback within BIM workflows, has several limitations, which must be acknowledged to contextualize the scope and applicability of the findings.

- Exploratory Scope and Sample Size: the study was exploratory in nature, involving a limited sample size (n = 10) with no control group, which restricts statistical generalizability.
- Participant Profile: the participant cohort consisted solely of architecture students, which may not reflect the responses of professional practitioners or diverse user groups.
- Short-Duration and Simulated Exposure: EEG data were recorded during brief simulated spatial experiences and may not represent long-term or ecologically valid interactions with built environments.
- Controlled Lab Setting: EEG recordings were conducted with participants seated in a controlled environment to reduce motion artifacts. This does not fully replicate real-world engagement scenarios, such as dynamic movement within immersive spaces.
- Subjective Feedback Bias: While the study triangulated EEG data with qualitative self-reports, subjective feedback may be prone to perceptual and reporting bias, despite reflexive measures adopted during thematic analysis.
- Framework Development Stage: The BIM-VR-EEG integration remains in the proof-of-concept phase and has not yet been tested in real-world BIM execution workflows or project team collaborations.

5.3. *Future Research Directions*

Future work will expand sample diversity, integrate multimodal sensing, and conduct longitudinal studies in real-world design settings. Developing BIM plug-ins for real-time

EEG integration will support seamless, iterative evaluation within design tools. The significant aspects of further study are as follows:

- Increase in sample diversity and size for statistical validation;
- Introducing multimodal physiological sensing (e.g., GSR, HRV, eye tracking) for richer feedback integration;
- Conducting longitudinal studies across diverse typologies and real-world studio workflows;
- Developing BIM plug-ins for real-time EEG data integration, enabling seamless iterative evaluation in design tools.

This study offers a compelling proof-of-concept that EEG-informed VR-BIM workflows can capture experiential data in meaningful ways. It lays a strong foundation for affect-responsive architectural design and signals a shift toward a more empathetic, user-aligned, and cognitively informed future in the built environment.

Supplementary Materials: The following supporting information can be downloaded at: https://www.mdpi.com/article/10.3390/infrastructures10060127/s1. Semi-structured Interviews.

Author Contributions: B.R.: Conceptualization, Methodology, Investigation, Formal analysis, and Data visualization and Writing—original draft. R.S.P.: Conceptualization, Methodology, Investigation, Validation, Writing—Review, and Supervision. G.R.K.: Formal analysis, Investigation, and Writing—Review. J.T.: Formal analysis, Investigation, and Writing—Review. R.S.: Methodology, Formal analysis, Investigation, and Writing—Review. All authors have read and agreed to the published version of the manuscript.

Funding: The authors gratefully acknowledge the financial support received from SRM Institute of Science and Technology (SRMIST), Kattankulathur Campus, for conducting this research. This work was funded under the Selective Excellence Research Initiative (SERI) 2024, which enabled the successful execution of experimental studies and data analysis integral to this project.

Informed Consent Statement: Informed consent was obtained from the participants.

Data Availability Statement: All the data utilized for the research are available within the article itself. The semi-structured interview report will be made available upon request.

Acknowledgments: The authors would like to acknowledge the invaluable support of students from the Department of Biomedical Engineering, SRM Institute of Science and Technology (SRMIST), for their assistance setting up the experimental protocol and facilitating participant engagement for this research. We give special thanks to Dhanush Kumar, Ismail Mohmed, and Varagunarajan K for their dedicated contributions and technical support throughout the study.

Conflicts of Interest: The authors declare no conflicts of interest.

References

1. Kanyilmaz, A.; Hoi Dang, V.; Kondratenko, A. How does conceptual design impact the cost and carbon footprint of structures? *Structures* **2023**, *58*, 105102. [CrossRef]
2. Wang, Y.; Liu, A.; Tao, F.; Nee, A.Y.C. Chapter 2—Digital twin driven conceptual design. In *Digital Twin Driven Smart Design*; Tao, F., Liu, A., Hu, T., Nee, A.Y.C., Eds.; Academic Press: Cambridge, MA, USA, 2020; pp. 33–66. [CrossRef]
3. Hettithanthri, U.; Hansen, P.; Munasinghe, H. Exploring the architectural design process assisted in conventional design studio: A systematic literature review. *Int. J. Technol. Des. Educ.* **2023**, *33*, 1835–1859. [CrossRef]
4. Pietrzyk, K. Wicked Problems in Architectural Research: The Role of Research by Design. *ARENA J. Archit. Res.* **2022**, *7*, 3. [CrossRef]
5. Hocking, V.T.; Brown, V.A.; Harris, J.A. Tackling wicked problems through collective design. *Intell. Build. Int.* **2016**, *8*, 24–36. [CrossRef]
6. Tabassum, R.R.; Park, J. Development of a Building Evaluation Framework for Biophilic Design in Architecture. *Buildings* **2024**, *14*, 3254. [CrossRef]

7. Min, I.A.; Noguchi, R.A. The architecture design and evaluation process: A decision support framework for conducting and evaluating architecture studies. In Proceedings of the 2016 IEEE Aerospace Conference, Big Sky, MT, USA, 5–12 March 2016; pp. 1–14. [CrossRef]
8. Crilly, N. *Design Thinking and Other Approaches: How Different Disciplines See, Think and Act*, 1st ed.; Cambridge University Press: Cambridge, UK, 2024. [CrossRef]
9. Peng, F. Design Thinking: From Empathy to Evaluation. In *Foundations of Robotics: A Multidisciplinary Approach with Python and ROS*; Herath, D., St-Onge, D., Eds.; Springer Nature: Singapore, 2022; pp. 63–81. [CrossRef]
10. Kwon, J.; Choi, Y.; Hwang, Y. Enterprise Design Thinking: An Investigation on User-Centered Design Processes in Large Corporations. *Designs* **2021**, *5*, 43. [CrossRef]
11. Eissa, D. Concept generation in the architectural design process: A suggested hybrid model of vertical and lateral thinking approaches. *Think. Ski. Creat.* **2019**, *33*, 100589. [CrossRef]
12. Li, Y.; Chen, H.; Yu, P.; Yang, L. A Review of Artificial Intelligence in Enhancing Architectural Design Efficiency. *Appl. Sci.* **2025**, *15*, 1476. [CrossRef]
13. Brozovsky, J.; Labonnote, N.; Vigren, O. Digital technologies in architecture, engineering, and construction. *Autom. Constr.* **2024**, *158*, 105212. [CrossRef]
14. Seghier, T.E.; Khosakitchalert, C.; Liu, Z.; Ohueri, C.C.; Lim, Y.-W.; Bin Zainazlan, A.F. From BIM to computational BIM: A systematic review of visual programming application in building research. *Ain Shams Eng. J.* **2025**, *16*, 103173. [CrossRef]
15. Altwassi, E.J.; Aysu, E.; Ercoskun, K.; Abu Raed, A. From Design to Management: Exploring BIM's Role across Project Lifecycles, Dimensions, Data, and Uses, with Emphasis on Facility Management. *Buildings* **2024**, *14*, 611. [CrossRef]
16. Bryde, D.; Broquetas, M.; Volm, J.M. The project benefits of Building Information Modelling (BIM). *Int. J. Proj. Manag.* **2013**, *31*, 971–980. [CrossRef]
17. Raitviir, C.-R.; Lill, I. Conceptual Framework of Information Flow Synchronization Throughout the Building Lifecycle. *Buildings* **2024**, *14*, 2207. [CrossRef]
18. Ali, K.N.; Alhajlah, H.H.; Kassem, M.A. Collaboration and Risk in Building Information Modelling (BIM): A Systematic Literature Review. *Buildings* **2022**, *12*, 571. [CrossRef]
19. Bitaraf, I.; Salimpour, A.; Elmi, P.; Shirzadi Javid, A.A. Improved Building Information Modeling Based Method for Prioritizing Clash Detection in the Building Construction Design Phase. *Buildings* **2024**, *14*, 3611. [CrossRef]
20. Borkowski, A.S. Evolution of BIM: Epistemology, genesis and division into periods. *J. Inf. Technol. Constr. ITcon* **2023**, *28*, 646–661. [CrossRef]
21. Hyarat, E.; Hyarat, T.; Al Kuisi, M. Barriers to the Implementation of Building Information Modeling among Jordanian AEC Companies. *Buildings* **2022**, *12*, 150. [CrossRef]
22. Wu, P.; Jin, R.; Xu, Y.; Lin, F.; Dong, Y.; Pan, Z. The analysis of barriers to bim implementation for industrialized building construction: A China study. *J. Civ. Eng. Manag.* **2021**, *27*, 1–13. [CrossRef]
23. Martin, H.; Watson, C.; Brooks, T. Synergizing BIM and RIBA in architectural practice—Technology workflow efficiencies, challenges, and insights. *Archit. Eng. Des. Manag.* **2025**, *21*, 35–59. [CrossRef]
24. Aladağ, H.; Demirdöğen, G.; Demirbağ, A.T.; Işık, Z. Understanding the perception differences on BIM adoption factors across the professions of AEC industry. *Ain Shams Eng. J.* **2023**, *14*, 102545. [CrossRef]
25. Coates, P.; Arayici, Y.; Koskela, L.; Kagioglou, M.; Usher, C.; O'Reilly, K. The Limitations of BIM in the Architectural Process. In Proceedings of the ICSU 2010: First International Conference on Sustainable Urbanization 2010, Hong Kong, China, 15–17 December 2010; pp. 2–10. Available online: http://eprints.hud.ac.uk/id/eprint/19678/ (accessed on 20 May 2025).
26. Andersson, R.; Eidenskog, M. Beyond barriers—Exploring resistance towards BIM through a knowledge infrastructure framework. *Constr. Manag. Econ.* **2023**, *41*, 926–941. [CrossRef]
27. Lu, W.; Peng, Y.; Shen, Q.; Li, H. Generic Model for Measuring Benefits of BIM as a Learning Tool in Construction Tasks. *J. Constr. Eng. Manag.* **2013**, *139*, 195–203. [CrossRef]
28. Furuta, S.; Nakazato, J.; Tsukada, M. Web-Based BIM Platform for Building Digital Twin. In Proceedings of the 2023 IEEE 3rd International Conference on Digital Twins and Parallel Intelligence (DTPI), Orlando, FL, USA, 7–9 November 2023; pp. 1–6. [CrossRef]
29. Hagedorn, P.; Block, M.; Zentgraf, S.; Sigalov, K.; König, M. Toolchains for Interoperable BIM Workflows in a Web-Based Integration Platform. *Appl. Sci.* **2022**, *12*, 5959. [CrossRef]
30. Abanda, F.H.; Balu, B.; Adukpo, S.E.; Akintola, A. Decoding ISO 19650 Through Process Modelling for Information Management and Stakeholder Communication in BIM. *Buildings* **2025**, *15*, 431. [CrossRef]
31. Ashworth, S.; Dillinger, M.; Körkemeyer, K. BIM guidance to optimise the operational phase: Defining information requirements based on ISO 19650. *Facilities* **2022**, *41*, 337–356. [CrossRef]
32. Godager, B.; Mohn, K.; Merschbrock, C.; Klakegg, O.J.; Huang, L. Towards an improved framework for enterprise BIM: The role of ISO 19650. *J. Inf. Technol. Constr. ITcon* **2022**, *27*, 1075–1103. [CrossRef]

33. East, E. An Overview of the u. s. National Building Information Model Standard (Nbims). *Comput. Civ. Eng.* **2007**, 59–66. [CrossRef]
34. Nawari, N.O.; Sgambelluri, M. The Role of National BIM Standard in Structural Design. In *Structures Congress*; ASCE: Reston, VA, USA, 2012; pp. 1660–1671. [CrossRef]
35. Ibrahim, M.Y.; Al-Kazzaz, D.A. A comparative analysis of BIM standards and guidelines between UK and USA. *J. Phys. Conf. Ser.* **2021**, *1973*, 012176. [CrossRef]
36. Bui, N.; Merschbrock, C.; Munkvold, B.E.; Hjelseth, E. Role of An Innovation Community in Supporting Bim Deployment: The Case of Buildingsmart Norway. In *WIT Transactions on The Built Environment*; WIT Press: Seville, Spain, 2019. [CrossRef]
37. Borrmann, A.; Amann, J.; Chipman, T.; Hyvärinen, J.; Liebich, T.; Muhič, S.; Plume, J.; Mol, L.; Scarponcini, P. IFC Infra Overall Architecture Project Documentation and Guidelines 2017. Available online: https://www.buildingsmart.org/wp-content/uploads/2017/07/08_bSI_OverallArchitecure_Guidelines_final.pdf (accessed on 17 May 2025).
38. Lama, S.; Fu, C. BIM and Design Briefing Integration: Feasibility Study and Conceptual System Framework Development. *J. Constr. Res.* **2021**, *2*. Available online: https://journals.bilpubgroup.com/index.php/jcr-b/article/download/2960/2783 (accessed on 20 May 2025).
39. Jelić, A. Designing "pre-reflective" architecture: Implications of neurophenomenology for architectural design and thinking. *Ambiances* **2015**, *1*. [CrossRef]
40. Kaklauskas, A.; Ubarte, I.; Tupenaite, L.; Raupys, D. Video neuro-advertising recommender model for affective BIM. In Proceedings of the 2018 7th International Conference on Computers Communications and Control (ICCCC), Oradea, Romania, 8–12 May 2018; pp. 246–251. [CrossRef]
41. Kim, J.I.; Li, S.; Chen, X.; Keung, C.; Suh, M.; Kim, T.W. Evaluation framework for BIM-based VR applications in design phase. *J. Comput. Des. Eng.* **2021**, *8*, 910–922. [CrossRef]
42. Johansson, M.; Roupé, M. Real-world applications of BIM and immersive VR in construction. *Autom. Constr.* **2024**, *158*, 105233. [CrossRef]
43. Shim, J.-H.; Ham, N.-H.; Kim, J.-J. Collaborative BIM-Based Construction Coordination Progress Modeling Using Extended Collaborative Process Modeling (exCPM). *Buildings* **2024**, *14*, 358. [CrossRef]
44. Onungwa, I.; Olugu-Uduma, N.; Shelden, D.R. Cloud BIM Technology as a Means of Collaboration and Project Integration in Smart Cities. *SAGE Open* **2021**, *11*, 21582440211033250. [CrossRef]
45. Chahrour, R.; Hafeez, M.A.; Ahmad, A.M.; Sulieman, H.I.; Dawood, H.; Rodriguez-Trejo, S.; Kassem, M.; Naji, K.K.; Dawood, N. Cost-benefit analysis of BIM-enabled design clash detection and resolution. *Constr. Manag. Econ.* **2021**, *39*, 55–72. [CrossRef]
46. Akponeware, A.O.; Adamu, Z.A. Clash Detection or Clash Avoidance? An Investigation into Coordination Problems in 3D BIM. *Buildings* **2017**, *7*, 75. [CrossRef]
47. Kim, H.-S.; Kim, S.-K.; Kang, L.-S. BIM performance assessment system using a K-means clustering algorithm. *J. Asian Archit. Build. Eng.* **2021**, *20*, 78–87. [CrossRef]
48. Jin, R.; Zhong, B.; Ma, L.; Hashemi, A.; Ding, L. Integrating BIM with building performance analysis in project life-cycle. *Autom. Constr.* **2019**, *106*, 102861. [CrossRef]
49. Bailey, G.; Kammler, O.; Weiser, R.; Fuchkina, E.; Schneider, S. Performing Immersive Virtual Environment User Studies with VREVAL. In *Co-Creating the Future: Inclusion in and Through Design, Proceedings of the 40th Conference on Education and Research in Computer Aided Architectural Design in Europe (eCAADe 2022), Ghent, Belgium, 13–16 September 2022*; Pak, B., Wurzer, G., Stouffs, R., Eds.; SCIX Open publishing: Washington, DC, USA, 2022; Volume 2, pp. 437–446. [CrossRef]
50. Hermund, A. Human Computer Interacting Through a Game Engine: Qualifying Inclusive Design in Architecture. In Proceedings of the Human-Computer Interaction. Theory, Methods and Tools; Kurosu, M., Ed.; Springer International Publishing: Cham, Switzerland, 2021; pp. 30–49. [CrossRef]
51. Graham, K.; Chow, L.; Fai, S. From BIM to VR: Defining a level of detail to guide virtual reality narratives. *J. Inf. Technol. Constr.* **2019**, *24*, 553–568. [CrossRef]
52. Podkosova, I.; Reisinger, J.; Kaufmann, H.; Kovacic, I. BIMFlexi-VR: A Virtual Reality Framework for Early-Stage Collaboration in Flexible Industrial Building Design. *Front. Virtual Real.* **2022**, *3*, 782169. [CrossRef]
53. Renganathan, B.; Shanthi Priya, R.; Senthil, R. Pre-Occupancy Evaluation of Affective Experiences in Building Information Modeling Through SLR and an e-Delphi Survey. *Technologies* **2024**, *13*, 14. [CrossRef]
54. Gómez-Sirvent, J.L.; Fernández-Sotos, D.; López de la Rosa, F.; Fernández-Caballero, A. Building information modeling and affective occupancy evaluation: A scoping review. *J. Ambient Intell. Smart Environ.* **2023**, 1–12. [CrossRef]
55. Carrera-Rivera, A.; Ochoa, W.; Larrinaga, F.; Lasa, G. How-to conduct a systematic literature review: A quick guide for computer science research. *MethodsX* **2022**, *9*, 101895. [CrossRef] [PubMed]
56. Moher, D.; Liberati, A.; Tetzlaff, J.; Altman, D.G.; the PRISMA Group. Preferred Reporting Items for Systematic Reviews and Meta-Analyses: The PRISMA Statement. *Ann. Intern. Med.* **2009**, *151*, 264–269. [CrossRef] [PubMed]

57. Lee, J.H.; Ostwald, M.J.; Arasteh, S.; Oldfield, P. BIM-Enabled Design Collaboration Processes in Remote Architectural Practice and Education in Australia. *J. Archit. Eng.* **2023**, *29*, 05022012. [CrossRef]
58. Magaldi, D.; Berler, M. Semi-structured Interviews. In *Encyclopedia of Personality and Individual Differences*; Zeigler-Hill, V., Shackelford, T.K., Eds.; Springer International Publishing: Cham, Switzerland, 2020; pp. 4825–4830. [CrossRef]
59. Adams, W.C. Conducting Semi-Structured Interviews. In *Handbook of Practical Program Evaluation*; John Wiley & Sons, Ltd.: Hoboken, NJ, USA, 2015; pp. 492–505. [CrossRef]
60. Kvale, S. *Interviews: An Introduction to Qualitative Research Interviewing*; Sage Publications: Thousand Oaks, CA, USA, 1996.
61. Aburamadan, R.; Trillo, C. Applying design science approach to architectural design development. *Front. Archit. Res.* **2020**, *9*, 216–235. [CrossRef]
62. Runeson, P.; Engström, E.; Storey, M.-A. The Design Science Paradigm as a Frame for Empirical Software Engineering. In *Contemporary Empirical Methods in Software Engineering*; Felderer, M., Travassos, G.H., Eds.; Springer International Publishing: Cham, Switzerland, 2020; pp. 127–147. [CrossRef]
63. Patton, M.Q. Two Decades of Developments in Qualitative Inquiry: A Personal, Experiential Perspective. *Qual. Soc. Work* **2002**, *1*, 261–283. [CrossRef]
64. Adeoye-Olatunde, O.A.; Olenik, N.L. Citation for: Research and scholarly methods: Semi-structured interviews. *Am. Coll. Clin. Pharm.* **2021**, *4*, 1358–1367. Available online: https://accpjournals.onlinelibrary.wiley.com/action/showCitFormats?doi=10.1002/jac5.1441&mobileUi=0. (accessed on 20 April 2025). [CrossRef]
65. Corbin, J.M.; Strauss, A. Grounded theory research: Procedures, canons, and evaluative criteria. *Qual. Sociol.* **1990**, *13*, 3–21. [CrossRef]
66. Gioia, D.A.; Corley, K.G.; Hamilton, A.L. Seeking Qualitative Rigor in Inductive Research: Notes on the Gioia Methodology. *Organ. Res. Methods* **2013**, *16*, 15–31. [CrossRef]
67. Spradley, J.P. *The Ethnographic Interview*; Wadsworth: Belmont, CA, USA, 2011.
68. McCracken, G. *The Long Interview*; SAGE Publications, Inc.: Thousand Oaks, CA, USA, 1988. [CrossRef]
69. Guest, G.; Bunce, A.; Johnson, L. How Many Interviews Are Enough?: An Experiment with Data Saturation and Variability. *Field Methods* **2006**, *18*, 59–82. [CrossRef]
70. Palinkas, L.A.; Horwitz, S.M.; Green, C.A.; Wisdom, J.P.; Duan, N.; Hoagwood, K. Purposeful sampling for qualitative data collection and analysis in mixed method implementation research. *Adm. Policy Ment. Health* **2015**, *42*, 533–544. [CrossRef] [PubMed]
71. Rahimi, S.; Khatooni, M. Saturation in qualitative research: An evolutionary concept analysis. *Int. J. Nurs. Stud. Adv.* **2024**, *6*, 100174. [CrossRef] [PubMed]
72. Hennink, M.; Kaiser, B.N. Sample sizes for saturation in qualitative research: A systematic review of empirical tests. *Soc. Sci. Med.* **2022**, *292*, 114523. [CrossRef]
73. Olmos-Vega, F.M.; Stalmeijer, R.E.; Varpio, L.; Kahlke, R. A practical guide to reflexivity in qualitative research: AMEE Guide No. 149. *Med. Teach.* **2022**, *45*, 241–251. [CrossRef]
74. Braun, V.; Clarke, V. Using thematic analysis in psychology. *Qual. Res. Psychol.* **2006**, *3*, 77–101. [CrossRef]
75. Abdurrahman, A.; Gustomo, A.; Prasetio, E.A.; Rustiadi, S. Designing an Open Innovation Framework for Digital Transformation Based on Systematic Literature Review. *J. Inf. Syst. Eng. Bus. Intell.* **2022**, *8*, 100–108. [CrossRef]
76. Calancie, L.; Frerichs, L.; Davis, M.M.; Sullivan, E.; White, A.M.; Cilenti, D.; Corbie-Smith, G.; Hassmiller Lich, K. Consolidated Framework for Collaboration Research derived from a systematic review of theories, models, frameworks and principles for cross-sector collaboration. *PLoS ONE* **2021**, *16*, e0244501. [CrossRef]
77. McMeekin, N.; Wu, O.; Germeni, E.; Briggs, A. How methodological frameworks are being developed: Evidence from a scoping review. *BMC Med. Res. Methodol.* **2020**, *20*, 173. [CrossRef]
78. Malmqvist, J.; Hellberg, K.; Möllås, G.; Rose, R.; Shevlin, M. Conducting the Pilot Study: A Neglected Part of the Research Process? Methodological Findings Supporting the Importance of Piloting in Qualitative Research Studies. *Int. J. Qual. Methods* **2019**, *18*, 1609406919878341. [CrossRef]
79. Reynolds, P.S. *A Guide to Sample Size for Animal-Based Studies*, 1st ed.; Wiley: Hoboken, NJ, USA, 2023. [CrossRef]
80. Zangeneh Soroush, M.; Zeng, Y. EEG-based study of design creativity: A review on research design, experiments, and analysis. *Front. Behav. Neurosci.* **2024**, *18*, 1331396. [CrossRef] [PubMed]
81. Kim, N.; Chung, S.; Kim, D.I. Exploring EEG-based Design Studies: A Systematic Review. *Arch. Des. Res.* **2022**, *35*, 91–113. [CrossRef]
82. Soufineyestani, M.; Dowling, D.; Khan, A. Electroencephalography (EEG) Technology Applications and Available Devices. *Appl. Sci.* **2020**, *10*, 7453. [CrossRef]
83. Cho, J.E.; Kang, S.Y.; Hong, Y.Y.; Jun, H.J. Predicting Architectural Space Preferences Using EEG-Based Emotion Analysis: A CNN-LSTM Approach. *Appl. Sci.* **2025**, *15*, 4217. [CrossRef]

84. Taherysayah, F.; Malathouni, C.; Liang, H.-N.; Westermann, C. Virtual reality and electroencephalography in architectural design: A systematic review of empirical studies. *J. Build. Eng.* **2024**, *85*, 108611. [CrossRef]
85. Eudave, L.; Vourvopoulos, A. Multimodal mapping of spatial attention for unilateral spatial neglect in VR: A proof of concept study using eye-tracking and mobile EEG. *Virtual Real.* **2025**, *29*, 24. [CrossRef]
86. Şekerci, Y.; Kahraman, M.U.; Özturan, Ö.; Çelik, E.; Ayan, S.Ş. Neurocognitive responses to spatial design behaviors and tools among interior architecture students: A pilot study. *Sci. Rep.* **2024**, *14*, 4454. [CrossRef]
87. Pandey, R.; Sorathia, K. A Pilot Study on Hand-Referenced Menu User Interfaces for Head-Mounted Display Virtual Reality. In Proceedings of the 27th International Academic Mindtrek Conference, Tampere, Finland, 8–11 October 2024; Association for Computing Machinery: New York, NY, USA, 2024; pp. 260–263. [CrossRef]
88. Paes, D.; Irizarry, J. Virtual Reality Technology Applied in the Building Design Process: Considerations on Human Factors and Cognitive Processes. *Adv. Intell. Syst. Comput.* **2016**, *485*, 3–15. [CrossRef]
89. Amini Gougeh, R.; Falk, T.H. Enhancing motor imagery detection efficacy using multisensory virtual reality priming. *Front. Neuroergon.* **2023**, *4*, 1080200. [CrossRef]
90. Choi, H.S.; Bruyns, G.; Zhang, W.; Cheng, T.; Sharma, S. Spatial Cognition and Three-Dimensional Vertical Urban Design Guidelines—Cognitive Measurement and Modelling for Human Centre Design. *Urban Sci.* **2023**, *7*, 125. [CrossRef]
91. Spark, J.; Rowe, E.; Alvarez-Jimenez, M.; Bell, I.; Byrne, L.; Dzafic, I.; Ellinghaus, C.; Lavoie, S.; Lum, J.; McLean, B.; et al. Integrating Virtual Reality, Neurofeedback, and Cognitive Behavioral Therapy for Auditory Verbal Hallucinations (Hybrid): Protocol of a Pilot, Unblinded, Single-Arm Interventional Study. *JMIR Res. Protoc.* **2025**, *14*, e63405. [CrossRef] [PubMed]
92. Kim, S.; Park, H.; Choo, S. Effects of Changes to Architectural Elements on Human Relaxation-Arousal Responses: Based on VR and EEG. *Int. J. Environ. Res. Public Health* **2021**, *18*, 4305. [CrossRef]
93. Hamadah, Q. The Polymorphic Diagram. In *Respecting Fragile Places, Proceedings of the 29th Conference on Education and Research in Computer Aided Architectural Design in Europe (eCAADe 2011), Ljubljana, Slovenia, 21–24 September 2011*; University of Ljubljana, Faculty of Architecture: Ljubljana, Slovenia, 2011; pp. 318–324. [CrossRef]
94. Kim, H.; Grobler, F. Building Ontology to Support Reasoning in Early Design. In Proceedings of the Computing in Civil Engineering (2007), Pittsburgh, PA, USA, 24–27 July 2007; American Society of Civil Engineers: Pittsburgh, PA, USA, 2007; pp. 151–158. [CrossRef]
95. Ziegler, C.; Bielski, J.; Langenhan, C. This Is How We Do It—Observing architects sketch to inform effective recognition of semantic building information. In Proceedings of the eCAADe 2021: Towards a New, Configurable Architecture, Novi Sad, Serbia, 8–10 September 2021; pp. 393–402. [CrossRef]
96. Cavieres, A.; Bhatia, U.; Joshi, P.; Zhao, F.; Ram, A. CBArch: A Case-Based Reasoning Framework for Conceptual Design of Commercial Buildings. In *Artificial Intelligence and Sustainable Design—Papers from the AAAI 2011 Spring Symposium (SS-11-02); Proceedings of the AAAI Spring Symposium, Palo Alto, CA, USA, 21–23 March 2011*; AAAI Press: Palo Alto, CA, USA, 2011.
97. Alassaf, N.; Clayton, M.J. The Use of Diagrammatic Reasoning to Aid Conceptual Design in Building Information Modeling (BIM). In Proceedings of the eCAADe 2021: Towards a New, Configurable Architecture, Novi Sad, Serbia, 8–10 September 2021; pp. 39–48. [CrossRef]
98. Çavuşoğlu, Ö.H.; Çağdaş, G. Why Do We Need Building Information Modeling (BIM) in Conceptual Design Phase? In *Computer-Aided Architectural Design. Future Trajectories*; Çağdaş, G., Özkar, M., Gül, L.F., Gürer, E., Eds.; Communications in Computer and Information Science; Springer Singapore: Singapore, 2017; Volume 724, pp. 121–135. [CrossRef]
99. Çavuşoğlu, Ö.H. Building Information Modeling Tools: Opportunities for Early Stages of Architectural Design. In Proceedings of the CAADRIA 2015: Emerging Experience in Past, Present and Future of Digital Architecture, Daegu, Republic of Korea, 20–23 May 2015; pp. 427–436. [CrossRef]
100. Donath, D.; Lobos, D. Massing Study Support. In Proceedings of the eCAADe 2008: Architecture "In Computro"—Integrating Methods and Techniques, Antwerp, Belgium, 17–19 September 2008; pp. 101–108. [CrossRef]
101. Wu, N.; Shih, S.-G. A BIM Inspired Supporting Platform for Architectural Design. *Comput.-Aided Des. Appl.* **2015**, *12*, 327–337. [CrossRef]
102. Abdelhameed, W. BIM from Conceptual Model to Construction. In Proceedings of the 2018 International Conference on Innovation and Intelligence for Informatics, Computing, and Technologies (3ICT), Sakhier, Bahrain, 18–20 November 2018; pp. 1–4. [CrossRef]
103. Abdelhameed, W. Is BIM Too Detailed for Conceptual Designing? A Case Study. In *Proceedings of the Future Technologies Conference (FTC) 2021, Volume 3*; Arai, K., Ed.; Lecture Notes in Networks and Systems; Springer International Publishing: Cham, Switzerland, 2022; Volume 360, pp. 686–701. [CrossRef]
104. Abdelhameed, W. Conceptual Design of Architecture Revisited. In Proceedings of the Smart Cities Symposium 2018, Zallaq, Bahrain, 22–23 April 2018; Institution of Engineering and Technology: Zallaq, Bahrain, 2018; pp. 1–4. [CrossRef]

105. Renev, I.; Chechurin, L.; Perlova, E. Early design stage automation in Architecture-Engineering-Construction (AEC) projects. In Proceedings of the eCAADe 2017: ShoCK—Sharing of Computable Knowledge, Rome, Italy, 20–22 September 2017; pp. 373–382. [CrossRef]
106. London, K.; Singh, V.; Taylor, C.; Gu, N.; Brankovic, L. Building Information Modelling project decision support framework. In Proceedings of the 24th Annual Conference of the Association of Researchers in Construction Management (ARCOM 2008), Cardiff, UK, 1–3 September 2008; Volume 2.
107. Khamis, A.A.; Ibrahim, S.A.; Khateb, M.A.; Abdel-Fatah, H.; Barakat, M.A. Introducing the Architecture Parametric Design Procedure: From Concept to Execution. *IOP Conf. Ser. Earth Environ. Sci.* **2022**, *1056*, 012004. [CrossRef]
108. Park, H.-J. Evolution + Bim: The Utilization of Building Information Modelling at an Early Design Stage. In Proceedings of the CAADRIA 2008: Beyond Computer-Aided Design, Chiang Mai, Thailand, 9–12 April 2008; pp. 552–559. [CrossRef]
109. Abrishami, S.; Goulding, J.; Pour Rahimian, F.; Ganah, A. Integration of BIM and generative design to exploit AEC conceptual design innovation. *Electron. J. Inf. Technol. Constr.* **2014**, *19*, 350–359.
110. Abrishami, S.; Goulding, J.; Pour Rahimian, F.; Ganah, A. Virtual generative BIM workspace for maximising AEC conceptual design innovation: A paradigm of future opportunities. *Constr. Innov.* **2015**, *15*, 24–41. [CrossRef]
111. Abrishami, S.; Goulding, J.; Rahimian, F.P.; Ganah, A.; Sawhney, A. G-BIM Framework and Development Process for Integrated AEC Design Automation. *Procedia Eng.* **2014**, *85*, 10–17. [CrossRef]
112. Abrishami, S.; Goulding, J.; Rahimian, F. Generative BIM workspace for AEC conceptual design automation: Prototype development. *Eng. Constr. Archit. Manag.* **2020**, *28*, 482–509. [CrossRef]
113. Rafiq, M.Y.; Rustell, M.J. Building Information Modeling Steered by Evolutionary Computing. *J. Comput. Civ. Eng.* **2013**, *28*, 130201212714004. [CrossRef]
114. Kochański, Ł.; Borkowski, A.S. Automating the conceptual design of residental areas using visual and generative programming. *J. Eng. Des.* **2024**, *35*, 195–216. [CrossRef]
115. Rostamiasl, V.; Jrade, A. Integrating Building Information Modeling and Virtual Reality to Facilitate the Implementation of Universal Design for Facilities at the Conceptual Stage. In *Canadian Society of Civil Engineering Annual Conference 2021. CSCE 2021*; Springer: Singapore, 2022; pp. 651–662. [CrossRef]
116. Pereira, V.A.; Paio, A. The architectural design process: An intuitive modeling in BIM's parametric design. In Proceedings of the eCAADe 2014: Fusion, Newcastle upon Tyne, UK, 10–12 September 2014; pp. 405–410. [CrossRef]
117. Rostamiasl, V.; Jrade, A. Integrating Building Information Modeling (BIM), Universal Design (UD) Standards, and Age-In-Place Requirements at the Conceptual Design Stage of Houses. In *Proceedings of the Canadian Society for Civil Engineering Annual Conference 2023, Volume 5*; Desjardins, S., Poitras, G.J., Nik-Bakht, M., Eds.; Lecture Notes in Civil Engineering; Springer Nature: Cham, Switzerland, 2024; Volume 499, pp. 201–216. [CrossRef]
118. Souza, M.P.D.; Fabricio, M.M. BIM and virtual reality for pre-design evaluation of building performance. *Archit. Eng. Des. Manag.* **2024**, *20*, 1260–1279. [CrossRef]
119. Rostamiasl, V.; Jrade, A. A cloud-based integration of Building Information Modeling and Virtual Reality through game engine to facilitate the design of Age-in-Place homes at the conceptual stage. *J. Inf. Technol. Constr. ITcon* **2024**, *29*, 377–399. [CrossRef]
120. Olivera, E.L.; Booth, C.; Agyekum, K.; Al-Tarazi, D. A phenomenological inquiry of Building Information Modelling macro-adoption in Uruguay. *Proc. Inst. Civ. Eng. Manag. Procure. Law* **2023**, *177*, 137–149. [CrossRef]
121. Zaker, R.; Coloma, E. Virtual reality-integrated workflow in BIM-enabled projects collaboration and design review: A case study. *Vis. Eng.* **2018**, *6*, 4. [CrossRef]
122. Park, C.; Lee, Y.G. Developing a Human Behavior Simulation Technology in the Real-Time Manner Based on BIM. In *Proceedings of the HCI International 2016—Posters' Extended Abstracts*; Stephanidis, C., Ed.; Springer International Publishing: Cham, Switzerland, 2016; pp. 519–523. [CrossRef]
123. Podkosova, I.; Reisinger, J.; Zahlbruckner, M.A.; Kovacic, I.; Kaufmann, H. Evaluation of Virtual Reality for Early-Stage Structure and Production Planning for Industrial Buildings. In Proceedings of the 2023 IEEE International Symposium on Mixed and Augmented Reality Adjunct (ISMAR-Adjunct), Sydney, Australia, 16–20 October 2023; pp. 159–166. [CrossRef]
124. Hattab, M.A.; Hamzeh, F. Information Flow Comparison Between Traditional and BIM-Based Projects in the Design Phase. In Proceedings of the 21st Annual Conference of the International Group for Lean Construction, Fortaleza, Brazil, 29 July–2 August 2013. [CrossRef]
125. El-Khouly, T.; Abdelhalim, O. Preserving conceptual design integrity: Strategies for enhancing interoperability in architectural digital design workflows. *Sci. Rep.* **2024**, *14*, 30595. [CrossRef]
126. Shehzad, H.M.F.; Ibrahim, R.B.; Yusof, A.F.; Khaidzir, K.A.M.; Iqbal, M.; Razzaq, S. The role of interoperability dimensions in building information modelling. *Comput. Ind.* **2021**, *129*, 103444. [CrossRef]
127. Zahedi, A.; Petzold, F. *Utilization of Simulation Tools in Early Design Phases Through Adaptive Detailing Strategies*; Tsinghua University: Beijing, China, 2018; pp. 11–20. [CrossRef]

128. Rinaldi, S.; Bittenbinder, F.; Liu, C.; Bellagente, P.; Tagliabue, L.C.; Ciribini, A.L.C. Bi-directional interactions between users and cognitive buildings by means of smartphone app. In Proceedings of the 2016 IEEE International Smart Cities Conference (ISC2), Trento, Italy, 12–15 September 2016; pp. 1–6. [CrossRef]
129. Langenhan, C.; Bielski, J.; Ziegler, C.; Eisenstadt, V.; Althoff, K.-D.; Dengel, A. Cross-Disciplinary Semantic Building Fingerprints, AI Knowledge Graphs to Store Topological Building Information Derived From Semantic Building Models (BIM) to Apply Methods of Artificial Intelligence (AI). In Proceedings of the CAADRIA 2023: Human-Centric, Navrangpura, India, 18–24 March 2023; pp. 129–138. [CrossRef]
130. Leon, M.; Laing, R. Cloud and Computer Mediated Collaboration in the Early Architectural Design Stages: A Study of Early Design Stage Collaboration Related to BIM and the Cloud. In Proceedings of the 2013 IEEE 5th International Conference on Cloud Computing Technology and Science, Bristol, UK, 2–5 December 2013; Volume 2, pp. 94–99. [CrossRef]
131. Baudoux, G.; Leclercq, P. Usability of BIM in Preliminary Design: A Study of the Relevance of the Different Representations Allowed by BIM. In *Proceedings of the Cooperative Design, Visualization, and Engineering*; Luo, Y., Ed.; Springer International Publishing: Cham, Switzerland, 2022; pp. 305–316. [CrossRef]
132. Wu, K.; Tang, S. BIM-Assisted Workflow Enhancement for Architecture Preliminary Design. *Buildings* **2022**, *12*, 601. [CrossRef]
133. Abualdenien, J.; Schneider-Marin, P.; Zahedi, A.; Harter, H.; Exner, H.; Steiner, D.; Singh, M.M.; Borrmann, A.; Lang, W.; Petzold, F.; et al. Consistent management and evaluation of building models in the early design stages. *J. Inf. Technol. Constr. ITcon* **2020**, *25*, 212–232. [CrossRef]
134. Jabi, W.; Chatzivasileiadi, A. Topologic: Exploring Spatial Reasoning Through Geometry, Topology, and Semantics. In *Formal Methods in Architecture*; Eloy, S., Leite Viana, D., Morais, F., Vieira Vaz, J., Eds.; Advances in Science, Technology & Innovation; Springer International Publishing: Cham, Switzerland, 2021; pp. 277–285. [CrossRef]
135. Donato, V. Towards design process validation integrating graph theory into BIM. *Archit. Eng. Des. Manag.* **2017**, *13*, 22–38. [CrossRef]
136. Meagher, M.; Langley, P. TopoBIM: Web-Based Spatial Topology for Early Design Participation. In *Complexity & Simplicity—Proceedings of the 34th Conference on Education and Research in Computer Aided Architectural Design in Europe (eCAADe 2016), Oulu, Finland, 22–26 August 2016*; Herneoja, A., Österlund, T., Markkanen, P., Eds.; Volume 2, pp. 663–672. [CrossRef]
137. Wu, T.; Jeng, T. Reforming Design Studios: Experiments in Integrating BIM, Parametric Design, Digital Fabrication, and Interactive Technology. In *Digital Physicality—Proceedings of the 30th Conference on Education and Research in Computer Aided Architectural Design in Europe (eCAADe 2012), Prague, Czech Republic, 12–14 September 2012*; Achten, H., Pavlicek, J., Hulin, J., Matejovska, D., Eds.; Czech Technical University: Prague, Czech Republic, 2012; Volume 1, pp. 49–54. [CrossRef]
138. Chai, W.; Zhang, C.; Wang, Z. Analysis on the Application of BIM Technology in the Interior Decoration Design of Public Buildings. In Proceedings of the 2022 International Conference on Big Data, Information and Computer Network (BDICN), Sanya, China, 20–22 January 2022; pp. 428–431. [CrossRef]
139. Krystallis, I.; Demian, P.; Price, A.D.F. Supporting Future-Proof Healthcare Design by Narrowing the Design Space of Solutions Using Building Information Modelling. In Proceedings of the 29th Annual ARCOM Conference, Reading, UK, 2–4 September 2013.
140. Lobos, D. BIM Supported Building Envelopes and Space Layout Based on a Case Study in South America. Ph.D. Thesis, Bauhaus University Weimar Press, Weimar, Germany, 2011. [CrossRef]
141. Lungu, M. The Coding Manual for Qualitative Researchers. *Am. J. Qual. Res.* **2022**, *6*, 232–237. [CrossRef]
142. Auernhammer, J.; Bruno, J.; Booras, A.; Claire, M.; Daniel, H.; Saggar, M. NeuroDesign: Greater than the sum of its parts. In *Design Thinking Research*; Springer: Cham, Switzerland, 2023.
143. Auernhammer, J.; Sonalkar, N.; Saggar, M. NeuroDesign: From Neuroscience Research to Design Thinking Practice. In *Design Thinking Research*; Springer: Cham, Switzerland, 2021; pp. 347–355. [CrossRef]
144. Bacevice, P.A.; Ducao, A. Use of biometric data and EEG to assess architectural quality of two office spaces: A pilot experiment. *Intell. Build. Int.* **2022**, *14*, 433–454. [CrossRef]
145. Dey, A.; Chatourn, A.; Billinghurst, M. Exploration of an EEG-Based Cognitively Adaptive Training System in Virtual Reality. In Proceedings of the 2019 IEEE Conference on Virtual Reality and 3D User Interfaces (VR), Osaka, Japan, 23–27 March 2019; pp. 220–226. [CrossRef]
146. Choi, Y.; Kim, M.; Chun, C. Measurement of occupants' stress based on electroencephalograms (EEG) in twelve combined environments. *Build. Environ.* **2015**, *88*, 65–72. [CrossRef]
147. Azevedo, A.S.; Jorge, J.; Campos, P. Combining EEG Data with Place and Plausibility Responses as an Approach to Measuring Presence in Outdoor Virtual Environments. *Presence Teleoperators Virtual Environ.* **2014**, *23*, 354–368. [CrossRef]
148. Nayak, C.S.; Anilkumar, A.C. EEG Normal Waveforms. In *StatPearls [Internet]*; StatPearls Publishing LLC: Treasure Island, FL, USA, 2023. Available online: https://www.ncbi.nlm.nih.gov/sites/books/NBK539805/ (accessed on 21 April 2025).
149. Williams, N.S.; McArthur, G.M.; de Wit, B.; Ibrahim, G.; Badcock, N.A. A validation of Emotiv EPOC Flex saline for EEG and ERP research. *PeerJ* **2020**, *8*, e9713. [CrossRef]

150. Kirch, W. Pearson's Correlation Coefficient. In *Encyclopedia of Public Health*; Kirch, W., Ed.; Springer Netherlands: Dordrecht, The Netherlands, 2008; pp. 1090–1091. [CrossRef]
151. Schober, P.; Boer, C.; Schwarte, L.A. Correlation Coefficients: Appropriate Use and Interpretation. *Anesth. Analg.* **2018**, *126*, 1763–1768. [CrossRef] [PubMed]
152. Cahn, B.R.; Polich, J. Meditation states and traits: EEG, ERP, and neuroimaging studies. *Psychol. Bull.* **2006**, *132*, 180–211. [CrossRef]
153. Aftanas, L.I.; Golocheikine, S.A. Human anterior and frontal midline theta and lower alpha reflect emotionally positive state and internalized attention: High-resolution EEG investigation of meditation. *Neurosci. Lett.* **2001**, *310*, 57–60. [CrossRef]
154. Klimesch, W. EEG alpha and theta oscillations reflect cognitive and memory performance: A review and analysis. *Brain Res. Brain Res. Rev.* **1999**, *29*, 169–195. [CrossRef]
155. Herrmann, C.S.; Strüber, D.; Helfrich, R.F.; Engel, A.K. EEG oscillations: From correlation to causality. *Int. J. Psychophysiol.* **2016**, *103*, 12–21. [CrossRef]
156. Vecchiato, G.; Tieri, G.; Jelic, A.; De Matteis, F.; Maglione, A.G.; Babiloni, F. Electroencephalographic Correlates of Sensorimotor Integration and Embodiment during the Appreciation of Virtual Architectural Environments. *Front. Psychol.* **2015**, *6*, 1944. [CrossRef]
157. Başar, E.; Başar-Eroglu, C.; Karakaş, S.; Schürmann, M. Gamma, alpha, delta, and theta oscillations govern cognitive processes. *Int. J. Psychophysiol. Off. J. Int. Organ. Psychophysiol.* **2001**, *39*, 241–248. [CrossRef] [PubMed]
158. Başar, E.; Schürmann, M.; Başar-Eroglu, C.; Karakaş, S. Alpha oscillations in brain functioning: An integrative theory. *Int. J. Psychophysiol. Off. J. Int. Organ. Psychophysiol.* **1997**, *26*, 5–29. [CrossRef] [PubMed]
159. Knyazev, G.G. Motivation, emotion, and their inhibitory control mirrored in brain oscillations. *Neurosci. Biobehav. Rev.* **2007**, *31*, 377–395. [CrossRef] [PubMed]
160. Tarrant, J.; Viczko, J.; Cope, H. Virtual Reality for Anxiety Reduction Demonstrated by Quantitative EEG: A Pilot Study. *Front. Psychol.* **2018**, *9*, 1280. [CrossRef]
161. Jelić, A.; Tieri, G.; De Matteis, F.; Babiloni, F.; Vecchiato, G. The Enactive Approach to Architectural Experience: A Neurophysiological Perspective on Embodiment, Motivation, and Affordances. *Front. Psychol.* **2016**, *7*, 481. [CrossRef]
162. Chiossi, F.; Ou, C.; Gerhardt, C.; Putze, F.; Mayer, S. Designing and Evaluating an Adaptive Virtual Reality System using EEG Frequencies to Balance Internal and External Attention States. *Int. J. Hum. Comput. Stud.* **2025**, *196*, 103433. [CrossRef]
163. Xia, G.; Henry, P.; Queiroz, F.; Westland, S. Effects of object colour stimuli on human brain activities and subjective feelings in physical environment and virtual reality. In Proceedings of the International Colour Association AIC Conference 2019, AIC 2019, Colour and Landscapes, Buenos Aires, Argentina, 14–17 October 2019.
164. Banaei, M.; Hatami, J.; Yazdanfar, A.; Gramann, K. Walking through Architectural Spaces: The Impact of Interior Forms on Human Brain Dynamics. *Front. Hum. Neurosci.* **2017**, *11*, 477. [CrossRef]
165. Assem, H.; Khodeir, L.; Fathy, F. Designing for human wellbeing: The integration of neuroarchitecture in design -A systematic review. *Ain Shams Eng. J.* **2023**, *14*, 102102. [CrossRef]
166. Bower, I.; Tucker, R.; Enticott, P. Impact of built environment design on emotion measured via neurophysiological correlates and subjective indicators: A systematic review. *J. Environ. Psychol.* **2019**, *66*, 101344. [CrossRef]
167. Panya, D.S.; Kim, T.; Gu, H.; Choo, S. Development of a Real-time BIM-VR Multi-Collaborative Design Environment. In Proceedings of the eCAADe 2023: Digital Design Reconsidered 2023, Graz, Austria, 20–23 September 2023; Volume 2, pp. 733–739.
168. Wong, M.O.; Du, J.; Zhang, Z.Q.; Liu, Y.Q.; Chen, S.M.; Lee, S.H. An experience-based interactive lighting design approach using BIM and VR: A case study. *IOP Conf. Ser. Earth Environ. Sci.* **2019**, *238*, 012006. [CrossRef]
169. Hermund, A.; Klint, L.; Bundgaard, T. BIM with VR for architectural simulations Building Information Models in Virtual Reality as an architectural and urban designtool. In Proceedings of the ACE 2018 Singapore: ACE 2018—Global Science and Technology Forum, Singapore, 14–15 May 2018.
170. Kochanowska, M.; Gagliardi, W.R.; Ball, J. The Double Diamond Model: In Pursuit of Simplicity and Flexibility. In *Perspectives on Design II*; Raposo, D., Neves, J., Silva, J., Eds.; Springer Series in Design and Innovation; Springer International Publishing: Cham, Switzerland, 2022; Volume 16, pp. 19–32. [CrossRef]
171. The Double Diamond—Design Council. Available online: https://www.designcouncil.org.uk/our-resources/the-double-diamond/ (accessed on 19 September 2024).
172. *The Field Guide to Human-Centered Design: Design Kit*, 1st ed.; IDEO, Ed.; Design Kit: San Francisco, CA, USA, 2015.
173. Abazari, T.; Potvin, A.; Demers, C.M.; Gosselin, L. A biophilic wellbeing framework for positive indoor-outdoor connections in energy-efficient Arctic buildings. *Build. Environ.* **2022**, *226*, 109773. [CrossRef]

174. Asim, F.; Chani, P.S.; Shree, V.; Rai, S. Restoring the mind: A neuropsychological investigation of university campus built environment aspects for student well-being. *Build. Environ.* **2023**, *244*, 110810. [CrossRef]
175. Zahoor, K.; Bawany, N.Z. ALF—Ambient Assisted Living for Healthcare Framework Based on IoMT and Big Data. In *Innovations in Smart Cities Applications Volume 6*; Ben Ahmed, M., Boudhir, A.A., Santos, D., Dionisio, R., Benaya, N., Eds.; Lecture Notes in Networks and Systems; Springer International Publishing: Cham, Switzerland, 2023; Volume 629, pp. 291–298. [CrossRef]

infrastructures

MDPI

Article

Blockchain Technology Adoption for Sustainable Construction Procurement Management: A Multi-Pronged Artificial Intelligence-Based Approach

Atul Kumar Singh [1,*], Saeed Reza Mohandes [2,*], Pshtiwan Shakor [3], Clara Cheung [2], Mehrdad Arashpour [4], Callum Kidd [2] and V. R. Prasath Kumar [5]

[1] Department of Civil Engineering, Chandigarh University, Mohali 140413, Punjab, India
[2] Department of Civil Engineering and Management, School of Engineering, The University of Manchester, Manchester M13 9PL, UK; clara.cheung@manchester.ac.uk (C.C.); c.kidd@manchester.ac.uk (C.K.)
[3] Civil Engineering Department, Al-Qalam University College, Kirkuk, Iraq; pshtiwan.shakor@alqalam.edu.iq
[4] Department of Civil Engineering, Monash University, Melbourne, VIC 3800, Australia; mehrdad.arashpour@monash.edu
[5] Department of Civil Engineering, College of Engineering and Technology, SRM Institute of Science and Technology, Chengalpattu 603203, Tamil Nadu, India; prasathv@srmist.edu.in
* Correspondence: atulkumarsingh22031996@gmail.com (A.K.S.); saeedreza.mohandes@manchester.ac.uk (S.R.M.)

Abstract: While blockchain technology (BT) has gained attention in the construction industry, limited research has focused on its application in sustainable construction procurement management (SCPM). Addressing this gap, the present study investigates the key drivers influencing BT adoption in SCPM using a hybrid methodological approach. This study includes a systematic review of academic and grey literature, expert consultations, and quantitative analysis using advanced fuzzy-based algorithms, k-means clustering, and social network analysis (SNA). Data were collected through an online survey distributed to professionals experienced in SCPM and blockchain implementation. The Fuzzy DEMATEL results identify "high quality", "decentralization and data security", and "cost of the overall project" as the most critical drivers. Meanwhile, SNA highlights "stability of the system", "overall performance of the project", and "customer satisfaction" as the most influential nodes within the network. These insights provide actionable guidance for industry stakeholders aiming to advance SCPM through blockchain integration and contribute to theoretical advancements by proposing novel analytical frameworks.

Keywords: blockchain technology; sustainability; procurement management; industry 4.0; fuzzy logic; artificial intelligence

1. Introduction

The construction industry (CI) has been recognized as one of the most environmentally unfriendly industries, with a substantial environmental footprint stemming from the extraction and processing of raw materials, transportation, and waste disposal [1]. These activities contribute significantly to greenhouse gas emissions, pollution, and ecosystem degradation. However, the integration of Industry 4.0 technologies such as 3D printing, robotics, drones, BIM, and IoT has created new opportunities for adopting more sustainable construction practices [2–4]. Among these, blockchain technology (BT) has emerged as a promising solution to enhance transparency, accountability, and operational efficiency within construction supply chains. Its application enables better tracking of raw materials and logistics, helping to curb illegal or unsustainable practices such as mining in restricted areas and deforestation. Furthermore, BT can facilitate efficient resource management, minimize construction waste, and support environmentally responsible decision-making [5].

Past research has increasingly emphasized BT's potential to promote sustainability within the construction industry. For instance, studies by Zhong et al. [6] and Perera et al. [7] demonstrate how BT can improve traceability of sustainable materials, foster trust among stakeholders, and reduce procurement fraud. These findings support the argument that BT is not merely a digital tool but a strategic enabler of sustainable construction transformation. Despite such advances, the integration of sustainability goals into procurement practices remains underexplored.

Originally conceptualized by Satoshi Nakamoto in 2008 with the release of Bitcoin [8], BT has since expanded beyond financial systems into various sectors, including agriculture, logistics, healthcare, and construction. Recent studies have examined BT's potential to improve transparency, efficiency, and trust in construction procurement, particularly within the scope of sustainable construction procurement management (SCPM) [9–12]. SCPM refers to procurement processes that incorporate sustainability principles—such as environmental stewardship, social responsibility, and economic viability—into decision-making.

Unlike conventional procurement, SCPM introduces additional expectations such as environmental certifications, responsible sourcing, and lifecycle impact assessments. Although BT has been widely acknowledged for improving transparency and reducing transaction costs [13,14], prior studies have primarily focused on general procurement applications. Consequently, the literature offers limited insights into how BT can be strategically leveraged to meet the distinct sustainability objectives of SCPM, leaving a crucial research gap that this study aims to fill.

To address this gap, this study aims to explore and evaluate the key drivers influencing the adoption of blockchain technology in SCPM. Specifically, this study formulates the following research objectives (ROs), designed in a logical progression.

The novelty of this research lies in its comprehensive approach to addressing the identified research gap. To fill this gap, this study explores the following research objectives (RO):

(RO1): To conduct a comprehensive literature review to identify a set of potential drivers for adopting BT in SCPM.

(RO2): To identify the relative influence of these identified drivers using the modified Fuzzy Decision-Making Trial and Evaluation Laboratory (FDEMATEL) method, thereby determining which drivers have the most impact in the SCPM context.

(RO3): To analyze the relational structure and prominence of these influential drivers through Social Network Analysis (SNA), thereby identifying those with the greatest strategic potential for facilitating BT adoption.

These objectives are sequential: first, identifying a comprehensive list of drivers (RO1), then quantifying their influence (RO2), and finally mapping their interrelationships and strategic significance (RO3). This layered approach ensures that the analysis is both data-driven and context-specific to SCPM, providing a comprehensive understanding of the subject matter. The novelty of this study lies in its integrative methodology, which combines FDEMATEL and SNA. These two advanced analytical tools have not been extensively applied in this domain to unravel the interdependencies and centrality of BT adoption drivers in sustainable procurement. Moreover, by focusing on SCPM rather than general procurement, this study advances theoretical understanding and provides actionable insights for aligning emerging technologies with sustainability goals in the construction sector. The findings are expected to support both academic inquiry and practical policy formulation by offering a framework for decision-makers and practitioners to prioritize interventions that foster BT-enabled sustainable procurement systems in the construction industry.

2. Literature Review

2.1. BT and SCPM

Blockchain Technology (BT) enhances transparency, efficiency, and trust in information sharing through its decentralized, encrypted ledger system. BT has four key characteristics. First, it allows businesses to share data seamlessly across networks, making it particularly valuable for multi-organizational settings, such as financial consortia and procurement systems, because it is distributed and synchronized among all participants [15]. Second, BT incorporates smart contracts, which are pre-defined agreements stored on the blockchain before execution. These contracts are automated computer protocols designed to facilitate, verify, and enforce terms digitally, enabling trustworthy transactions without the need for third-party intermediaries. Depending on the protocol, specific operations such as payments are automatically executed. Smart contracts can also define non-monetary conditions, providing stakeholders with assurance that all actions within the network adhere to agreed-upon rules [16]. Third, BT operates on a peer-to-peer (P2P) network where consensus among participants is required for validating transactions, thereby preventing inaccuracies or fraudulent records. Lastly, the immutability of blockchain data ensures that validated transactions cannot be altered [17].

Traditional construction procurement primarily focuses on the "iron triangle" of cost, time, and quality. In contrast, sustainable construction procurement management (SCPM) integrates environmental, social, and economic dimensions into procurement decisions. SCPM emphasizes reducing environmental impacts through practices such as lifecycle assessments, the use of low-carbon materials, and minimizing construction waste. It also promotes social equity (e.g., ensuring fair labor practices and supporting local employment) and long-term economic resilience. Unlike conventional procurement, which often prioritizes immediate project deliverables, SCPM demands transparency, traceability, and accountability across the entire supply chain. BT supports SCPM by enabling end-to-end material traceability, verification of sustainability claims, and real-time performance monitoring, which are essential for achieving SCPM's holistic objectives.

Any asset recorded on a blockchain can be traced to its origin, journey, and final status, offering complete provenance tracking. BT can operate as either public (e.g., Bitcoin) or private (permissioned) networks, with the key difference being who is allowed to participate in the network. Public blockchains are open to anyone and typically use incentive mechanisms to encourage participation, as seen with Bitcoin [18]. However, public blockchains require high computational power due to consensus mechanisms like proof-of-work (PoW). Alternatives such as Proof of Stake (PoS) or Proof of Authority (PoA) may be used depending on the network design. Private blockchains, on the other hand, require invitations and validation before participants can join, making them particularly suitable for business supply chains (SCs) where controlled access and confidentiality are critical [19]. The adoption of BT can significantly expand business horizons by enabling decentralized, "trustless" transactions, removing intermediaries, and facilitating coordination among diverse supply chain members. For example, Maersk, a global logistics leader, reported saving billions of dollars in maritime container management through its partnership with IBM on BT [20].

For BT to operate effectively in procurement or broader economic systems, trust must be established among participants in generating, storing, and sharing critical records. In traditional settings, industries such as banking, healthcare, and education rely heavily on third-party entities for record-keeping and maintenance. In contrast, BT-based digital systems reduce the risk of human error or corruption by ensuring secure, tamper-proof record-keeping [21]. Within procurement, particularly SCPM—BT, enhances coordination and collaboration across various stakeholders engaged in planning, monitoring, and executing projects [22].

While BT offers multiple advantages, it also brings challenges. On the positive side, it enables procurement to achieve traditional objectives—such as cost reduction, risk mitigation, and quality control while improving sustainability. For example, logistics services that are critical for delivering customer value can be streamlined and authenticated using BT-supported quality assurance mechanisms. Although conventional quality audits remain necessary, BT accelerates processes and makes transactional records more robust and reliable [12,23]. BT also provides a solid foundation for resolving global procurement challenges through its ability to enhance transparency, traceability, and security. Its architecture, based on cryptographically timestamped and immutable records, ensures that every transaction event is verifiable [24]. These features not only improve the trustworthiness of smart contracts but also facilitate stronger stakeholder relationships. Despite these benefits, many SC networks have encountered technical, organizational, behavioral, and policy-related barriers during the adoption of BT [25].

Recent research has explored the intersection of BT and sustainability. Studies indicate that BT's tracking and tracing capabilities can support more sustainable procurement by improving resource monitoring, waste reduction, and compliance with environmental standards [26]. BT has applications in sustainable water management, agriculture procurement, and product decarbonization, reflecting its growing potential in industries committed to reducing their environmental footprint. Companies are becoming increasingly aware of the need for traceability and sustainability, not only for regulatory compliance but also to maintain public trust and build stronger relationships with environmentally conscious consumers. These considerations often take precedence over traditional cost-cutting measures. However, sustainability and cost efficiency are sometimes perceived as conflicting priorities, making SCPM a complex yet crucial focus area [27].

2.2. Past Studies on BT Adoption in SCPM

The existing literature highlights the transformative potential of blockchain technology (BT) in addressing procurement challenges such as improving transparency, ensuring integrity and compliance, resolving disputes, and managing stakeholder expectations. By maintaining immutable records of product origins, procurement members, and operational processes on blockchain ledgers, BT can enhance traceability, authenticity, and chain-of-custody management [26]. Numerous studies emphasize the positive impact of BT procurement sustainability; however, research on the specific challenges of integrating BT into sustainable procurement programs—such as resource constraints, inter-organizational coordination, and technological limitations—remains limited [23]. This study aims to address these gaps by examining the drivers of BT adoption in SCPM, offering actionable insights to help organizations navigate the complexities of integration and enhance sustainability practices in construction procurement.

Much of the existing research explores firms' intentions to adopt BT rather than identifying the critical success factors (CSFs) influencing implementation. Transaction Cost Theory (TCT) is frequently employed to evaluate the impact of BT procurement, with studies highlighting its ability to reduce opportunistic behavior, minimize uncertainty, and lower transaction costs [28]. Several studies have employed Decision-Making Trial and Evaluation Laboratory (DEMATEL) techniques to investigate factors influencing the adoption of BT in agriculture and other supply chains (SCs). These studies primarily assessed interrelationships among adoption barriers in industrial and service sectors.

However, such research often focuses on cryptocurrency-related challenges in public blockchain contexts, with limited attention to private BT applications or SCPM-specific issues, reducing their relevance to construction procurement [29].

BT's potential to enhance supply chain performance by improving quality, reducing costs, and increasing transparency has been widely recognized. In manufacturing, BT combined with radio frequency identification (RFID) has been proposed for tracking agricultural goods during early adoption stages. Parmentola et al. [30] emphasized that BT can create accurate, tamper-proof, and secure systems that improve sustainability.

However, reviews of digital supply chain (SC) technologies often overlook BT entirely, suggesting a lack of academic focus on its role in procurement transformation. Studies on SC traceability have examined both the practical applications of blockchain and the challenges associated with its implementation in trade-related SCs [31].

BT has been extensively discussed as a tool for enhancing data security, access control, and records management. However, much of the literature overlooks the exploration of how blockchain-based structural frameworks and architectures can be implemented effectively. Research has highlighted the potential of BT in sustainable and innovative cities, considering environmental, economic, and social governance perspectives [29]. Popular platforms such as Ethereum and Hyperledger Fabric are widely cited for blockchain deployment. In supply chain contexts, BT is considered a critical technology for achieving efficient logistics operations through approaches such as mean-variance optimization [32].

Similarly, studies in the gem industry emphasize BT's ability to ensure ethical sourcing and sustainability, particularly in sectors facing reputational risks due to unethical practices. In agri-food supply chains (ASCs), blockchain-based traceability has been extensively studied. For instance, BT has been applied to Indian food-retail SCs to monitor product authenticity [33], while the integration of IoT, HACCP, and blockchain has been proposed to improve real-time food safety and transparency [34]. A blockchain-based quality management framework was also suggested to address issues of centralized trust, information asymmetry, and inspection costs in SCs [35].

Sharma et al. [36] developed a blockchain-enabled IoT supply chain management system to monitor environmental parameters, including humidity, soil moisture, and air temperature, in real-time, thereby achieving high levels of transparency and security in procurement operations. Their system also leveraged barcodes and QR codes to enhance traceability in the mango and pork value chains [23]. Moreover, BT's potential to contribute to the United Nations Sustainable Development Goals (SDGs) has been explored, with experts emphasizing its ability to improve transparency, build trust, and ensure secure information sharing in procurement [37]. Some studies even propose that BT could allow firms to develop proprietary cryptocurrencies and protect sensitive data [38].

Despite its promise, BT adoption in construction and SCPM faces significant barriers. The absence of clear norms and regulatory guidelines is a major impediment, preventing seamless integration into existing systems in the highly regulated construction industry [39,40]. Establishing standardized protocols is necessary to boost organizational confidence and avoid legal or governmental conflicts [41]. From a technical perspective, construction projects necessitate complex, multi-party data exchanges, which raises concerns about trust in decentralized systems and the protection of sensitive data [42]. Studies also emphasize the need for stronger security and privacy frameworks to ensure effective adoption [43].

Furthermore, the high implementation costs of BT pose challenges for small- and medium-sized enterprises (SMEs). There is also a widespread lack of understanding of BT's operational and strategic potential, combined with the absence of well-defined technical standards [11]. Therefore, further research is needed to address these gaps, particularly in identifying SCPM-specific drivers, barriers, and success strategies to enable widespread adoption and maximize BT's potential to enhance transparency, sustainability, and cost efficiency [44].

2.3. Research Gaps

The existing literature highlights the transformative potential of BT in enhancing transparency, traceability, and operational efficiency across various sectors, including supply chain management, agriculture, and logistics [45,46]. In the construction domain, studies suggest that BT can reduce transactional friction, mitigate information asymmetry, and ensure the immutability of records [7,46]. These capabilities are particularly relevant to construction procurement, which often suffers from fragmented processes and a lack of trust among stakeholders.

However, several important gaps remain in the scholarly discourse:

- While BT has been widely discussed in general supply chains [47], there have been limited empirical studies specifically addressing its application within SCPM. Existing research often centers around public blockchains or cryptocurrency-focused systems, with insufficient emphasis on the potential of private or consortium blockchains tailored for construction procurement with sustainability mandates [6,48].
- Although studies have examined BT adoption in construction [49,50], many focus on perceived benefits or general intentions. Few have systematically identified and evaluated the critical drivers that influence adoption in sustainability-driven procurement environments, where environmental, social, and governance (ESG) goals play a central role [51,52].
- The current literature lacks robust analytical models that employ Multi-Criteria Decision-Making (MCDM) techniques, such as Fuzzy Delphi, AHP, or DEMATEL, to prioritize BT adoption factors in SCPM. Most contributions remain at a conceptual level or offer generalized roadmaps, without presenting structured frameworks that can support practical implementation and policymaking [53,54].

Based on these observed gaps, this study aims to provide a grounded and methodologically robust framework for identifying and evaluating key drivers of BT adoption in SCPM. By combining expert input with integrated MCDM techniques, this research aspires to inform both theoretical understanding and practical decision-making in sustainable construction procurement contexts.

3. Research Methodology

The research methodology for this study is structured into three sequential phases, as illustrated in the research flow chart (Figure 1). The first phase involves identifying the drivers for the adoption of BT in SCPM through a comprehensive literature review and expert consultations. The second phase focuses on data collection to validate and refine the identified drivers. The third phase involves data analysis using a hybrid approach combining fuzzy DEMATEL, k-means clustering, and social network analysis (SNA) to evaluate the drivers and their interrelationships. This phased approach ensures a systematic pathway from conceptualization to empirical analysis.

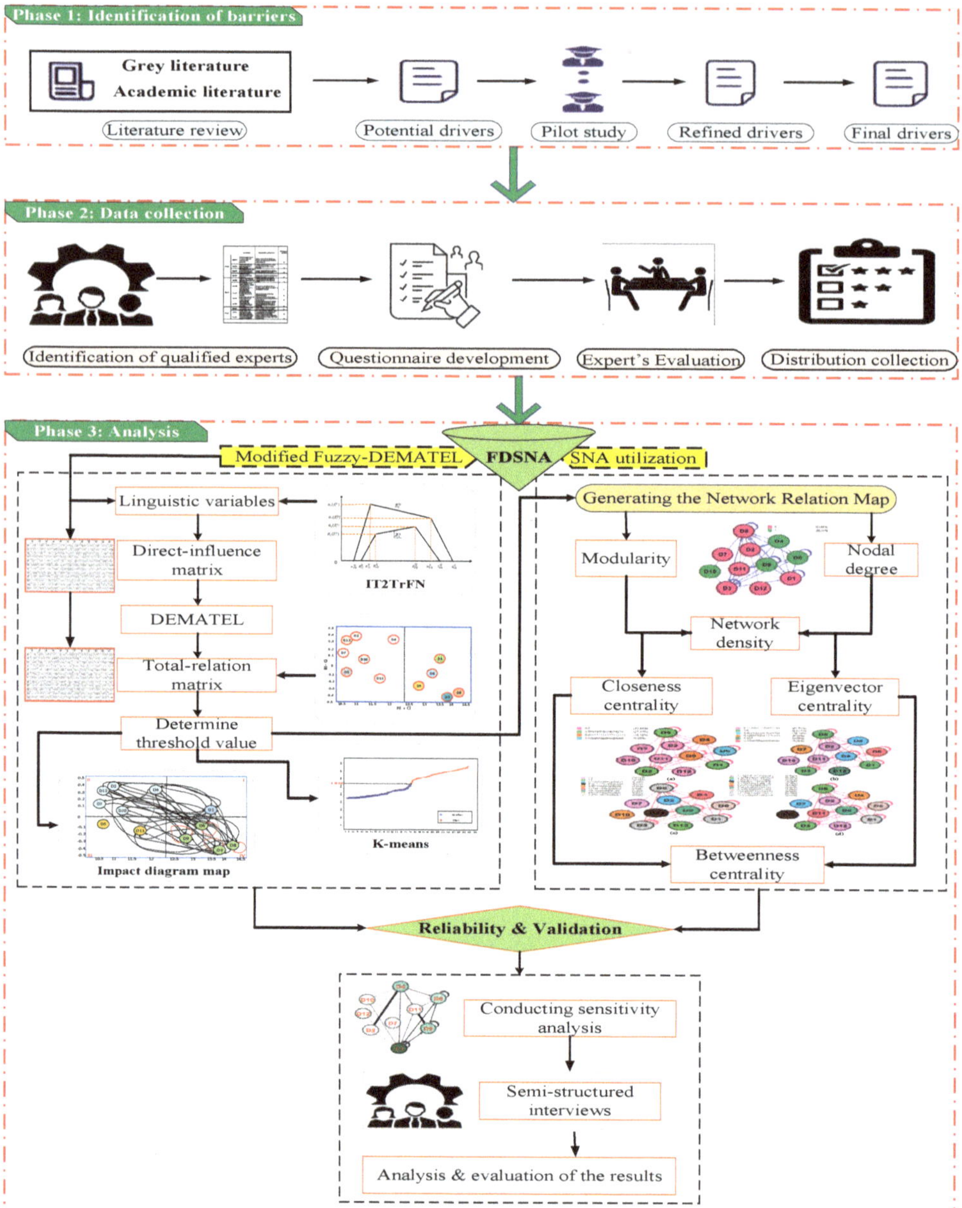

Figure 1. Research framework.

3.1. Phase 1: Identification of Drivers

This section outlines the process of identifying and validating the drivers for BT adoption in SCPM. Initially, a comprehensive review of academic literature and grey literature (including government, industry, and project reports) was conducted to extract a preliminary list of 32 potential drivers [55]. To ensure robust validation, a panel of 13 experts was formed, with selection criteria requiring each expert to have at least 3 years of direct experience in construction procurement and a professional background in the construction industry or academia [56]. The number 13 was chosen following recommendations from expert elicitation studies, which suggest that panels of 10–15 participants are sufficient to ensure diversity and reliable feedback while maintaining feasibility for in-depth discussions. An expert discussion was conducted via Zoom Meeting to evaluate and refine the identified drivers [9]. During the session, experts assessed the relevance and clarity of the 32 drivers and proposed organizing them into broader thematic categories. For example, the driver "Stability of the system" was expanded to include related elements such as transparency, security, elimination of intermediaries, and trustworthiness [11,57]. Based on iterative consensus, the 32 drivers were consolidated into 12 overarching categories to enhance clarity and usability. This refined set of 12 drivers forms the foundation for subsequent analysis in this study (Table 1).

Table 1. List of drivers with description.

Drivers	Code	Description	References
Stability of the system	D1	The four variables in assessing system robustness are transparency, security, removing intermediaries, and trustworthiness. After integrating BT into the SCPM, all of these characteristics made the SCPM and system more robust and sustainable, due to the transparency in information flow, stability of data, and peer-to-peer transactions that did not involve a third party.	[2–5,58]
Cost of the overall project	D2	Three variables are combined to form the overall cost, energy, and project costs. The overall SCPM cost encompasses all financial investments, including documentation fees, stationery expenses, human resources costs, electricity costs, facility costs, and documentation incurred during production, among other expenses.	[15–18]
The overall performance of the project	D3	The project's overall performance is measured by the efficiency, effectiveness, and speed with which the job is performed, characterized by low response times, high standardization, and reduced complexity of the work. A composite of five variables is built by examining the correlations between efficiency, speed, automation, simplification of current paradigms, and sharing demand in SCPM.	[9,37,59–62]
Decentralization and data security	D4	By experts under the name data safety and decentralization, a group of four variables is grouped. All people involved directly or indirectly in the system need records and information regarding standardization, production, and supply. It is only on BT-integrated SCPM that it is possible to hack, change, control, or lose data for any reason.	[12,21,63–65]
Adaptability	D5	Regarding adaptability, three variables—traceability, visibility, and identifying issues—form an ordinary driver. Using IoT/Industry 4.0, adaptability means tracking causes, goods location, accidents, and fraud between a SCPM's manufacturing and end-use processes.	[24–27]

Table 1. *Cont.*

Drivers	Code	Description	References
Policies and laws	D6	A significant amount of time and effort is required to document laws and policies. Documentation plays a crucial role in all contracts, but public ledger technology provides transparency, speeds up work, checks for corruption in government, and also helps identify scams within any organization. The importance of data records for any legal action cannot be overstated. This technology does not allow for the deletion or modification of data, ensuring that there can be no fraud. Laws and government policies have been taken into account in the design to ensure accessibility.	[23,26,28,29,61]
The system with innovative features	D7	With the intelligent system, smart contracts are implemented, invoicing is simplified, and inventory levels are improved. By eliminating fraud in documentation, ensuring timely tax payments, and ensuring the timely delivery of goods from the shipyard area, the system eliminates documentation fraud.	[29–32]
Satisfaction of customers	D8	Feedback and customer centricity are the basis of customer satisfaction. The customer is satisfied when they provide positive feedback and respond positively. To achieve this, it is necessary to deliver the right product with the correct information at the right time, place, and in the right hands and commit to providing periodic service after the sale.	[23,33,34,36–38]
The system with high reliability	D9	Four key drivers determine the Reliability of Systems: Scalability, Reliability, Durability, and the ability to withstand data loss. A record of the raw, semi-finished, and finished material is kept at every BT location. By maintaining accurate information about goods, IT saves time and money.	[19,20,23,34,59]
Detailed documentation	D10	Auditable, accounting, and ecosystem documentation are part of the documentation. In BT, integration is preferred for several reasons, including smooth auditing processes, simplified financial systems, and seamless currency flow.	[3,12,23,33,34]
A data management system	D11	A data management system controls access to data from end-users, simultaneously manipulates transactions from a single account, eliminates human error in documentation and other tasks, and enables real-time information flow. Three variables are considered in data management: data quality, flow and control of information, and access control.	[4,5,37,38,58]
High-quality	D12	Quality assurance and quality fairness are drivers that affect quality. Any irregularity in the process, transportation, raw material specification, or the final product is considered a quality issue. The availability of complete information and the elimination of human error are key.	[2,23,66,67]

3.2. Phase 2: Data Collection

The second phase involved a broader expert group to evaluate and assign weights to the 12 drivers. The study population consisted of professionals with substantial expertise in BT and SCPM. While the Phase 1 panel (13 experts) refined the driver list, Phase 2 relied on a separate, larger group of 27 experts to assess the relative importance of each driver. This clarification directly addresses the reviewer's query: the same 13 experts were not reused but were complemented by a more diverse pool of academics and practitioners for this phase. The sampling method began with a pilot study involving four SCPM specialists to refine the questionnaire and ensure its relevance to the local context (see Appendix C). Following this, 40 experts were invited to participate in the survey, from which 27 were selected [57,68]. This group comprised 16 academics actively involved in blockchain and SCPM research, as well as 11 practitioners with management or technical roles in the construction and procurement industries (see Table 2). The selected experts had over 3 years of experience in their respective fields, with academics averaging 16.14 years of experience (standard deviation: 9.11 years) and practitioners averaging 19.08 years (standard deviation: 7.12 years). Before the survey, all participants underwent a briefing session on BT and SCPM concepts to ensure a uniform understanding of the study's objectives and criteria. The weights assigned by experts reflect the perceived significance of each driver relative to SCPM goals such as cost efficiency, transparency, and sustainability. To ensure all participants had a solid understanding of the study's context, they received training on BT and SCPM before completing the questionnaires. This approach aimed to enhance the reliability of the data and ensure a thorough understanding of the research objectives [69–72].

Table 2. Details of the experts participating in this study.

Categories	Size of the Company	Occupations	Degree	Years of Experience	
				Construction	Procurement
Practitioner	More than 3000 employees	Technical specialist	M.Eng	More than 15	Between 6 and 10
	More than 5000 employees	Project Manager	M.Eng	More than 15	Between 6 and 10
	More than 5000 employees	Technical specialist	M.Eng	More than 15	Between 6 and 10
	More than 4000 employees	Legal Manager	M.Eng	More than 15	Between 6 and 10
	More than 5000 employees	Senior engineer	M.Eng	More than 15	Between 6 and 10
	More than 10,000 employees	Quality Manager	B.Eng	More than 15	Between 1 and 5
	More than 5000 employees	Project Supervisor	B.Eng	Between 6 and 10	Between 6 and 10
	More than 5000 employees	Technical specialist	B.Eng	Between 11 and 15	Between 6 and 10
	More than 10,000 employees	Project Supervisor	M.Eng	More than 15	Between 6 and 10
	More than 5000 employees	Project Supervisor	M.Eng	Between 11 and 15	Between 6 and 10
	More than 5000 employees	Technical specialist	B.Eng	More than 15	Between 6 and 10
	More than 5000 employees	Project Manager	M.Eng	Beween 11 and 15	Between 6 and 10
	More than 5000 employees	Consultant	M.Eng	Between 11 and 15	Between 6 and 10
Academic	More than 1000 employees	Professor	PhD	More than 15	Between 1 and 5
	More than 5000 employees	Professor	PhD	More than 15	Between 6 and 10
	More than 2000 employees	Professor	PhD	More than 15	Between 6 and 10
	More than 5000 employees	Professor	PhD	More than 15	Between 6 and 10

Table 2. *Cont.*

Categories	Size of the Company	Occupations	Degree	Years of Experience	
				Construction	Procurement
Academic	More than 1000 employees	Professor	PhD	More than 15	Between 6 and 10
	More than 7000 employees	Assistant Professor	PhD	Between 6 and 10	Between 1 and 5
	More than 6000 employees	Associate Professor	PhD	Between 11 and 15	Between 1 and 5
	More than 5000 employees	Professor	PhD	More than 15	Between 6 and 10
	More than 5000 employees	Associate Professor	PhD	Between 11 and 15	Between 1 and 5
	More than 2000 employees	Assistant Professor	PhD	Between 6 and 10	Between 1 and 5
	More than 2000 employees	Assistant Professor	PhD	Between 6 and 10	Between 1 and 5
	More than 5000 employees	Associate Professor	PhD	Between 11 and 15	More than 10
	More than 5000 employees	Assistant Professor	PhD	Between 6 and 10	Between 1 and 5
	More than 7000 employees	Assistant Professor	PhD	Between 6 and 10	Between 1 and 5

The research was conducted in Tamil Nadu, a large state located in the southern part of India. In recent years, people from rural areas have migrated to cities due to their moderate climate. Many construction projects have been launched to meet this increased demand. In this study, this region is an ideal case study since it has a high number of construction projects, and there are several reported fatalities. The researchers conducted additional interviews with several experts who had not been previously selected to validate the findings obtained through the advanced fuzzy-based algorithm, k-means clustering, and social network analysis technique. To enhance the reliability of findings, this study was conducted in Tamil Nadu, a region with high construction activity and reported procurement challenges. Nine additional experts (distinct from the earlier groups) were interviewed in a semi-structured format to validate the identified drivers and assigned weights using advanced fuzzy-based algorithms, k-means clustering, and social network analysis. Their profiles, which include experience levels exceeding 15 years in civil engineering and procurement, are provided in Table 3.

Table 3. Description of the experts for validation.

Interviewees	Degree	Years of Experience	Occupations
1	M.Eng	21	Civil Engineering
2	B.Eng	26	Civil Engineering
3	B.Eng	18	Civil Engineering
4	M.Eng	24	Civil Engineering
5	B.Eng	19	Civil Engineering
6	B.Eng	17	Civil Engineering
7	M.Eng	20	Civil Engineering
8	M.Eng	25	Civil Engineering
9	B.Eng	16	Civil Engineering

3.3. Phase 3: Data Analysis

3.3.1. Modified Fuzzy DEMATEL Method

Traditional DEMATEL cannot directly convert qualitative judgments in the linguistic variables into crisp numbers due to the inherent vagueness of human thought. This makes assigning an accurate and reasonable threshold value challenging because experts' experiences determine it. Two aspects of the DEMATEL method are improved using IT2TrFNs: (1) Experts' evaluations are represented through IT2TrFNs, allowing for more efficient information transfer. (2) K-means clustering is used instead of subjective judgment to optimize the threshold value. The modification of the DEMATEL technique is derived from the IT2TrFN theory. A detailed explanation of the modified fuzzy DEMATEL approach (proposed by [73]) and the use of Trapezoidal Fuzzy Number Type-II Interval is provided in Appendix C, along with the complete computational framework. The following steps are involved in the modified fuzzy DEMATEL.

Step 1. Establishing expert-driven matrices of direct impact.

To make impact judgments, decision-makers should conceptualize drivers as semantic values. Assuming they are responsible for analyzing the relationship between drivers [73], decision-makers can provide the impact matrix directly as semantic values [74]:

$$D^k = \begin{bmatrix} d_{11} & d_{12} & \cdots & d_{1n} \\ d_{21} & d_{22} & \cdots & d_{2n} \\ \vdots & \vdots & \vdots & \vdots \\ d_{n1} & d_{n2} & \cdots & d_{nn} \end{bmatrix} \quad (1)$$

The mapping between semantic values and IT2TrFN is complete when $aij = LVij\ Pij$ describes the effect of driver i on driver j, where $LVij$ represents the semantic value and Pij represents the decision-maker's membership. Celik et al. [75] present the mapping relations in Table 4. The following is the matrix of direct influence:

$$\overset{\approx k}{D} = \begin{bmatrix} e_{11} & e_{12} & \cdots & e_{1n} \\ e_{21} & e_{22} & \cdots & e_{2n} \\ \vdots & \vdots & \vdots & \vdots \\ e_{n1} & e_{n2} & \cdots & e_{nn} \end{bmatrix} \quad (2)$$

Table 4. Linguistic variables are presented along with their corresponding IT2TrFNs.

Linguistic Terms	IT2TrFNs
No influence (NI)	(0.1,0.1,0.1,0.1;1,1,0.1,0.1,0.1,0.1;1,1)
Very low influence (VL)	(0.1,0.2,0.4,0.5;1,1,0.12,0.22,0.38,0.48;0.8,0.8)
Low influence (L)	(0.3,0.4,0.6,0.7;1,1,0.32,0.42,0.58,0.68;0.8,0.8)
High influence (H)	(0.5,0.6,0.8,0.9;1,1,0.52,0.62,0.78,0.88;0.8,0.8)
Very high influence (VH)	(0.7,0.8,0.9,0.9;1,1,0.72,0.82,0.8,0.9;0.8,0.8)

The overall averaged matrix should be calculated based on a determined weight for each expert. Due to the varying backgrounds of each expert, including their professional titles and job experience, this is more challenging in practice. The ordered weighted averaging operation (OWD) [76] was considered for this study. A fuzzy number of interval type-II can be expressed in the following manner with the OWA operator [77]:

$$OWD\left(e_{ij}^{1}, e_{ij}^{2}, \cdots, e_{ij}^{k}\right) = \bigoplus_{k=1}^{K} \left(\omega_k e_{ij}^{k}\right) \tag{3}$$

An expert k decides to weigh the weight of driver i based on their decision; bk ij represents the influence of driver i on driver j due to that expert's decision.

Step 2. Creating the matrix of direct influences that are normalized.

Criteria are selected based on the maximum value:

$$\text{Maxvar} = max\left(\underset{1 \le i \le n}{max} \sum_{i=1}^{n} e_{ij}, \underset{1 \le j \le n}{max} \sum_{j=1}^{n} e_{ij} \right) \tag{4}$$

As a result, let us look at the normative matrix directly:

$$M = \left(\frac{y_{ij}}{Maxvar} \right)_{n \times n} \tag{5}$$

Step 3. The total-relation matrix can be calculated by using the formula below.

$$T = \left(M + M^2 + M^3 + \cdots + M^k\right) = \sum_{k=1}^{\infty} M^k = M(I - M)^{-1} \tag{6}$$

$$T - \left[c_{ij}\right]_{n \times n} i, j \in \{1, 2, 3, \cdots, n\} \tag{7}$$

When there are n units in I, IT2TrFN is used to express C_{ij}.

Step 4. Estimating influence severity and prominence.

The total-relation matrix shows the degree of influence for each row and column. Calculations are based on the following formulas:

$$D_i - \sum_{j=1}^{n} c_{ij}, (i = 1, 2, 3, \cdots, n) \tag{8}$$

$$R_i = \sum_{j=1}^{n} c_{ji}, (i = 1, 2, 3, \cdots, n) \tag{9}$$

Step 5. By using K-Means clustering, confirm thresholds.

Using the k-means clustering algorithm, CIJ is determined. The clustering of all C_{ij} is set to 2 when the parameter of cluster number is set to 2. The objective function is then calculated by using Equation (1). Once both center point sets are equal, the algorithm terminates, which means $|SSEn - SSEn - 1| < \varepsilon$ (ε is selected). If this is not the case, the distance calculation should be repeated with the updated centroids.

Step 6. A map of impact digraphs is constructed.

Follow step 4 to calculate $Di + Ri$ to determine the centrality of the evaluation driver list, which indicates the importance of each factor; $Di - Ri$ for determining the cause importance of an element. Obstacle elements with a positive cause degree are classified as cause elements. Classification will be made if the result element group does not meet the criteria. The weight of a driver indicates its importance within the driver system. To calculate the results, follow the formula below.

$$W_i = \frac{\sqrt{D_i^2 + R_i^2}}{\sum_{i=1}^{n} \sqrt{D_i^2 + R_i^2}} \tag{10}$$

The selection of the modified fuzzy DEMATEL approach using IT2TrFN was motivated by its ability to handle higher-order uncertainties and capture the vagueness inherent in expert judgments, which traditional Type-I fuzzy sets cannot adequately represent. IT2TrFN effectively models both primary and secondary uncertainties, providing a more reliable representation of subjective evaluations [78–81]. Furthermore, k-means clustering was integrated to objectively determine the threshold for constructing the NRM, replacing the conventional reliance on subjective expert opinion. By grouping influence values (Cij) into statistically distinct clusters, k-means ensures that the threshold reflects the natural separation within the data. To validate the robustness of this selection, a sensitivity analysis was conducted by testing alternative cluster numbers (K = 2, 3), which yielded consistent results. Additional diagnostics, such as within-cluster sum of squares (WCSS) and silhouette scores, confirmed the quality of the clustering. This hybrid approach thus enhances both the precision and objectivity of the causal mapping process.

3.3.2. Social Network Analysis

The analysis of social networks using network and graph theory, known as social network analysis (SNA), examines various social networks that impact their actors differently [82]. As reported by Sodhro et al. [83], the nodes and edges within social networks have predefined content. Depending on how the edges interact, actors can have "directed" or "undirected" relationships. Through several metrics, SNA reveals complex relationships between nodes. SNA is used in this study in the following way:

Step 1: NRM (Network Relation Map) generation: This diagram provides a clear understanding of the problem by simplifying a complex system into manageable segments. A threshold is established during the development of the NRM to reduce complexity and identify significant obstacles. Thus, only relationships with a threshold value greater than the T-relation matrix must be mapped to the graph. There are two options to determine the threshold for creating the NRM. In the total relationship matrix, the average can be calculated based on the opinions of expert consultants [84]. If a relationship was more significant than the threshold, the relationship between the main drivers was filtered. The network and the critical links between the drivers are then imported into Gephi (version 0.10) for examination and visualization.

Step 2. Analysis: The complex interrelationships among potential drivers are explored using several metrics:

(1) Network density: Divide the total number of connections by the maximum number of connections to obtain the network density. The potential density of drivers can vary from 0 (no connections) to 1 (all possible connections), depending on the number of connections. Node networks become denser and more cohesive as the motus value increases. A dense network allows for better communication than a sparse network.

(2) Modularity: Groups, clusters, or communities are defined as modules by their modularity [85]. Modularity is used to optimize network methods for detecting community structures. Internal components are complex in clusters with high modularity, and there are dense connections between potential drivers. In a network with high modularity, there is a significant degree of connectivity between potential drivers.

(3) Nodal weighted degree: Drivers are sized according to how many connections (edges) they have with other drivers, forming the 12 potential drivers. Adding the weights of all the edges of the drivers determines their weighted degree. A network's weighted degree of nodes can give more significance to inward edges than outward edges. Specifically, the inward edges are weighted based on their total weight. A net weighted degree is calculated by subtracting an in-degree score from an out-degree score. Weighed edges show that a driver with the greatest impact on its neighbor has a greater likelihood of having a negative effect on it than a driver with a smaller impact.

(4) Betweenness centrality: As a driver appears on the shortest path between other drivers, it is said to have betweenness centrality. As indicated by this indicator, some of the 12 potential drivers in the network function as "bridges". A driver in a network with

high betweenness centrality holds significant power and control over the network, as more information passes through it, assuming that data is transmitted along the shortest path.

(5) Closeness centrality: Measurement of the closeness centrality of the network can be conducted using the distance centrality indicator. Based on this measure, one can compute the average distance between all drivers in the network. When two drivers are located close to each other, they can easily communicate and influence one another without involving multiple intermediaries.

(6) Eigenvector centrality: Using the number of connections within the network, one can determine an eigenvector driver's centrality. High-scoring drivers require higher scores on the eigen centrality analysis than those that do not.

4. Results

4.1. Analysis Results

Table 5 presents the results of the fuzzy DEMATEL analysis, which utilizes the following Equations (6)–(10) to compute each driver's prominence ($Ri + Ci$), net influence ($Ri - Ci$), and relative importance (weight). Here, Ri represents the total influence a driver exerts on others, while Ci represents the total influence it receives from others. The values of $Ri + Ci$ indicate the centrality of each driver in the system, whereas $Ri - Ci$ distinguishes cause factors (positive values) from effect factors (negative values). The weight column reflects the normalized importance of each driver within the network, derived through the defuzzification process applied to the fuzzy influence matrix.

Table 5. Values of Ri, Ci, $Ri + Ci$, and $Ri - Ci$ for the driver.

	Ri	*Ci*	*Ri + Ci*	*Ri − Ci*	Weight
D1	6.801543	6.720673	13.52222	0.08087	0.093236
D2	5.631987	5.284317	10.9163	0.347671	0.083007
D3	6.707044	7.107573	13.81462	−0.40053	0.114555
D4	6.315886	5.939657	12.25554	0.376229	0.114775
D5	5.093696	5.136026	10.22972	−0.04233	0.108224
D6	6.459117	6.58895	13.04807	−0.12983	0.154793
D7	5.358921	5.22465	10.58357	0.134271	0.14855
D8	6.87887	7.254008	14.13288	−0.37514	0.232977
D9	6.344795	6.570017	12.91481	−0.22522	0.277563
D10	5.745256	5.714912	11.46017	0.030344	0.34093
D11	5.595075	5.77906	11.37413	−0.18398	0.513407
D12	5.583879	5.196229	10.78011	0.38765	1.00000

The comprehensive influence factors obtained from the fuzzy DEMATEL total-relation matrix were further classified using the k-means clustering algorithm to differentiate between significant and negligible influences. This step refines the fuzzy DEMATEL output by objectively determining the thresholds for "effect" and "no effect" relationships, rather than relying on subjective expert judgment. The clustering analysis identified the upper and lower bounds of the "no impact" range as [0.0729, 0.1719] and [0.1745, 0.3035], as illustrated in Figure A1. Based on this analysis, a threshold value of 0.50358 (Figure 2) was derived, where 88 total-relation indices below this threshold were classified as "no effect", and 63 indices above this value were classified as "effect". This threshold-based classification provides the foundation for constructing the Network Relation Map (NRM), which is subsequently analyzed through Social Network Analysis (SNA) to uncover the interrelationships and centrality of drivers.

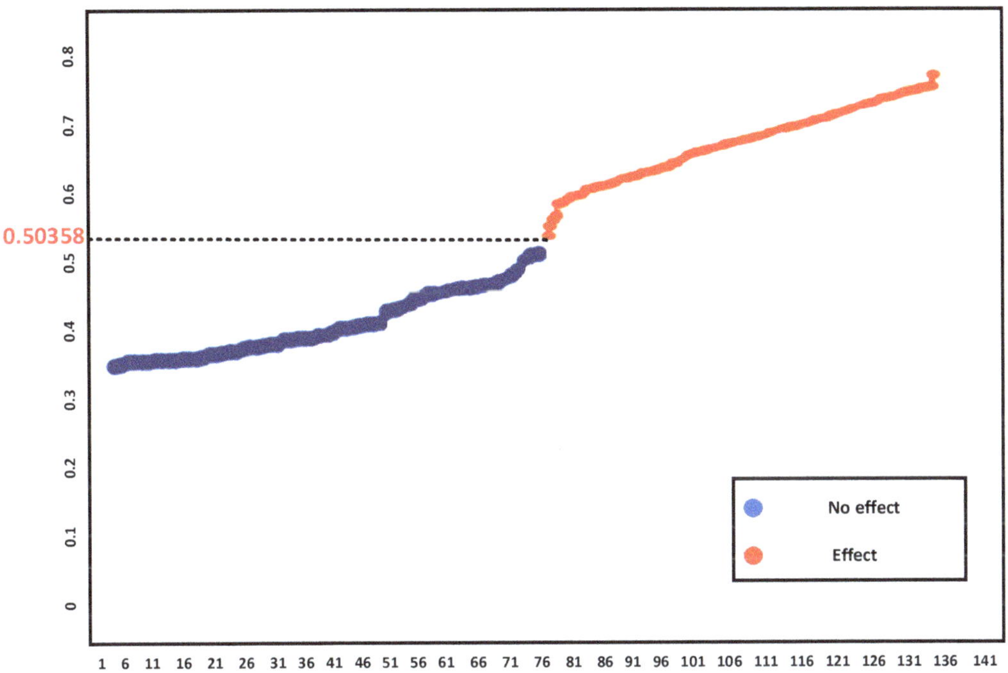

Figure 2. Total-relation index classification by the k-means clustering algorithm.

Figure A2 illustrates how the investment drivers interact with one another intuitively. A digraph map is presented in Figure 3, where the horizontal axis represents the centrality degree (i.e., the overall importance of each driver in the network), and the vertical axis represents the cause degree (i.e., the extent to which a driver influences others). Drivers located in Quadrant I are considered cause factors, meaning they exert a strong influence on the system and should be prioritized for intervention. Conversely, drivers in Quadrants III and IV are effect factors, which are more reactive and influenced by others. The priority for strategic focus follows the order: Quadrant I (most important), then Quadrant II, Quadrant IV, and finally Quadrant III.

To further explore these relationships, the Network Relationship Map (NRM) is constructed using Gephi, as shown in Figure 4, based on a threshold value of 0.50358, derived from the average total influence score. This filtering ensures that only the most meaningful interactions are included. The network consists of 12 drivers and 213 connections, with each node representing a driver and arrows indicating influence pathways. The network's density is calculated to be 0.818, suggesting that most drivers are directly connected. The average path length of 1.27 and a network diameter of 2 imply that communication or influence flows easily between drivers, which is ideal for coordinated implementation strategies in firms.

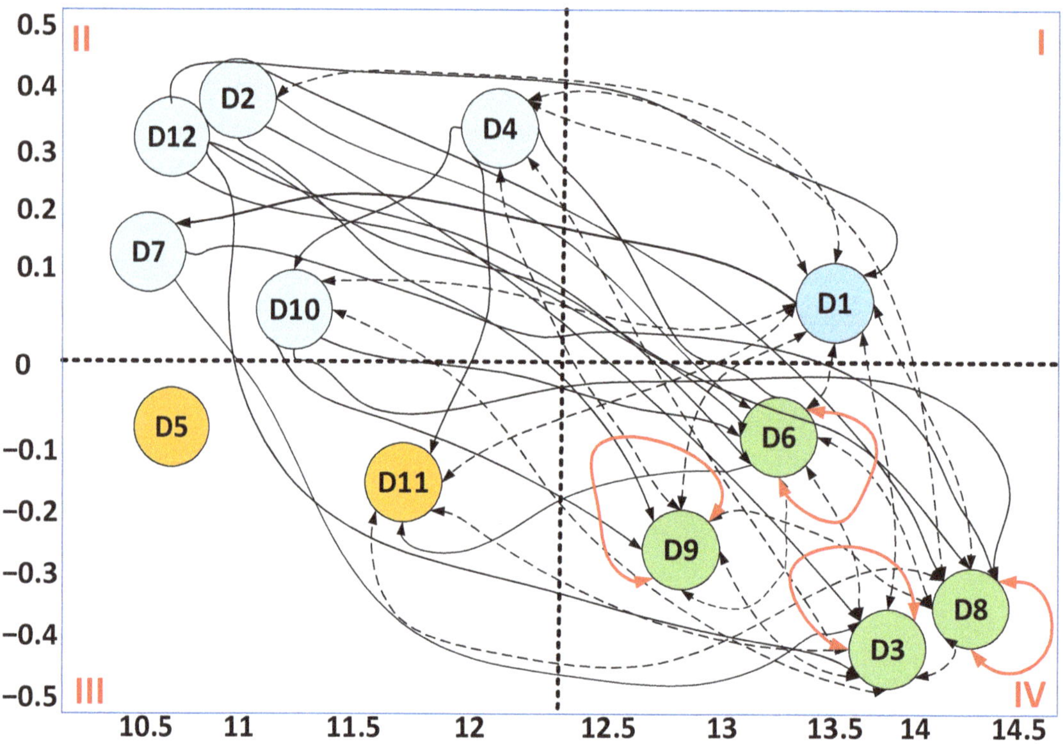

Figure 3. Diagraph map of impact (note: dashed-line arrows indicate two-way relations; solid-line arrows indicate one-way relations).

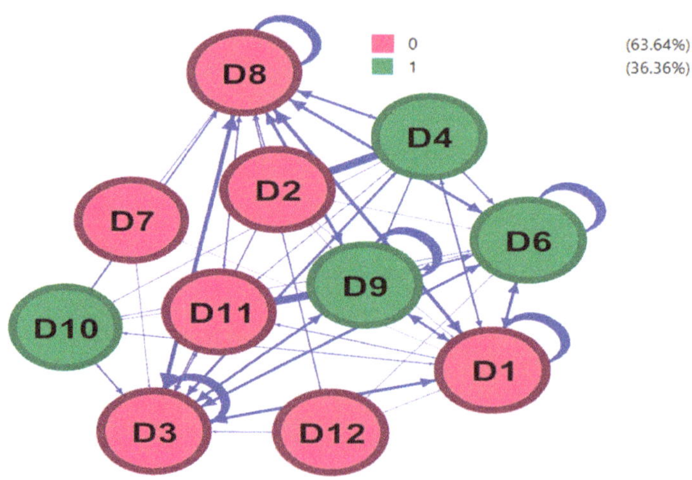

Figure 4. Clusters of the 12 drivers.

Furthermore, the modularity value of 0.013, derived from community detection algorithms, suggests low clustering, indicating that the drivers form a highly integrated network rather than fragmented subgroups. In practice, this suggests that Indian construction firms adopting blockchain may encounter challenges in forming cohesive implementation teams or silos. However, the low modularity also implies strong communication and alignment across various drivers, which could be leveraged to implement blockchain in a more unified and strategic manner across departments and functions.

Several measures can be used to analyze the impact of each driver on the entire NRM, including betweenness centrality, closeness centrality, and eigenvector centrality. The scores for these measures are listed in Table 5, while the driver network maps for each measure are displayed in Figure 5a–d. As indicated by their highest closeness centrality scores, the drivers with the shortest path to other drivers are D1, D3, and D8. These drivers are at the center of the network, as demonstrated in Figure 5b, with the highest betweenness centrality scores indicating their control over the relationships among drivers. In addition, D3, D8, and D1 had the highest eigenvector centrality scores, as shown in Table 6 and Figure 5c, due to their connections with highly influential nodes. It is essential to note that although D5 does not directly impact other drivers, its connections and susceptibility to the impact of other drivers make it crucial to the entire network map. Indian construction firms have utilized BT to hinder the development of clusters within their networks, as illustrated in Figure 5, where colors denote groups of clusters.

Figure 5. The NRM is created by considering the following drivers: (**a**) the betweenness centrality, (**b**) the closeness centrality, (**c**) the weighted degree, and (**d**) the eigenvector centrality status.

Table 6. Ranking of drivers based on different measures.

Drivers	Closeness	Rank	Betweenness	Rank	Eigenvector	Rank
D1	1.0000	1	0.9074	1	0.97632	2
D2	0.7143	4	0.0000	5	0.00000	8
D3	1.0000	1	0.0907	2	1.00000	1
D4	0.8333	3	0.0111	4	0.58782	5
D5	0.0000	7	0.0000	5	0.00000	8
D6	0.9091	2	0.0389	3	0.87358	3
D7	0.5882	6	0.0000	5	0.14895	7
D8	1.0000	1	0.0907	2	1.00000	1
D9	0.8333	3	0.0111	4	0.87205	3
D10	0.7143	4	0.0000	5	0.39117	6
D11	0.7143	4	0.0000	5	0.67801	4
D12	0.6250	5	0.0000	5	0.00000	8

The nodal degree has traditionally been measured by measuring the number of direct ties between nodes in social network analysis (SNA) [10]. However, these studies did not consider the relative weight of each node's influence on the entire network. To address this limitation, FDSNA was developed, which leverages the driver's influence weight calculated by DEMATEL in SNA analysis. By doing so, FDSNA can compute both the nodal and the weighted nodal degrees.

A weighted nodal degree is derived by summing up the weighted in-degrees and weighted out-degrees of the nodal drivers in Figure 5d. Those drivers with stronger connections, such as D1, D3, and D8, are centrally placed, while those with weaker ones are located at the sides of the graph. The thickness of the arrows represents the influence of the interconnections between the drivers. A net weight degree and a net degree value are also presented in Table 7. The effect factor D3, however, is considered to have a high net degree value despite its high net degree value. The net weight degrees of D3, D8, and D1 are the highest in Table 7, whereas the three nodal degrees are D3, D8, and D1. Accordingly, higher net nodal degrees are not necessarily associated with more significant net influences on others. Therefore, the FDSNA approach is given preference over the SNA approach.

Table 7. A description of the net weight degree status.

Label	Weight Degree	Rank	Weighted Out-Degree	Net Weight	Rank	Out-Degree	Degree	Net Degree	Rank
D1	5.7553	3	5.34147	11.0968	3	9	19	10	2
D2	0.0000	10	2.64692	2.6469	9	5	5	0	9
D3	6.6165	2	4.81892	11.4355	1	8	19	11	1
D4	2.1848	7	4.04860	6.2334	6	7	11	4	4
D5	0.0000	10	0.00000	0.0000	12	0	0	0	9
D6	5.1674	4	3.58322	8.7507	4	6	15	9	3
D7	0.5060	9	1.04772	1.5537	11	2	3	1	8
D8	6.7520	1	4.36891	11.1209	2	7	18	11	1
D9	4.6335	5	3.53429	8.1678	5	6	14	8	4

Table 7. Cont.

Label	Weight Degree	Rank	Weighted Out-Degree	Net Weight	Rank	Out-Degree	Degree	Net Degree	Rank
D10	1.5858	8	2.72950	4.3153	7	5	8	3	7
D11	2.6659	6	1.64290	4.3088	8	3	8	5	5
D12	0.0000	10	2.10489	2.1049	10	4	4	0	9

A network can be analyzed using SNA metrics and statistics to identify key influencers. The SNA metrics closeness and betweenness centrality, eigenvector centrality, and weighted degree give more weight to drivers scoring high in these metrics. Based on the results of SNA metrics, Table 8 displays the top three drivers for five SNA indicators. Five times more frequently than the remaining top three SNA metrics, D3 and D8 are the most critical.

Table 8. Key drivers for the five SNA indicators.

Driver ID	Key Drivers	SNA Indicator
D1	Stability of the system	Betweenness centrality
D3	The overall performance of the project	Betweenness centrality
D8	Satisfaction of customers	Betweenness centrality
D1	Stability of the system	Closeness centrality
D3	The overall performance of the project	Closeness centrality
D8	Satisfaction of customers	Closeness centrality
D3	The overall performance of the project	Eigenvector centrality
D8	Satisfaction of customers	Eigenvector centrality
D1	Stability of the system	Eigenvector centrality
D8	Satisfaction of customers	Weighted degree
D3	The overall performance of the project	Weighted degree
D1	Stability of the system	Weighted degree
D3	The overall performance of the project	Net weighted degree
D8	Satisfaction of customers	Net weighted degree
D1	Stability of the system	Net weighted degree

4.2. Sensitivity Analysis

A sensitivity analysis was conducted to test the proposed NRM's strength. To assess the network's sensitivity to the removal of crucial drivers, D3 and D8 were excluded as the main drivers, resulting in a new network consisting of nine nodes and 29 edges, as depicted in Figure 6. After thoroughly analyzing the new network, three notable observations emerged. Firstly, the new network exhibits a reduced level of complexity, boasting a density of 0.403, considerably lower than the original network's density of 0.818. Secondly, there was an increase in the network diameter and average path length, which now stand at 2 and 1.17, respectively, in contrast to the original network's 2 and 1.27. Table 9 illustrates the significant impact on the SNA metrics for the entire network resulting from the exclusion of D3 and D8. Consequently, several nodes can have a negative impact on the new network. Based on the results of this research, it appears that removing D3 and D8, which displayed the strongest interconnections with other drivers, led to reduced network complexity and more significant challenges when sharing information.

This underscores the effectiveness and robustness of the proposed FDSNA approach in visualizing the intricate interrelationships between drivers and identifying key drivers.

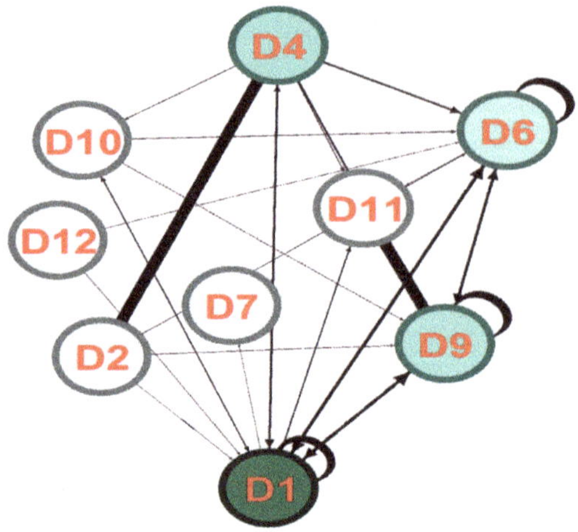

Figure 6. Results of the sensitivity analysis.

Table 9. Sensitivity analysis of the SNA new metrics score.

Drivers	Closeness New	Original	Betweenness Centrality New	Original	Net Weighted Degree New	Original	Eigenvector Centrality New	Original
D1	1.000000	1.0000	0.345238	0.9074	8.41425	11.0968	1.000000	0.97632
D2	0.636364	0.7143	0.000000	0.0000	1.54058	2.6469	0.000000	0.00000
D4	0.857143	0.8333	0.014881	0.0111	3.84807	6.2334	0.452504	0.58782
D5	0.000000	0.0000	0.000000	0.0000	0.00000	0.0000	0.000000	0.00000
D6	0.666667	0.9091	0.041667	0.0389	6.14765	8.7507	0.863759	0.87358
D7	0.000000	0.5882	0.000000	0.0000	0.50602	1.5537	0.243604	0.14895
D9	0.666667	0.8333	0.026786	0.0111	5.59116	8.1678	0.861323	0.87205
D10	0.666667	0.7143	0.000000	0.0000	2.63279	4.3153	0.352672	0.39117
D11	0.545455	0.7143	0.000000	0.0000	2.10011	4.3088	0.564159	0.67801
D12	0.583333	0.6250	0.000000	0.0000	1.02211	2.1049	0.000000	0.00000

4.3. Reliability and Validity

Several measures were employed to assess the reliability of the responses. To determine the dependability of the FDSNA results, raw numerical data were used, following the suggestion of Durdyev et al. [86], rather than relying on linguistic factors provided by specialists. The consistency of responses was deemed acceptable if the calculated value exceeded 0.70 for all experts. The experts were requested to re-evaluate their responses if this threshold was not met. On average, the selected experts achieved a consistency value of 0.7479, indicating that the results were reliable. The CR value of each pairwise evaluation had to be greater than 0.1 to guarantee the reliability of the FDSNA findings. In cases where the added CR value was less than 0.1, the expert was required to complete a new survey. The average CR values for all experts who evaluated the FDSNA results were above the threshold value of 0.093, indicating significant differences between their responses.

This investigation aimed to verify construction engineering and management forecasts in line with Durdyev et al. [87]'s conclusions. Four validation techniques were employed to achieve this objective. Firstly, competent experts participated in identifying obstacles, as suggested by Mohandes et al. [70], to ensure excellent internal validity. Secondly, qualified specialists were involved throughout the research, resulting in robust face validity. Thirdly, four expert panels reviewed and tested the survey questionnaires, enhancing the study's face validity. Lastly, the study objectives were assessed based on the framework provided to ensure the research goals were achieved.

A semi-structured interview was conducted with an experienced expert to obtain external validation for this study. According to Table 10, the validation process is viewed from the perspective of the selected specialists. As a result, the generalization of the findings to a broader context was enabled after comparing the results between the primary and validation studies.

Table 10. Validation of results based on expert interviews.

Interviews	Experts' Involvement in Validation									Aggregation of Responses
Drivers	Expert 1	Expert 2	Expert 3	Expert 4	Expert 5	Expert 6	Expert 7	Expert 8	Expert 9	
D1	2	5	3	4	5	3	3	2	2	3.222
D2	3	5	2	5	5	2	1	2	1	2.889
D3	5	5	5	3	4	3	4	4	3	4.000
D4	1	5	2	2	5	3	3	2	4	3.000
D5	4	5	4	5	4	4	2	1	3	3.556
D6	5	5	3	3	5	4	4	3	2	3.778
D7	3	5	5	3	4	5	3	5	1	3.778
D8	3	4	4	5	4	2	5	5	5	4.111
D9	2	5	5	4	2	3	4	5	5	3.889
D10	2	2	4	2	1	3	5	5	4	3.111
D11	3	1	3	3	2	4	3	5	3	3.000
D12	3	2	3	3	4	4	5	4	5	3.667

5. Discussion

A new technique, referred to as modified FDSNA, was proposed in this research to identify significant connections between drivers and establish an NRM. The findings indicate that not only did the SNA strategy corroborate the outcomes obtained from DEMATEL, but it also provided supplementary insights to decision-makers and managers to effectively identify potential drivers for the successful execution of BT in the Indian SCPM.

A DEMATEL analysis was conducted to measure the direct/indirect influence of drivers (Table 1) and categorize them as cause or effect factors (Figure 1). Based on the causal diagram, D3, D6, and D8 are the most likely drivers since they interact most with other drivers. In addition, they have a high rate and a positive value, making them net drivers of causation. Moreover, D12, D4, and D2 have the highest positive values and the greatest influence on other drivers, indicating that they are the most causal drivers. According to these results, CI experts in India believe that BT adoption requires attention to "the overall performance of the project", "customer satisfaction", and "stability of the system". The market is still in its infancy, and there is no straightforward recipe for success yet. Numerous companies fail to realize returns on their investments as they overlook the significance and feasibility of blockchain solutions, as there exists no methodical assessment of the value at stake [88–90].

In this context, what are the most effective ways for companies to evaluate if blockchain is a viable option for strategic purposes, warranting significant investments? Several blockchain functions must be utilized for SCPM to be more trustworthy, transparent, resource-efficient, and traceable. Several functions contribute to SCPM adoption. By integrating IoT, BDA, and cloud computing technologies, these blockchain technologies can be remedied [22,66,91]. Moreover, governments have adopted oppressive policies towards Bitcoin, thereby hindering the development of other blockchain-based solutions [92].

SCPM/BT integration and procurement collaboration are the most prominent challenges related to specific procurement challenges. It is possible to overcome these obstacles by developing a collaborative ecosystem. To establish proficient governance frameworks for adopting blockchain, it is imperative to identify suitable collaborators [93]. The protection of sensitive and proprietary information requires clear disclosure policies. Promoting adoption by sharing less sensitive details on effective sustainability techniques is advisable, rather than disclosing negative or critical sustainability practices. Collaborating and exchanging information on social and environmental practices, progress, and developmental information may also prove beneficial. Positive experiences from sharing SCPM information can lead to more positive practices and collaborations between companies, ultimately bringing about greater competitive advantages.

The relationships between the drivers were revealed using the NRM constructed with SNA based on the recognition of influential correlations between them (Table A1). Analyzing the NRM using different SNA metrics revealed the following remarkable findings:

1. Strong relationships were found among the three clusters of identified drivers, facilitating the flow of information through the network. As a result, treating a single driver can have an immediate impact on the entire NRM. BT cannot be successfully used within Indian construction companies if one driver is selected and solved without considering the effect of other drivers on that driver.
2. The flow of information and communication in a network is influenced by drivers such as "stability of the system", "overall performance of the project", and "customer satisfaction". These factors are most influential on the flow of information in a network. Due to this, it is crucial to understand aspects with a high degree of closeness because they play an imperative role in facilitating the timely availability of competitive information [94]. This means that the three drivers mentioned above are crucial to the rapid transformation of the entire driver network.
3. There are several key drivers with the highest weighted degree, including "stability of the system", "the overall performance of the project", and "customer satisfaction". These drivers interact the most with other drivers and have the largest net weight and influence over them. It is evident from the causal diagram that these three drivers have the most significant impact on other drivers.
4. SNA analysis showed that although the "adaptability" and "high-quality" drivers scored the lowest among the four SNA static factors, they remain significant drivers with strong eigenvector centrality; therefore, they can influence the entire NRM. "Adaptability" and "quality" are key drivers since they are inextricably linked to decisive factors. As revealed by the NRM, addressing and treating all other drivers wisely can increase "adaptability" and "high-quality".
5. Table 8 summarizes the main drivers for the successful implementation of BT in Indian construction companies, including "stability of the system", "overall project performance", and "customer satisfaction".

Blockchain has been less prevalent in the CI due to its naive nature [90]. BT adoption in the construction industry is driven by three key success factors: system stability, overall project performance, and customer satisfaction. A survey conducted by Deloitte and PwC in 2018 also confirmed this study's findings. In addition to these drivers, blockchain can support SCPM by promoting the use of more environmentally friendly materials and practices [14]. In addition to enabling procurement teams to make informed decisions about which materials to purchase, blockchain can also be used to track the carbon footprint of various materials and products. It can also be used to create digital certificates of origin and quality for materials, promoting the use of sustainably sourced materials and reducing the potential for fraud and misrepresentation [15].

Despite the many potential benefits of BT in SCPM, some challenges and barriers must be addressed. One of the key challenges is the need for interoperability between different blockchain platforms and systems. This can be a particular issue in the CI, where many different stakeholders and systems are involved in the procurement process [62]. Another challenge is the need for standardization and regulation of BT in the CI [23]. Ensuring that all parties use the technology consistently and responsibly without clear standards and regulations can be challenging. This raises concerns regarding data privacy and security issues.

Alternatively, all nations could sign a cooperative international agreement. International trade connects most procurements. Therefore, blockchain adoption can facilitate international adoption since most procurements are not geographically isolated [9]. Moreover, using blockchain will help construction stakeholders build trust, which can mitigate any uncertainty surrounding its use. India's Andhra Pradesh government acquired land for the state's new capital, Amaravati, to develop it. This is a great example. To increase the adoption rate of blockchain technology in the construction industry, the government can play a crucial role by implementing suitable measures. Along with addressing privacy, security, and interoperability concerns, blockchain systems must overcome stakeholder controversies. Addressing these issues can enhance the user experience, improving consumer satisfaction [59,61].

Lastly, it is worth noting that government agencies and their affiliated organizations actively endorse such endeavors. Public-private partnerships (PPPs) between the government and the private sector are expected to accelerate the adoption of blockchain in the construction industry [25]. Since several developing countries share comparable conditions with India, this study's findings may apply to other nations without requiring modifications based on their respective contexts and requirements.

Overall, the use of BT in SCPM can revolutionize the industry's operations. By promoting greater transparency, accountability, efficiency, and sustainability, blockchain can help to create a more responsible and environmentally friendly CI. However, it is essential that the challenges and barriers to adoption are addressed and that the technology is used responsibly and ethically.

Implications for Practice and Policy

The findings from fuzzy DEMATEL and SNA reveal that drivers such as overall project performance (D3), customer satisfaction (D6), and system stability (D8) hold central positions within the network, exerting significant influence on other factors. For firms, this highlights the importance of prioritizing performance-driven blockchain applications that enhance transparency, streamline processes, and facilitate real-time data exchange. These drivers should be treated as critical focus areas when integrating blockchain into construction procurement workflows.

This study offers actionable insights for construction project managers, procurement professionals, and policymakers seeking to implement blockchain in sustainable construction procurement management (SCPM). The ranked drivers allow organizations to allocate resources strategically, design targeted training programs, develop supportive regulatory frameworks, and adopt blockchain-based tools aligned with specific sustainability goals. For example, drivers such as system stability and data management emphasize the importance of secure and transparent procurement operations, while policies and laws highlight the role of governance in enabling smoother implementation. These findings bridge the gap between theory and practical applications of digital procurement transformation.

From a managerial standpoint, firms can optimize blockchain adoption by first focusing on system stabilization (D8), followed by enhancing performance (D3), and improving client engagement and satisfaction (D6). This structured approach fosters adaptability, cost efficiency, and long-term digital maturity. Additionally, training and capacity-building initiatives should align with these central drivers to maximize their effectiveness.

For policymakers, the findings suggest a targeted approach to regulation and incentives. Policy measures should prioritize transparent procurement protocols, support the development of digital infrastructure, and promote blockchain-based pilot projects with measurable performance outcomes. Furthermore, regulatory bodies could establish national digital standards and data-sharing frameworks to enhance systemic stability and facilitate the scalable implementation of blockchain technology in the construction sector.

6. Conclusions

This study investigates the drivers promoting the adoption of BT within SCPM using a hybrid methodological approach. Several methods were utilized in this study, including a systematic literature review, expert interviews, and a questionnaire survey. According to the data collected from the pool of qualified experts, the following conclusions can be drawn:

1. The drivers that act as stumbling blocks to adopting BT-based technologies were identified through a comprehensive literature review and expert interviews.
2. Analyzing the effect and intensity of interrelationships among drivers as a function of their causal interrelationships using the modified FDEMATEL method, the study results indicate that decision-makers must deal effectively with drivers by addressing potential "causal" drivers.
3. Created an NRM of potential drivers and determined their complex relationships using SNA. The most potential drivers for BT adoption in SCPM are "stability of the system", "overall project performance", and "customer satisfaction".

This study contributes to the body of knowledge and practice by providing detailed, novel insights that benefit stakeholders in the construction industry, policymakers, and those involved in sustainability practices. Examining the drivers of BT towards promoting SCPM can potentially revolutionize how the CI operates. By leveraging blockchain's unique features, such as transparency, immutability, and decentralization, sustainable procurement practices can be effectively implemented, reducing environmental impact and increasing social responsibility. While examining the potential drivers of BT in promoting SCPM, some limitations become evident. One such limitation is the lack of clear guidelines and standards for blockchain integration into construction procurement practices. Additionally, there is a need for further research into the financial implications and cost–benefit analysis of implementing BT in construction procurement. Also, comprehensive case studies will be conducted to assess real-world financial impacts, develop clear guidelines for seamless integration, and explore regulatory challenges to overcome legal barriers.

Furthermore, research should address barriers to blockchain adoption in SCPM by analyzing socio-economic, technical, and organizational hurdles. Developing blockchain-based tools tailored to construction procurement systems is crucial, as is exploring the broader potential of blockchain in project and supply chain management to optimize workflows and enhance transparency. By focusing on these areas, future research can provide actionable insights for industry stakeholders and policymakers, facilitating the adoption of blockchain technology and advancing sustainability practices in the construction sector.

Author Contributions: Conceptualization, A.K.S. and S.R.M.; Data curation, A.K.S. and V.R.P.K.; Formal analysis, A.K.S.; Investigation, A.K.S.; Methodology, A.K.S. and S.R.M.; Project administration, S.R.M. and M.A.; Resources, P.S. and C.K.; Software, P.S.; Supervision, S.R.M. and V.R.P.K.; Validation, S.R.M. and C.C.; Writing—original draft, A.K.S., S.R.M., P.S. and V.R.P.K.; Writing—review and editing, A.K.S., S.R.M., P.S., C.C., M.A. and C.K. All authors have read and agreed to the published version of the manuscript.

Funding: This research received no external funding.

Acknowledgments: We appreciate the contributions of experts in blockchain technology and sustainable construction procurement management, actively ensuring the survey's comprehensive foundation.

Conflicts of Interest: The authors declare no conflict of interest.

Appendix A

This appendix comprises three invaluable tables: Table A1 showcases a matrix that normalizes direct influences, while Table A2 presents a matrix of total relationships. Lastly, Table A3 elucidates the influence relationships for the construction network relation map.

Table A1. Matrix normalizing direct influences.

Drivers	D1	D2	D3	D4	D5	D6	D7	D8	D9	D10	D11	D12
D1	1.550637	0.494392	0.676472	0.55752	0.477099	0.635608	0.506017	0.693607	0.639211	0.538975	0.543418	0.488587
D2	0.510831	1.356655	0.552925	0.469006	0.420286	0.512895	0.408399	0.553412	0.516858	0.447933	0.453823	0.428966
D4	0.647963	0.4886	0.576108	0.550039	0.459983	0.635327	0.472641	0.704186	0.633554	0.52914	0.542604	0.466899
D5	0.583205	0.456752	0.626832	1.44955	0.447396	0.579505	0.459817	0.644215	0.572374	0.517687	0.524791	0.453763
D6	0.471377	0.3941	0.491027	0.435406	1.313745	0.443896	0.381754	0.502049	0.450448	0.406439	0.418881	0.384574
D7	0.623938	0.476413	0.650388	0.526864	0.451429	1.513981	0.456224	0.668061	0.609097	0.501166	0.517754	0.463804
D9	0.493992	0.392909	0.520716	0.444065	0.402508	0.480994	1.335391	0.526996	0.487231	0.442977	0.428138	0.403003
D10	0.664488	0.501779	0.708183	0.56422	0.494287	0.649222	0.494248	1.601897	0.643516	0.539576	0.537379	0.480074
D11	0.612839	0.466955	0.643667	0.513004	0.440147	0.605077	0.465483	0.655943	1.503768	0.482427	0.49977	0.455716
D12	0.535598	0.418495	0.573169	0.485252	0.412343	0.525368	0.415376	0.5802	0.515165	1.393589	0.472225	0.418475
D1	0.514149	0.414434	0.557952	0.476893	0.404528	0.496627	0.405884	0.570797	0.49938	0.462008	1.387718	0.404704
D2	0.511656	0.422834	0.530134	0.467837	0.412276	0.510451	0.423415	0.552645	0.499415	0.452995	0.452558	1.347663

Table A2. Matrix of total relationships.

Drivers	D1	D2	D3	D4	D5	D6	D7	D8	D9	D10	D11	D12
D1	0.550637	0.494392	0.676472	0.55752	0.47710	0.635608	0.506017	0.693607	0.639211	0.538975	0.543418	0.488587
D2	0.510831	0.356655	0.552925	0.469006	0.420286	0.512895	0.40840	0.553412	0.516858	0.447933	0.453823	0.428966
D3	0.647963	0.48860	0.576108	0.550039	0.459983	0.635327	0.472641	0.704186	0.633554	0.52914	0.54260	0.46690
D4	0.58320	0.456752	0.626832	0.44955	0.44740	0.57950	0.459817	0.644215	0.572374	0.517687	0.524791	0.453763
D5	0.471377	0.39410	0.491027	0.435406	0.313745	0.44390	0.381754	0.502049	0.450448	0.406439	0.418881	0.384574
D6	0.623938	0.476413	0.650388	0.526864	0.451429	0.513981	0.456224	0.668061	0.60910	0.501166	0.517754	0.46380
D7	0.493992	0.392909	0.520716	0.444065	0.402508	0.480994	0.335391	0.52700	0.487231	0.442977	0.428138	0.40300
D8	0.664488	0.501779	0.708183	0.56422	0.494287	0.649222	0.494248	0.60190	0.643516	0.539576	0.537379	0.480074
D9	0.612839	0.466955	0.643667	0.51300	0.440147	0.605077	0.465483	0.655943	0.503768	0.482427	0.49977	0.455716
D10	0.53560	0.418495	0.573169	0.485252	0.412343	0.525368	0.415376	0.58020	0.515165	0.393589	0.472225	0.418475
D11	0.514149	0.414434	0.557952	0.476893	0.404528	0.496627	0.405884	0.57080	0.49938	0.462008	0.387718	0.40470
D12	0.511656	0.422834	0.530134	0.467837	0.412276	0.510451	0.423415	0.552645	0.499415	0.45300	0.452558	0.347663

Table A3. Influence relationships for the construction network relation map.

Drivers	D1	D2	D3	D4	D5	D6	D7	D8	D9	D10	D11	D12
D1	0.55064	0.00000	0.67647	0.55752	0.00000	0.63561	0.50602	0.69361	0.63921	0.53897	0.54342	0.00000
D2	0.51083	0.00000	0.55293	0.00000	0.00000	0.51289	0.00000	0.55341	0.51686	0.00000	0.00000	0.00000
D4	0.64796	0.00000	0.57611	0.55004	0.00000	0.63533	0.00000	0.70419	0.63355	0.52914	0.54260	0.00000
D5	0.58320	0.00000	0.62683	0.00000	0.00000	0.57950	0.00000	0.64422	0.57237	0.51769	0.52479	0.00000
D6	0.00000	0.00000	0.00000	0.00000	0.00000	0.00000	0.00000	0.00000	0.00000	0.00000	0.00000	0.00000
D7	0.62394	0.00000	0.65039	0.00000	0.00000	0.51398	0.00000	0.66806	0.60910	0.00000	0.51775	0.00000
D9	0.00000	0.00000	0.52072	0.00000	0.00000	0.00000	0.00000	0.52700	0.00000	0.00000	0.00000	0.00000
D10	0.66449	0.00000	0.70818	0.56422	0.00000	0.64922	0.00000	0.60190	0.64352	0.00000	0.53738	0.00000
D11	0.61284	0.00000	0.64367	0.51300	0.00000	0.60508	0.00000	0.65594	0.50377	0.00000	0.00000	0.00000
D12	0.53560	0.00000	0.57317	0.00000	0.00000	0.52537	0.00000	0.58020	0.51516	0.00000	0.00000	0.00000
D1	0.51415	0.00000	0.55795	0.00000	0.00000	0.00000	0.00000	0.57080	0.00000	0.00000	0.00000	0.00000
D2	0.51166	0.00000	0.53013	0.00000	0.00000	0.51045	0.00000	0.55264	0.00000	0.00000	0.00000	0.00000

Appendix B

This appendix comprises Figures A1 and A2, which provide a vivid depiction of both the degree of causation and centrality, as well as the intricate interplay between drivers and their corresponding levels of influence.

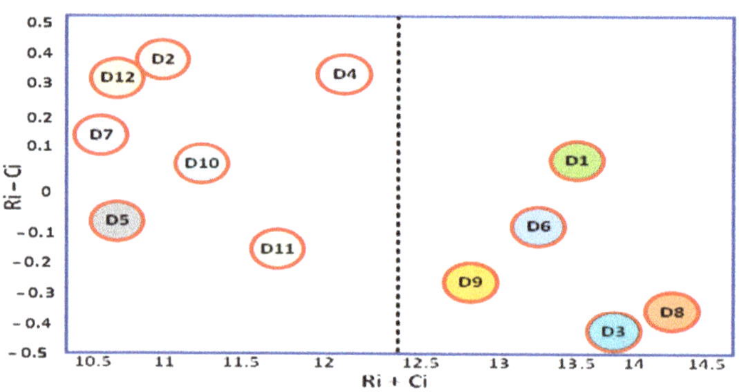

Figure A1. Visual representation of the cause degree and the center degree.

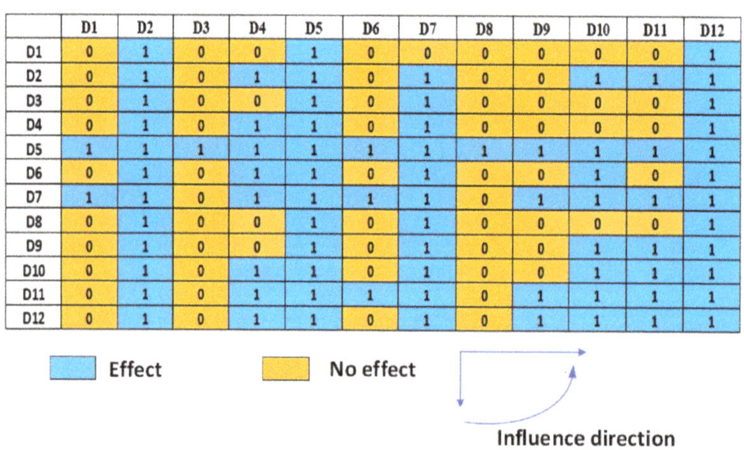

	D1	D2	D3	D4	D5	D6	D7	D8	D9	D10	D11	D12
D1	0	1	0	0	1	0	0	0	0	0	0	1
D2	0	1	0	1	1	0	1	0	0	1	1	1
D3	0	1	0	0	1	0	1	0	0	0	0	1
D4	0	1	0	1	1	0	1	0	0	0	0	1
D5	1	1	1	1	1	1	1	1	1	1	1	1
D6	0	1	0	1	1	0	1	0	0	1	0	1
D7	1	1	0	1	1	1	1	0	1	1	1	1
D8	0	1	0	0	1	0	1	0	0	0	0	1
D9	0	1	0	0	1	0	1	0	0	1	1	1
D10	0	1	0	1	1	0	1	0	0	1	1	1
D11	0	1	0	1	1	1	1	0	1	1	1	1
D12	0	1	0	1	1	0	1	0	1	1	1	1

Figure A2. Relationship between drivers and their influence.

Appendix C

"Development of Fuzzy-DEMATEL-based survey"

Section A: Experts need to answer the following questions:

How many construction projects are done using blockchain technology
Size of the company
Occupations
Degree
Years of experience

Section B: Questionnaire survey:

Table A4. Variables and codes to be used.

The Extent of the Impact	Code	Explanation
No influence	NI	There is no influence between the two drivers
Very low influence	VL	One driver has a very low influence on the other one
Low influence	L	One driver has a low influence on the other one
High influence	H	One driver has a high influence on the other one
Very high influence	VH	One driver has a very high influence on the other one

Examples:

- Based on your expertise, if you believe that D1 has no effect on D2. Please fill in NI in the box.
- Based on your expertise, if you believe that D3 has a "very high impact" on D4. Please fill in VH in the box.

With the above in mind, please fill in the boxes highlighted in YELLOW using the CODE shown in Table A5 based on variables in Table A6.

Table A5. Sample of variables to fill in the questionnaire.

Drivers	D1	D2	D3	D4	D5	D6	D7	D8	D9	D10	D11	D12
D1												
D2												
D3												
D4												
D5												
D6												
D7												
D8												
D9												
D10												
D11												
D12												

Table A6. List of drivers with description.

Drivers	Code	Description
Stability of the system	D1	Transparency, security, removing intermediaries, and trustworthiness are four variables considered in assessing system robustness. After the integration of BT into construction procurement management (CPM), all of these characteristics made the CPM and system more robust and sustainable due to the transparency in information flow, stability of data, and peer-to-peer transactions without involving a third party.
Cost of the overall project	D2	Three variables are combined to form the overall cost: cost, energy, and overall project costs. The overall CPM cost includes all financial investments like documentation fees, stationery expenditures, human resources costs, electricity costs, facility costs, documentation incurred during production, etc.
The overall performance of the project	D3	The overall performance of the project is the efficiency, effectiveness, and speed of doing the job effectively with low responses, high standardization, and reduced complexity of the work. A composite of five variables is built by examining the correlations between efficiency, speed, automation, simplification of current paradigms, and sharing demand in CPM.
Decentralization and data security	D4	By experts under the name data safety and decentralization, a group of four variables is grouped. All people involved directly or indirectly in the system need records and information regarding standardization, production, and supply. It is only on BT-integrated CPM that it is possible to hack, change, control, or lose data for any reason.
Adaptability	D5	Regarding adaptability, three variables, traceability, visibility, and identifying issues, form a common driver. Using IoT/Industry 4.0, adaptability means tracking causes, goods location, accidents, and fraud between a CPM's manufacturing and end-use processes.
Policies and laws	D6	A lot of time and effort is involved in documenting laws and policies. Documentation plays a crucial role in all contracts, but public ledger technology gives us transparency, speeds up work, checks for corruption in government, and also helps to find out scams in any organization. The importance of data records for any legal action cannot be overstated. This technology does not allow the deletion or modification of data, so there can be no fraud; laws and government policies have been factored together to form accessibility.
The system with innovative features	D7	With the smart system, smart contracts are implemented, invoicing is simplified, and inventory levels are improved. By eliminating any fraud in the documentation, paying taxes on time, and ensuring that goods are delivered on time from the shipyard area, the system removes any fraud in the documentation.

Table A6. *Cont.*

Drivers	Code	Description
Satisfaction of customers	D8	Feedback and customer centricity, based on expert opinion, is the basis of customer satisfaction. The customer is satisfied when they provide positive feedback and respond positively. In order to achieve this, it is necessary to deliver the right product with the correct information at the right time, at the right place, and in the right hands, and to commit to providing periodic service after the sale.
The system with high reliability	D9	Reliable systems are determined by four drivers: Scalability, Reliability, Durability, and the ability to lose data. A record of raw, semi-finished, and finished material is kept at every BT location. By keeping the correct information about goods, IT saves not only time but also money.
Detailed documentation	D10	Auditable, accounting, and ecosystem documentation are part of the documentation. In BT, integration is preferred for several reasons, such as smooth auditing processes, simplified financial systems, and smooth currency flow.
A data management system	D11	A data management system controls access to data from the end-users, manipulates transactions from a single account simultaneously, eliminates human error in the documentation and other tasks, and allows real-time information flow. Three variables are considered in data management: data quality, flow and control of information, and access control.
High-quality	D12	Quality assurance and quality fairness are drivers that affect quality. Any irregularity in the process, transportation, raw material specification, or the final product is considered poor quality, due to the lack of availability of complete information and the need to eliminate human error.

Section C: Additional questions

From your perspective, what would be the ways/strategies/suggestions/solutions (in terms of policy, regulation, engineering, application, etc.) for promoting the adoption of blockchain in the construction industry?

Suggestions/Strategies
1.
2.
3.
4.
5.

Appendix D

Appendix D.1. Fuzzy-DEMATEL

The modification of the DEMATEL technique is derived from the IT2TrFN theory. Within the next phase, we will introduce a modified fuzzy DEMATEL approach, which was proposed by [91,92]. This chapter concludes with a description of the analysis framework that has been constructed.

Appendix D.1.1. Trapezoidal Fuzzy Number Type-II Interval

Members of IT2TrFN describe a method in which personal information remains fuzzy, subjective, and uncertain despite the freedom and flexibility provided by the process.

Defining 1. *Fuzzy sets of type II can be expressed in two ways.*

$$\tilde{Z} = \{((a,b), b_{\tilde{Z}}(a,b)) \mid \forall x \in X, \forall b \in J_x \subseteq [0,1], 0 \leq b_{\tilde{Z}}(a,b) \leq 1\} \quad \text{(A1)}$$

$$\tilde{\tilde{Z}} = \frac{\int_{a \in X} \int_{b \in J_a} b_{\tilde{\sim}}(a,b)}{(a,b)dbda} \quad \text{(A2)}$$

where $b_{\underset{A}{\approx}}$ *Z has a type-II affiliation function,* $J_a \subseteq [0,1]$.

Defining 2. *The fuzzy set Z (a,b) that is described as Z (a,b) is an interval type-II fuzzy set, which takes the form of*

$$\tilde{\tilde{Z}} = \frac{\int_{a \in X} \int_{b \in J_x} 1}{(a,b)dbda} \quad \text{(A3)}$$

where Ja ⊆ [0, 1].

Defining 3. *Since the upper and lower bounds of IT2TrFN are trapezoidal fuzzy numbers shown in Figure A3, it is a trapezoidal fuzzy number. The following is the result:*

$$\tilde{\tilde{Z}} = \left(\tilde{Z}_i^U, \tilde{Z}_i^L\right) = \begin{bmatrix} \left(z_{i1}^U, z_{i2}^U, z_{i3}^U, z_{i4}^U; H_1\left(\tilde{Z}_i^U\right), H_2\left(\tilde{Z}_i^U\right)\right), \\ \left(z_{i1}^L, z_{i2}^L, z_{i3}^L, z_{i4}^L; H_1\left(\tilde{Z}_i^L\right), H_2\left(\tilde{Z}_i^L\right)\right) \end{bmatrix} \quad \text{(A4)}$$

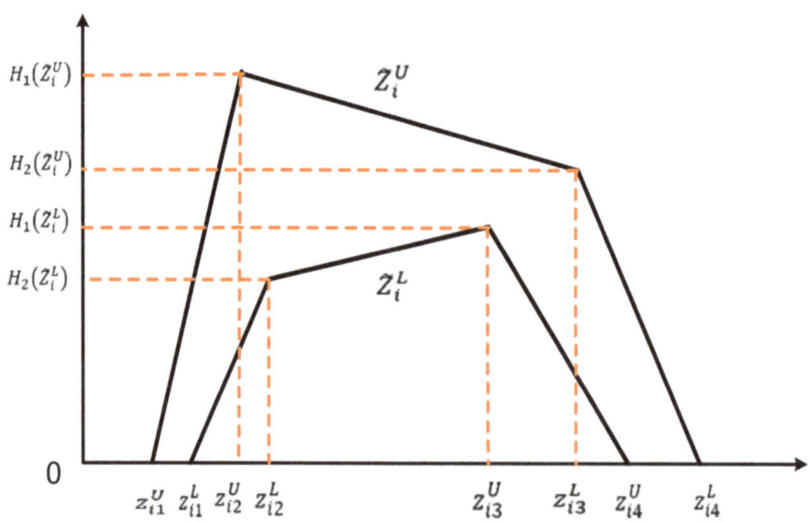

Figure A3. Schematic diagram of IT2TrFN.

The Hj(U i) becomes a member of the aU i(j + 1), and the Hj(Li) becomes a member of the aL i(j + 1), 1 + j + 2. $\overset{\sim U}{Z_i}$ $\overset{\sim L}{Z_i}$.

Sets of interval trapezoids with constant k are interval trapezoidal type-II fuzzy sets:

$$\ni \overset{\approx}{Z}_2 = \left(\overset{\sim U}{Z_1}, \overset{\sim L}{Z_1}\right) \oplus \left(\overset{\sim U}{Z_2}, \overset{\sim L}{Z_2}\right)$$

$$\left[\begin{array}{l}\left(z_{11}^U + z_{21}^U, z_{12}^U + z_{22}^U, z_{13}^U + z_{23}^U, z_{14}^U + z_{24}^U; min\left(H_1\left(\overset{\sim U}{Z_1}\right), H_1\left(\overset{\sim U}{Z_2}\right)\right), min\left(H_2\left(\overset{\sim U}{Z_1}\right), H_2\left(\overset{\sim U}{Z_2}\right)\right)\right), \\ \left(z_{11}^L + z_{21}^L, z_{12}^L + z_{22}^L, z_{13}^L + z_{23}^L, z_{14}^L + z_{24}^L; min\left(H_1\left(\overset{\sim L}{Z_1}\right), H_1\left(\overset{\sim L}{Z_2}\right)\right), min\left(H_2\left(\overset{\sim L}{Z_1}\right), H_2\left(\overset{\sim L}{Z_2}\right)\right)\right)\end{array}\right] \quad (A5)$$

$$X_1 \otimes \overset{\approx}{Z}_2 = \left(\overset{\sim U}{Z_1}, \overset{\sim L}{Z_1}\right) \otimes \left(\overset{\sim U}{Z_2}, \overset{\sim L}{Z_2}\right)$$

$$\left[\begin{array}{l}\left(z_{11}^U \times z_{21}^U, z_{12}^U \times z_{22}^U, z_{13}^U \times z_{23}^U, z_{14}^U \times z_{24}^U; min\left(H_1\left(\overset{\sim U}{Z_1}\right), H_1\left(\overset{\sim U}{Z_2}\right)\right), min\left(H_2\left(\overset{\sim U}{Z_1}\right), H_2\left(\overset{\sim U}{Z_2}\right)\right)\right), \\ \left(z_{11}^L \times z_{21}^L, z_{12}^L \times z_{22}^L, z_{13}^L \times z_{23}^L, z_{14}^L \times z_{24}^L; min\left(H_1\left(\overset{\sim L}{Z_1}\right), H_1\left(\overset{\sim L}{Z_2}\right)\right), min\left(H_2\left(\overset{\sim L}{Z_1}\right), H_2\left(\overset{\sim L}{Z_2}\right)\right)\right)\end{array}\right] \quad (A6)$$

$$k\overset{\approx}{Z}_1 = \left[\begin{array}{l}\left(kz_{11}^U, kz_{12}^U, kz_{13}^U, kz_{14}^U; H_1\left(\overset{\sim U}{Z_1}\right), H_2\left(\overset{\sim U}{Z_1}\right)\right), \\ \left(kz_{11}^L, kz_{12}^L, kz_{13}^L, kz_{14}^L; H_1\left(\overset{\sim L}{Z_1}\right), H_2\left(\overset{\sim L}{Z_1}\right)\right)\end{array}\right] \quad (A7)$$

According to Herrera-Viedma [74], IT2TrFN is defuzzed as follows:

$$\text{Defuzzified}\left(\overset{\approx}{Z_i}\right) = \frac{\frac{\left(z_{i4}^U - z_{i1}^U\right) + \left(H_1\left(\overset{\sim U}{Z_i}\right) * z_{i2}^U - z_{i1}^U\right) + \left(H_2\left(\overset{\sim U}{Z_i}\right) * z_{i3}^U - z_{i1}^U\right)}{4} + z_{i1}^U + \frac{\left(z_{i4}^L - z_{i1}^L\right) + \left(H_1\left(\overset{\sim L}{Z_i}\right) * z_{i2}^L - z_{i1}^L\right) + \left(H_2\left(\overset{\sim L}{Z_i}\right) * z_{i3}^L - z_{i1}^L\right)}{4} + z_{i1}^L}{2} \quad (A8)$$

Appendix D.1.2. Algorithm for Clustering Using K-Means

K-means clusters data based on their mean values in a non-hierarchical manner. Clustering data into K categories is achieved by optimizing the mean distance values iteratively. Its unsupervised machine-learning algorithm can improve the objectivity of threshold values.

Definition 1. *A sample data set of x_i would be compared to a cluster centroid set of c_j [93,94]. The distance between the sample data and the centroid is calculated using the Euclidean distance formula, which is derived from the Pythagorean theorem:*

$$dis(x_i, c_j) = \sqrt{(x_{i1} - c_{j1})^2 + (x_{i2} - c_{j2})^2 + \cdots + (x_{im} - c_{jm})^2} \quad (A9)$$

where $x_i = (x_{i1}, x_{i2}, \cdots, x_{im})$, $c_i = (c_{j1}, c_{j2}, \cdots, c_{jm})$ and M represents the dimension.

Definition 2. *A sample cluster's centroid is determined in each iteration by following the steps below [93,94]:*

$$c_j = \frac{1}{N(\varphi_j)} - \left\{\sum_{x_i \in \varphi_j} x_{i1}, \sum_{x_i \in \varphi_j} x_{i2}, \cdots, \sum_{x_i \in \varphi_j} x_{im}\right\} \quad (A10)$$

where $N(_{\varphi j})$ measures how many data points are within the cluster φj.

Definition 3. *Based on the definition of the clustering criterion function, when* $|SSEN + SSEN - 1| < \varepsilon$, *the clustering criterion function has converged [93,94]*

$$SSE = \sum_{j=1}^{K} \sum_{x_i \in \varphi_j} dis(x_i, c_j) \quad \text{(A11)}$$

where n denotes the number of iterations, the discriminant condition, etc.

References

1. Schiavoni, S.; D'Alessandro, F.; Bianchi, F.; Asdrubali, F. Insulation materials for the building sector: A review and comparative analysis. *Renew. Sustain. Energy Rev.* **2016**, *62*, 988–1011. [CrossRef]
2. Fang, H.; Wang, B.; Song, W. Analyzing the interrelationships among barriers to green procurement in photovoltaic industry: An integrated method. *J. Clean. Prod.* **2020**, *249*, 119408. [CrossRef]
3. Toktaş-Palut, P.; Baylav, E.; Teoman, S.; Altunbey, M. The impact of barriers and benefits of e-procurement on its adoption decision: An empirical analysis. *Int. J. Prod. Econ.* **2014**, *158*, 77–90. [CrossRef]
4. Yu, A.T.W.; Yevu, S.K.; Nani, G. Towards an integration framework for promoting electronic procurement and sustainable procurement in the construction industry: A systematic literature review. *J. Clean. Prod.* **2020**, *250*, 119493. [CrossRef]
5. Mansi, M. Sustainable procurement disclosure practices in central public sector enterprises: Evidence from India. *J. Purch. Supply Manag.* **2015**, *21*, 125–137. [CrossRef]
6. Zhong, B.; Pan, X.; Ding, L.; Chen, Q.; Hu, X. Blockchain-driven integration technology for the AEC industry. *Autom. Constr.* **2023**, *150*, 104791. [CrossRef]
7. Perera, S.; Nanayakkara, S.; Rodrigo, M.N.N.; Senaratne, S.; Weinand, R. Blockchain technology: Is it hype or real in the construction industry? *J. Ind. Inf. Integr.* **2020**, *17*, 100125. [CrossRef]
8. Ershadi, M.; Jefferies, M.; Davis, P.; Mojtahedi, M. Barriers to achieving sustainable construction project procurement in the private sector. *Clean. Eng. Technol.* **2021**, *3*, 100125. [CrossRef]
9. Waqar, A.; Khan, A.M.; Othman, I. Journal of Infrastructure Intelligence and Resilience Blockchain empowerment in construction supply chains: Enhancing efficiency and sustainability for an infrastructure development. *J. Infrastruct. Intell. Resil.* **2024**, *3*, 100065. [CrossRef]
10. Kumar, A.; Kumar, V.R.P.; Dehdasht, G.; Reza, S.; Manu, P.; Pour, F. Investigating the barriers to the adoption of blockchain technology in sustainable construction projects. *J. Clean. Prod.* **2023**, *403*, 136840. [CrossRef]
11. Maden, A.; Alptekin, E. Understanding the Blockchain Technology Adoption from Procurement Professionals' Perspective—An Analysis of the Technology Acceptance Model Using Intuitionistic Fuzzy Cognitive Maps. In *International Conference on Intelligent and Fuzzy Systems*; Springer International Publishing: Cham, Switzerland, 2021; Volume 1197, pp. 347–354. [CrossRef]
12. Li, J.; Kassem, M. Applications of distributed ledger technology (DLT) and Blockchain-enabled smart contracts in construction. *Autom. Constr.* **2021**, *132*, 103955. [CrossRef]
13. Saygili, M.; Mert, I.E.; Tokdemir, O.B. A decentralized structure to reduce and resolve construction disputes in a hybrid blockchain network. *Autom. Constr.* **2020**, *134*, 104056. [CrossRef]
14. Saberi, S.; Kouhizadeh, M.; Sarkis, J.; Shen, L. Blockchain technology and its relationships to sustainable supply chain management. *Int. J. Prod. Res.* **2019**, *57*, 2117–2135. [CrossRef]
15. Hamledari, H.; Fischer, M. Measuring the impact of blockchain and smart contracts on construction supply chain visibility. *Adv. Eng. Inform.* **2021**, *50*, 101444. [CrossRef]
16. Badi, S.; Ochieng, E.; Nasaj, M.; Papadaki, M. Technological, organisational and environmental determinants of smart contracts adoption: UK construction sector viewpoint. *Constr. Manag. Econ.* **2021**, *39*, 36–54. [CrossRef]
17. Ronaghi, M.H.; Mosakhani, M. The effects of blockchain technology adoption on business ethics and social sustainability: Evidence from the Middle East. *Environ. Dev. Sustain.* **2022**, *24*, 6834–6859. [CrossRef] [PubMed]
18. Tezel, A.; Papadonikolaki, E.; Yitmen, I.; Hilletofth, P. Preparing construction supply chains for blockchain technology: An investigation of its potential and future directions. *Front. Eng. Manag.* **2020**, *7*, 547–563. [CrossRef]
19. Sarker, I.; Datta, B. Re-designing the pension business processes for achieving technology-driven reforms through blockchain adoption: A proposed architecture. *Technol. Forecast. Soc. Change* **2022**, *174*, 121059. [CrossRef]
20. Akram, S.V.; Malik, P.K.; Singh, R.; Anita, G.; Tanwar, S. Adoption of blockchain technology in various realms: Opportunities and challenges. *Secur. Priv.* **2020**, *3*, e109. [CrossRef]
21. Pérez, A.T.E.; Rossit, D.A.; Tohmé, F.; Vásquez, Ó.C. Mass customized/personalized manufacturing in Industry 4.0 and blockchain: Research challenges, main problems, and the design of an information architecture. *Inf. Fusion* **2022**, *79*, 44–57. [CrossRef]

22. Rane, S.B.; Thakker, S.V. Green procurement process model based on blockchain–IoT integrated architecture for a sustainable business. *Manag. Environ. Qual. Int. J.* **2020**, *31*, 741–763. [CrossRef]
23. Nanayakkara, S.; Perera, S.; Senaratne, S.; Weerasuriya, G.T. Blockchain and Smart Contracts: A Solution for Payment Issuesa. *Informatics* **2021**, *8*, 36. [CrossRef]
24. Hamledari, H.; Fischer, M. Construction payment automation using blockchain-enabled smart contracts and robotic reality capture technologies. *Autom. Constr.* **2021**, *132*, 103926. [CrossRef]
25. Figueiredo, K.; Hammad, A.W.A.; Haddad, A.; Tam, V.W.Y. Assessing the usability of blockchain for sustainability: Extending key themes to the construction industry. *J. Clean. Prod.* **2022**, *343*, 131047. [CrossRef]
26. Smith, S.; O'rourke, L. Exploring the potential of Blockchain technology for the UK Construction industry—2019. *Sustainability* **2019**, *1*, 1–53.
27. Sadeghi, M.; Mahmoudi, A.; Deng, X. Adopting distributed ledger technology for the sustainable construction industry: Evaluating the barriers using Ordinal Priority Approach. *Environ. Sci. Pollut. Res.* **2022**, *29*, 10495–10520. [CrossRef]
28. Bai, C.; Quayson, M.; Sarkis, J. Analysis of Blockchain's enablers for improving sustainable supply chain transparency in Africa cocoa industry. *J. Clean. Prod.* **2022**, *358*, 131896. [CrossRef]
29. Parmentola, A.; Petrillo, A.; Tutore, I.; De Felice, F. Is blockchain able to enhance environmental sustainability? A systematic review and research agenda from the perspective of Sustainable Development Goals (SDGs). *Bus. Strategy Environ.* **2022**, *31*, 194–217. [CrossRef]
30. Sigalov, K.; Ye, X.; König, M.; Hagedorn, P.; Blum, F.; Severin, B.; Hettmer, M.; Hükinghaus, P.; Wölkerling, J.; Groß, D. Automated payment and contract management in the construction industry by integrating building information modeling and blockchain-based smart contracts. *Appl. Sci.* **2021**, *11*, 7653. [CrossRef]
31. Kouhizadeh, M.; Saberi, S.; Sarkis, J. Blockchain technology and the sustainable supply chain: Theoretically exploring adoption barriers. *Int. J. Prod. Econ.* **2021**, *231*, 107831. [CrossRef]
32. Luthra, S.; Sharma, M.; Kumar, A.; Joshi, S.; Collins, E.; Mangla, S. Overcoming barriers to cross-sector collaboration in circular supply chain management: A multi-method approach. *Transp. Res. Part E Logist. Transp. Rev.* **2022**, *157*, 102582. [CrossRef]
33. Surendra, V.; Singh, A.R.; Raut, R.D.; Hareesh, U. Resources, Conservation & Recycling Blockchain technology adoption barriers in the Indian agricultural supply chain: An integrated approach. *Resour. Conserv. Recycl.* **2020**, *161*, 104877. [CrossRef]
34. Zhong, B.; Wu, H.; Ding, L.; Luo, H.; Luo, Y.; Pan, X. Hyperledger fabric-based consortium blockchain for construction quality information management. *Front. Eng. Manag.* **2020**, *7*, 512–527. [CrossRef]
35. Sharma, D.K.; Kumar, A.; Bathla, G. Heavy vehicle defense procurement use cases and system design using blockchain technology. In *Autonomous and Connected Heavy Vehicle Technology*; Academic Press: Oxford, UK, 2022. [CrossRef]
36. Gürpinar, T.; Brüggenolte, M.; Meyer, D.; Ioannidis, P.A.; Henke, M. Blockchain technology in procurement—A systematic literature mapping. In Proceedings of the Scientific Track of the Blockchain Autumn School 2020, Mittweida, Germany, 28 September–2 October 2020, Volume 1, pp. 7–13. [CrossRef]
37. Kamble, S.S.; Gunasekaran, A.; Subramanian, N.; Ghadge, A.; Belhadi, A.; Venkatesh, M. Blockchain technology's impact on supply chain integration and sustainable supply chain performance: Evidence from the automotive industry. *Ann. Oper. Res.* **2023**, *327*, 575–600. [CrossRef]
38. Weerakoon, H.D.; Chandanie, H. Analysis of feasibility of blockchain technology for international trade related to Sri Lankan construction industry. In *Proceedings of the 9th World Construction Symposium, Sri Lanka [Online], 9–10 July 2021*; Sandanayake, Y.G., Gunatilake, S., Waidyasekara, K.G.A.S., Eds.; pp. 75–85. Available online: https://ciobwcs.com/papers/ (accessed on 21 May 2025). [CrossRef]
39. Waqar, A.; Qureshi, A.H.; Othman, I.; Saad, N.; Azab, M. Exploration of challenges to deployment of blockchain in small construction projects. *Ain Shams Eng. J.* **2024**, *15*, 102362. [CrossRef]
40. Malik, S.; Chadhar, M.; Chetty, M. Factors affecting the organizational adoption of blockchain technology: An Australian perspective. *Proc. Annu. Hawaii Int. Conf. Syst. Sci.* **2021**, *2020*, 5597–5606. [CrossRef]
41. Upadhyay, N. Demystifying blockchain: A critical analysis of challenges, applications and opportunities. *Int. J. Inf. Manag.* **2020**, *54*, 102120. [CrossRef]
42. Singh, A.K.; Mohandes, S.R.; Awuzie, B.O.; Omotayo, T.; Kumar, V.R.P.; Kidd, C. A roadmap for overcoming barriers to implementation of blockchain-enabled smart contracts in sustainable construction projects. *Smart Sustain. Built Environ.* **2024**. [CrossRef]
43. Nawari, N.O.; Ravindran, S. Blockchain technology and BIM process: Review and potential applications. *J. Inf. Technol. Constr.* **2019**, *24*, 209–238.
44. Khan, S.A.; Mubarik, M.S.; Kusi-Sarpong, S.; Gupta, H.; Zaman, S.I.; Mubarik, M. Blockchain technologies as enablers of supply chain mapping for sustainable supply chains. *Bus. Strategy Environ.* **2022**, *31*, 3742–3756. [CrossRef]
45. Shoaib, M.; Zhang, S.; Ali, H. A bibliometric study on blockchain-based supply chain: A theme analysis, adopted methodologies, and future research agenda. *Environ. Sci. Pollut. Res.* **2023**, *30*, 14029–14049. [CrossRef]

46. Li, R.; Song, T.; Mei, B.; Li, H.; Cheng, X.; Sun, L. Blockchain for Large-Scale Internet of Things Data Storage and Protection. *IEEE Trans. Serv. Comput.* **2019**, *12*, 762–771. [CrossRef]
47. Wang, Z.; Wang, T.; Hu, H.; Gong, J.; Ren, X.; Xiao, Q. Blockchain-based framework for improving supply chain traceability and information sharing in precast construction. *Autom. Constr.* **2020**, *111*, 103063. [CrossRef]
48. Li, C.Z.; Chen, Z.; Xue, F.; Kong, X.T.; Xiao, B.; Lai, X.; Zhao, Y. A blockchain- and IoT-based smart product-service system for the sustainability of prefabricated housing construction. *J. Clean. Prod.* **2021**, *286*, 125391. [CrossRef]
49. Msawil, M.; Greenwood, D.; Kassem, M. A Systematic evaluation of blockchain-enabled contract administration in construction projects. *Autom. Constr.* **2022**, *143*, 104553. [CrossRef]
50. Yoon, J.H.; Pishdad-Bozorgi, P. State-of-the-Art Review of Blockchain-Enabled Construction Supply Chain. *J. Constr. Eng. Manag.* **2022**, *148*, 03121008. [CrossRef]
51. Chen, W.; Wu, W.; Ouyang, Z.; Fu, Y.; Li, M.; Huang, G.Q. Event-based data authenticity analytics for IoT and blockchain-enabled ESG disclosure. *Comput. Ind. Eng.* **2024**, *190*, 109992. [CrossRef]
52. Wu, W.; Fu, Y.; Wang, Z.; Liu, X.; Niu, Y.; Li, B.; Huang, G.Q. Consortium blockchain-enabled smart ESG reporting platform with token-based incentives for corporate crowdsensing. *Comput. Ind. Eng.* **2022**, *172*, 108456. [CrossRef]
53. Erol, I.; Ar, I.M.; Peker, I.; Searcy, C. Alleviating the Impact of the Barriers to Circular Economy Adoption Through Blockchain: An Investigation Using an Integrated MCDM-based QFD With Hesitant Fuzzy Linguistic Term Sets. *Comput. Ind. Eng.* **2022**, *165*, 107962. [CrossRef]
54. Singh, A.K.; Kumar, V.R.P.; Dehdasht, G.; Mohandes, S.R.; Manu, P.; Rahimian, F.P. Investigating barriers to blockchain adoption in construction supply chain management: A fuzzy-based MCDM approach. *Technol. Forecast. Soc. Change* **2023**, *196*, 122849. [CrossRef]
55. Ali, U.; Kidd, C. Configuration Management maturation: An empirical investigation. *Proc. Inst. Mech. Eng. Part B J. Eng. Manuf.* **2015**, *229*, 321–327. [CrossRef]
56. Mohandes, S.R.; Zhang, X. Towards the development of a comprehensive hybrid fuzzy-based occupational risk assessment model for construction workers. *Saf. Sci.* **2019**, *115*, 294–309. [CrossRef]
57. Aziz, K.M.A.; Daoud, A.O.; Singh, A.K.; Alhusban, M. Integrating digital mapping technologies in urban development: Advancing sustainable and resilient infrastructure for SDG 9 achievement—A systematic review. *Alexandria Eng. J.* **2025**, *116*, 512–524. [CrossRef]
58. Ghaemi, H.; Ghaemi, H. Application of Structural Equation Modeling in Assessing the Relationship Between Stuttering Students' Cognitive and Metacognitive Strategies and Their Reading Comprehension Performance. *Lang. Test. Asia* **2011**, *1*, 7–32. [CrossRef]
59. Gustavsson, T.K.; Kadefors, A.; Lingegård, S.; Laedre, O.; Klakegg, O.J.; Olsson, N.; Larsson, J. Procurement research: Current state and future challenges in the Nordic countries. *Emerald Reach. Proc. Ser.* **2019**, *2*, 195–204. [CrossRef]
60. Loosemore, M.; Alkilani, S.Z.; Murphy, R. The institutional drivers of social procurement implementation in Australian construction projects. *Int. J. Proj. Manag.* **2021**, *39*, 750–761. [CrossRef]
61. Wang, Y.; Men, S.; Guo, T. Application of Blockchain Technology in Value Chain of Procurement in Manufacturing Enterprises. *Wirel. Commun. Mob. Comput.* **2021**, *2021*, 1674412. [CrossRef]
62. Akaba, T.I.; Norta, A.; Udokwu, C.; Draheim, D. A Framework for the Adoption of a Blockchain-Based E-Procurement System: A Case Study of Nigeria. *Responsible Des. Implement. Use Inf. Commun. Technol.* **2020**, *I3E 2020*, 3–14. [CrossRef]
63. Hunhevicz, J.J.; Motie, M.; Hall, D.M. Digital building twins and blockchain for performance-based (smart) contracts. *Autom. Constr.* **2022**, *133*, 103981. [CrossRef]
64. Al Azmi, N.; Sweis, G.; Sweis, R.; Sammour, F. Exploring Implementation of Blockchain for the Supply Chain Resilience and Sustainability of the Construction Industry in Saudi Arabia. *Sustainability* **2022**, *14*, 6427. [CrossRef]
65. Hughes, L.; Dwivedi, Y.K.; Misra, S.K.; Rana, N.P.; Raghavan, V.; Akella, V. Blockchain research, practice and policy: Applications, benefits, limitations, emerging research themes and research agenda. *Int. J. Inf. Manag.* **2019**, *49*, 114–129. [CrossRef]
66. Majeed, U.; Khan, L.U.; Yaqoob, I.; Kazmi, S.M.A.; Salah, K.; Hong, C.S. Blockchain for IoT-based smart cities: Recent advances, requirements, and future challenges. *J. Netw. Comput. Appl.* **2021**, *181*, 103007. [CrossRef]
67. Sanka, A.I.; Irfan, M.; Huang, I.; Cheung, R.C.C. A survey of breakthrough in blockchain technology: Adoptions, applications, challenges and future research. *Comput. Commun.* **2021**, *169*, 179–201. [CrossRef]
68. Sadeghi, H.; Zhang, X.; Mohandes, S.R. Developing an ensemble risk analysis framework for improving the safety of tower crane operations under coupled Fuzzy-based environment. *Saf. Sci.* **2023**, *158*, 105957. [CrossRef]
69. Farooque, M.; Jain, V.; Zhang, A.; Li, Z. Fuzzy DEMATEL analysis of barriers to Blockchain-based life cycle assessment in China. *Comput. Ind. Eng.* **2020**, *147*, 106684. [CrossRef]
70. Mohandes, S.R.; Sadeghi, H.; Fazeli, A.; Mahdiyar, A.; Hosseini, M.R.; Arashpour, M.; Zayed, T. Causal analysis of accidents on construction sites: A hybrid fuzzy Delphi and DEMATEL approach. *Saf. Sci.* **2022**, *151*, 105730. [CrossRef]
71. Wei, D.; Liu, H.; Shi, K. What are the key barriers for the further development of shale gas in China? A grey-DEMATEL approach. *Energy Rep.* **2019**, *5*, 298–304. [CrossRef]

72. Gan, X.; Chang, R.; Zuo, J.; Wen, T.; Zillante, G. Barriers to the transition towards off-site construction in China: An Interpretive structural modeling approach. *J. Clean. Prod.* **2018**, *197*, 8–18. [CrossRef]
73. Bello, A.O.; Eje, D.O.; Idris, A.; Semiu, M.A.; Khan, A.A. Drivers for the implementation of modular construction systems in the AEC industry of developing countries. *J. Eng. Des. Technol.* **2024**, *22*, 2043–2062. [CrossRef]
74. Xu, C.; Wu, Y.; Dai, S. What are the critical barriers to the development of hydrogen refueling stations in China? A modified fuzzy DEMATEL approach. *Energy Policy* **2020**, *142*, 111495. [CrossRef]
75. Celik, E.; Gumus, A.T.; Alegoz, M. A trapezoidal type-2 fuzzy MCDM method to identify and evaluate critical success factors for humanitarian relief logistics management. *J. Intell. Fuzzy Syst.* **2014**, *27*, 2847–2855. [CrossRef]
76. Dou, Y.; Sarkis, J.; Bai, C. Government green procurement: A Fuzzy-dematel analysis of barriers. *Stud. Fuzziness Soft Comput.* **2014**, *313*, 567–589. [CrossRef]
77. Wu, Y.; Liu, F.; Wu, J.; He, J.; Xu, M.; Zhou, J. Barrier identification and analysis framework to the development of offshore wind-to-hydrogen projects. *Energy* **2022**, *239*, 122077. [CrossRef]
78. Mendel, J.M. Advances in type-2 fuzzy sets and systems. *Inf. Sci.* **2007**, *177*, 84–110. [CrossRef]
79. Qin, J.; Liu, X. Multi-attribute group decision making using combined ranking value under interval type-2 fuzzy environment. *Inf. Sci.* **2015**, *297*, 293–315. [CrossRef]
80. Irfan, M.; Rauniyar, A.; Hu, J.; Singh, A.K.; Chandra, S.S. Modeling barriers to the adoption of metaverse in the construction industry: An application of fuzzy-DEMATEL approach. *Appl. Soft Comput.* **2024**, *167*, 112180. [CrossRef]
81. Rajabpour, E.; Fathi, M.R.; Torabi, M. Analysis of factors affecting the implementation of green human resource management using a hybrid fuzzy AHP and type-2 fuzzy DEMATEL approach. *Environ. Sci. Pollut. Res.* **2022**, *29*, 48720–48735. [CrossRef] [PubMed]
82. Yuan, M.; Li, Z.; Li, X.; Luo, X. Managing stakeholder-associated risks and their interactions in the life cycle of prefabricated building projects: A social network analysis approach. *J. Clean. Prod.* **2021**, *323*, 129102. [CrossRef]
83. Sodhro, A.H.; Pirbhulal, S.; Luo, Z.; de Albuquerque, V.H.C. Towards an optimal resource management for IoT based Green and sustainable smart cities. *J. Clean. Prod.* **2019**, *220*, 1167–1179. [CrossRef]
84. Yevu, S.K.; Yu, A.T.W.; Darko, A. Barriers to electronic procurement adoption in the construction industry: A systematic review and interrelationships. *Int. J. Constr. Manag.* **2023**, *23*, 964–978. [CrossRef]
85. Rasheed, A.; San, O.; Kvamsdal, T. Digital twin: Values, challenges and enablers from a modeling perspective. *IEEE Access* **2020**, *8*, 21980–22012. [CrossRef]
86. Durdyev, S.; Mohandes, S.R.; Tokbolat, S.; Sadeghi, H.; Zayed, T. Examining the OHS of green building construction projects: A hybrid fuzzy-based approach. *J. Clean. Prod.* **2022**, *338*, 130590. [CrossRef]
87. Hang, L.; Kim, D.-H. Optimal blockchain network construction methodology based on analysis of configurable components for enhancing Hyperledger Fabric performance. *Blockchain Res. Appl.* **2021**, *2*, 100009. [CrossRef]
88. Xu, Y.; Chong, H.Y.; Chi, M. Blockchain in the AECO industry: Current status, key topics, and future research agenda. *Autom. Constr.* **2022**, *134*, 104101. [CrossRef]
89. Björklund, V.; Vincze, T. Blockchain Smart Contracts, the New Rebar in the Construction Industry? Master's Thesis, Logistics and Transport Management, Mittweida University, Mittweida, Germany, 2 July 2019. Master Degree Project 2019:78.
90. Sheng, D.; Ding, L.; Zhong, B.; Love, P.E.D.; Luo, H.; Chen, J. Construction quality information management with blockchains. *Autom. Constr.* **2020**, *120*, 103373. [CrossRef]
91. Wu, Y.; Liao, Y.; Xu, M.; He, J.; Tao, Y.; Zhou, J.; Chen, W. Barriers identification, analysis and solutions to rural clean energy infrastructures development in China: Government perspective. *Sustain. Cities Soc.* **2022**, *86*, 104106. [CrossRef]
92. Herrera-Viedma, E. Fuzzy Sets and Fuzzy Logic in Multi-Criteria Decision Making. The 50Th Anniversary of Prof. Lotfi Zadeh'S Theory: Introduction. *Technol. Econ. Dev. Econ.* **2015**, *21*, 677–683. [CrossRef]
93. Ikotun, A.M.; Ezugwu, A.E.; Abualigah, L.; Abuhaija, B.; Heming, J. K-means clustering algorithms: A comprehensive review, variants analysis, and advances in the era of big data. *Inf. Sci.* **2023**, *622*, 178–210. [CrossRef]
94. Abernathy, A.; Celebi, M.E. The incremental online k-means clustering algorithm and its application to color quantization. *Expert Syst. Appl.* **2022**, *207*, 117927. [CrossRef]

infrastructures

Article

Evolution of Convolutional and Recurrent Artificial Neural Networks in the Context of BIM: Deep Insight and New Tool, Bimetria

Andrzej Szymon Borkowski [1,*], Łukasz Kochański [2] and Konrad Rukat [2]

1 Faculty of Geodesy and Cartography, Warsaw University of Technology, Politechniki Square 1, 00-661 Warsaw, Poland
2 Bimetria sp. z o.o., Warsaw, Poland; lukasz.kochanski@bimetria.com (Ł.K.); konrad.rukat@bimetria.com (K.R.)
* Correspondence: andrzej.borkowski@pw.edu.pl

Abstract

This paper discusses the evolution of convolutional (CNN) and recurrent (RNN) artificial neural networks in applications for Building Information Modeling (BIM). The paper outlines the milestones reached in the last two decades. The article organizes the current state of knowledge and technology in terms of three aspects: (1) computer visualization coupled with BIM models (detection, segmentation, and quality verification in images, videos, and point clouds), (2) sequence and time series modeling (prediction of costs, energy, work progress, risk), and (3) integration of deep learning results with the semantics and topology of Industry Foundation Class (IFC) models. The paper identifies the most used architectures, typical data pipelines (synthetic data from BIM models, transfer learning, mapping results to IFC elements) and practical limitations: lack of standardized benchmarks, high annotation costs, a domain gap between synthetic and real data, and discontinuous interoperability. We indicate directions for development: combining CNN/RNN with graph models and transformers for wider use of synthetic data and semi-/supervised learning, as well as explainability methods that increase trust in AECOO (Architecture, Engineering, Construction, Owners & Operators) processes. A practical case study presents a new application, Bimetria, which uses a hybrid CNN/OCR (Optical Character Recognition) solution to generate 3D models with estimates based on two-dimensional drawings. A deep review shows that although the importance of attention-based and graph-based architectures is growing, CNNs and RNNs remain an important part of the BIM process, especially in engineering tasks, where, in our experience and in the Bimetria case study, mature convolutional architectures offer a good balance between accuracy, stability and low latency. The paper also raises some fundamental questions to which we are still seeking answers. Thus, the article not only presents the innovative new Bimetria tool but also aims to stimulate discussion about the dynamic development of AI (Artificial Intelligence) in BIM.

Keywords: Building Information Modeling; CNN; RNN; OCR; IFC; AECOO; Artificial Intelligence; Bimetria

1. Introduction

Over the last two decades, Building Information Modeling (BIM) has emerged as the fundamental paradigm for the digitization of the broadly understood construction industry–AECOO (Architecture, Engineering, Construction, Owners & Operators) [1]. BIM combines geometric data, semantics (high-level relationships), and non-graphical data into coherent, shared models that support decision-making from the concept and design stage, through project implementation, to the maintenance and operation stage [2,3]. At the same time, there has been a strong evolution in artificial neural networks [4]. Convolutional Neural Networks (CNNs) have become the standard in spatial-visual tasks [5], and Recurrent Neural Networks (RNNs) in sequence and time series modeling [6]. The convergence of the relative maturity of BIM ecosystems with the maturity of CNNs/RNNs has triggered a wave of applications that promise automation, scalability, and greater efficiency of engineering processes. In recent years, the notion of "active BIM" has emerged to describe BIM environments augmented with optimization and decision-making capabilities that dynamically update project information during design and construction

Despite significant progress, the evolution of Artificial Intelligence (AI) is still ongoing, and new obstacles are being overcome [7]. However, there is a lack of standardized benchmarks dedicated to BIM tasks [8], the cost of annotation in native environments is high, and the differences between synthetic data (generated from models) and real data (from construction sites, scans, photos) make it difficult to transfer new solutions into practice [9]. In addition, the lack of seamless interoperability and different versions of IFC (Industry Foundation Class) standards complicates the replicability and comparability of results [10], and the limited "explainability" of AI models undermines confidence in applications critical to key costs or schedules [11], which are most important in the construction investment process. At the same time, the only source of truth, the so-called CDE (Common Data Environment), does not always function well in investment processes [12], which hinders the development of SaaS (Software as a Service). These barriers point to the need for work that will not only present existing solutions but also organize the evolution and ask questions to which we still do not know the answers.

The aim of this article was to conduct a synthetic review of the literature related to the evolution of CNNs and RNNs in BIM applications and to organize the current state of knowledge in three complementary perspectives: (1) computer visualization coupled with BIM models (detection, segmentation, and quality verification in images, videos, and point clouds), (2) sequence and time series modeling (prediction of costs, energy, work progress, and risk), and (3) integration of deep learning results with the semantics and topology of IFC models. In each of these areas, the dominant architectures, typical data flows (including the use of synthetic data and learning methods), and a set of evaluation metrics used in practice were identified. At the same time, limitations and gaps that hinder the large-scale implementation of BIM–AI were pointed out.

The article is not only review-oriented but also discussion-oriented. Several fundamental questions are posed that will determine the direction of research and practice in the coming years. The research questions are as follows:

1. When are classic CNNs/RNNs sufficient for BIM tasks, and when do models that explore relationships and attention (transformers) give an advantage?
2. How can model predictions be linked to IFC elements and relationships so that they are operationalizable in workflows and CDE systems?
3. What balance between synthetic and real data minimizes the domain gap while reducing the already high cost of annotation?
4. What XAI (Explainable AI) mechanisms are practically useful for engineers and how can their value be measured?
5. What are the minimum requirements (data schemas, APIs, metrics) needed to make results comparable and transferable between tools?

https://doi.org/10.3390/infrastructures11010006

The contribution of this work is threefold. (1) We provide a structured synthesis of scattered results on the evolution of CNN and RNN approaches for BIM-related tasks, highlighting typical data flows, model architectures and integration patterns. (2) We describe Bimetria as a concrete implementation of these ideas in an industrial setting, showing how AI-based perception can be linked with IFC structures and quantity takeoff in a repeatable way. (3) We adapt and extend the concept of "active BIM"–previously used mainly in the context of optimization-based construction management and logistics to an AI-perception-driven workflow in which CNN/RNN models automatically construct and update IFC-based building models and bills of quantities from 2D drawings. The structure of the article is as follows: Section 1 covers the Introduction, Section 2 presents theoretical background and terminology, as well as the origins and evolution of CNN/RNN. Section 3 discusses the new Bimetria (CNN) tool—its operation and pipeline. Section 4 discusses future research directions, identifies limitations, gaps, and challenges, and attempts to answer the fundamental questions posed. Section 5 presents recommendations and conclusions.

2. Evolution of CNN/RNN

2.1. Literature Search Protocol

To make the review reproducible, a transparent literature search protocol was applied. Two major bibliographic databases, Scopus and Web of Science, were queried, focusing on original research articles at the intersection of BIM and deep learning. The core query combined BIM-related terms with CNN/RNN-related terms, such as ("Building Information Modeling" OR BIM OR "Industry Foundation Classes" OR IFC) AND ("convolutional neural network" OR CNN OR "recurrent neural network" OR RNN OR "deep learning").

The time window was restricted to 2012–2025, corresponding to the period in which deep learning methods became practically relevant for BIM applications. The search was limited to peer-reviewed journal and conference papers in English. In the first screening stage, titles and abstracts were checked against the following inclusion criteria: (i) explicit use of BIM or IFCs in the problem formulation or data structures; (ii) use of CNNs, RNNs or closely related deep architectures; (iii) focus on AECOO-relevant tasks such as detection, segmentation, quality control, forecasting or model integration. Papers that only mentioned BIM tangentially, purely theoretical AI work without a BIM context, and non-archival publications such as position papers or theses were excluded.

In the second stage, full texts were assessed to confirm that BIM/IFC structures were effectively used and that quantitative evaluation was reported. Methodologically, the paper is based on an in-depth narrative review rather than a fully protocol-driven systematic review.

For each paper identified in the search, the primary task (e.g., object detection, semantic segmentation, defect detection, progress estimation, cost or energy forecasting), model family (e.g., 2D/3D CNN, U-Net/HR-Net, LSTM/GRU, graph neural networks, transformers), data type (raster drawings, 3D point clouds, IFC graphs, time series, video) and main evaluation metrics (e.g., IoU/Dice, mAP, accuracy/F1, MAE/RMSE) were noted. On this basis, papers were assigned to one or more of the three analytical lenses used in the article: computer visualization coupled with BIM models, sequence and time series modeling, and integration of deep learning outputs with IFC semantics and topology. The synthesis presented in Section 2 follows this structure.

2.2. Three Eras of Evolution

The first concepts of neural networks dominated until the 1990s, but computational limitations and the vanishing gradient problem hampered the development of deep architectures. Convolutions became standard in images (LeCun/LeNet-5 ideas) [13], but their

https://doi.org/10.3390/infrastructures11010006

potential only fully revealed itself in the era of GPUs and big data. In 2012, a major milestone was reached with the "AlexNet" network by Krizhevsky, Sutskever, and Hinton [14]. Their model outperformed the competition in ImageNet, demonstrating that deep CNNs radically improve image classification quality through scaling. Since then, we have seen an avalanche of similar architectures. The following can be considered further milestones in the development of CNNs and RNNs:

- 2013—word2vec by Tomas Mikolov et al. [15];
- 2014—seq2seq Sutskever, Vinyals, Le [16];
- 2016—AlphaGo (DeepMind) [17];
- 2017—Transformer ("Attention Is All You Need") as a milestone in the architecture behind GPT-x [18];
- 2018—GPT1—the first LLM from OpenAI [19];
- 2019—GPT2 — (described in the media as "too dangerous to release") [20];
- 2022—GPT3.5—public debut of ChatGPT on 30 November 2022 [21];
- 2023—GPT4—"Sparks of Artificial General Intelligence" [22];
- 2025—GPT5—with a built-in "thinking" mode, whose approach corresponds to the concept of System 2 (Kahneman) [23] used in the description of reasoning models.

Key events are mapped on the timeline in Figure 1.

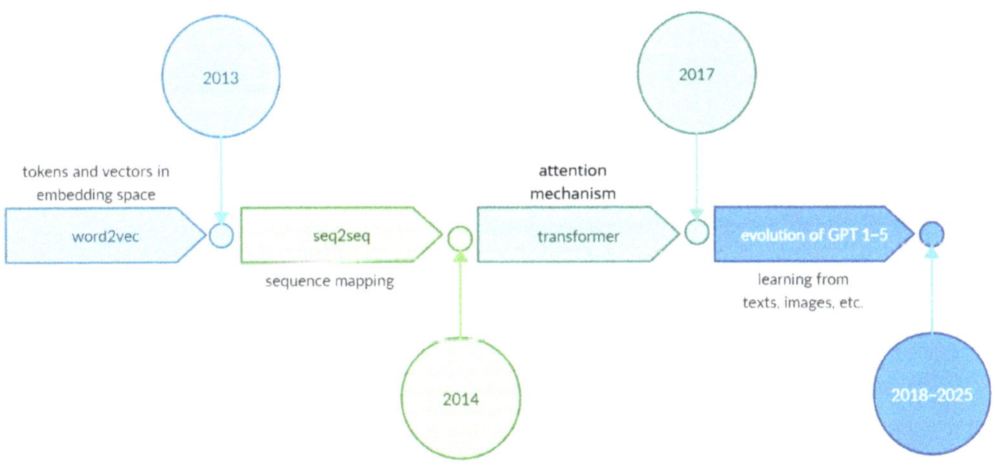

Figure 1. Timeline and milestones in the development of RNNs/CNNs (source: own work).

For BIM practice, the most direct implications of these milestones are the emergence of attention-based architectures and transformers for point clouds and graphs, enabling richer reasoning over IFC topology, as well as the use of large language models to support tasks such as code compliance checking, automated documentation and conversational access to project data. In this article, we focus on CNN/RNN architectures but briefly discuss how attention and transformer-based models can complement them in Sections 2 and 4.

As a result, BIM has access to CNN components that enable the detection and segmentation of structural elements, quality control in images or videos, and the integration of results with IFC models. At the same time, RNNs were also developing. For example, LSTM/GRU models (2014) established the standard for modeling time series and event sequences, and seq2seq (2014) showed how to map variable-length sequences to other sequences. The introduction of the attention mechanism (2014) shifted the emphasis to dynamic weighting of time steps, which improved performance and stability. For BIM, this

https://doi.org/10.3390/infrastructures11010006

translated into forecasting delays, costs, energy, and analysis of operational logs/records throughout the life cycle of a building.

With the advent of AlphaGo (a system combining deep CNN with reinforcement learning and tree search, which defeated champion Lee Sedol), it was proven that the combination of perception (CNN), memory/planning, and expert data scales to tasks with astronomical state space. AlphaGo Zero went even further, learning "from scratch" solely through self-play. Although Go has nothing to do with BIM, these milestones have changed the mental model of system design, as instead of solely supervised pipelines, the role of feedback, simulation, and decision policies is growing, which in turn translates into planning, scheduling, and active learning on synthetic data in AECOO.

The evolution can be seen as three eras in the development of BIM-AI (Figure 2):

1. 2D perception (2012–2016) including CNNs (AlexNet→ResNet, U-Net) and detection/segmentation of elements in images and videos, including the first linking to BIM.
2. Sequences and decisions (2014–2017) including LSTM/GRU with attention mechanisms and cost/deadline forecasting, as well as the integration of predictions with decisions and policies.
3. Geometry and relations (2017 to present) covering PointNet/GNN/sparse-3D solutions and the important HR-Net [24], including direct work on point clouds and IFC graphs, and the maturation of self-supervised methods.

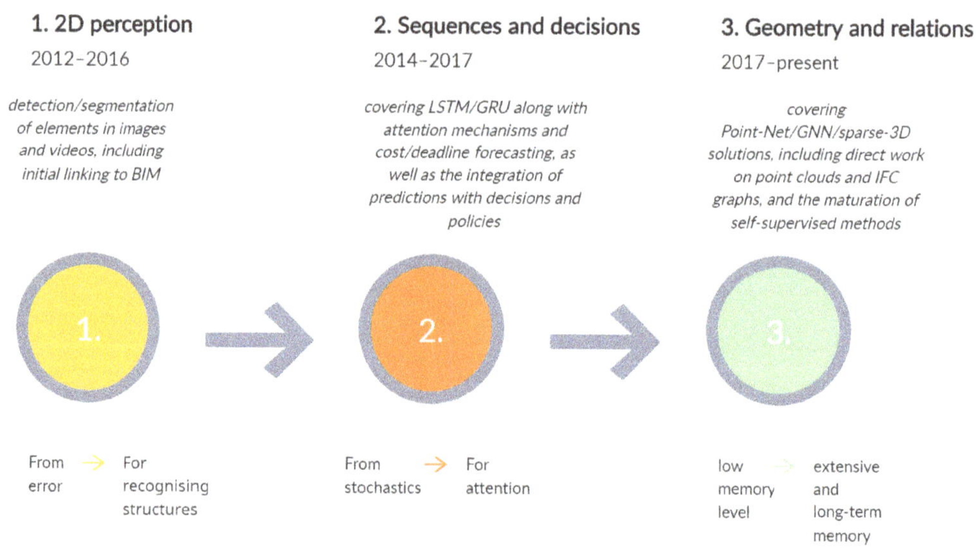

Figure 2. Three eras in the development of BIM–AI (source: own work).

In many current industrial deployments, CNNs remain the backbone choice for visual tasks, especially where stability and low latency are important, although systematic comparisons with transformer-based alternatives are still limited. RNNs (LSTM/GRU) are still practical for time series and event logs, although some tasks are being taken over by attention architectures. Geometric DL (GNN/3D) is crucial when predictions must be topologically and semantically consistent with IFC. Self-/semi-supervised + synthetics are the most effective response to benchmark deficiencies and high annotation costs.

https://doi.org/10.3390/infrastructures11010006

3. Example of Bimetria

This chapter presents a practical implementation of the previously discussed concepts of deep learning in the context of BIM, namely the Bimetria system. The aim is to show how the combination of convolutional neural networks (CNNs) and optical character recognition (OCR), supported by domain rules and a geometric reconstruction stage, can automate the transition from 2D drawings in the form of scans or PDF files to structured information about a building object and to an IFC model ready for use in AECOO processes. This example complements the literature review with a practical dimension, highlighting the trade-offs between accuracy and stability, limitations due to data quality and diversity, and architectural decisions dictated by engineering design requirements. Conceptually, Bimetria can be seen as a specific form of "active BIM". In existing research, active BIM usually refers to BIM environments enriched with optimization or decision-making capabilities that dynamically update project information, for example to optimize crane locations, site logistics or building energy performance [25–27]. In our case, the "activeness" of BIM is driven primarily by AI-based perception: CNN/RNN models interpret 2D drawings, construct and update an IFC-based model, and populate quantitative attributes that feed downstream cost estimation and model management. This shifts active BIM towards workflows where automated recognition and model generation are tightly coupled with BIM/IFC structures and quantity takeoff.

At a high level, Bimetria implements a processing chain that includes preliminary cleaning and normalization of drawing sheets, detection and segmentation of graphic elements such as symbols, lines, and dimensions using CNN models, extraction of descriptive content and dimensional numbers using OCR methods, merging of geometric and semantic information, reconstruction of three-dimensional elements along with their parameterization, mapping to IFCs and relations, followed by data integrity validation and export of results. The system in question takes a pragmatic approach that includes transfer learning, semi-supervised labeling, and consistency checking rules, which together allow for repeatable results in production conditions.

3.1. Input Data

The starting point for the Bimetria pipeline is a single 2D drawing sheet (Figure 3), which is the source of geometric and semantic information used to reconstruct the 3D model. A 2D drawing usually contains layers of a mixed nature: construction graphics, installation symbols, dimensions, text descriptions, legends, and title blocks. In practice, both vector PDF files and raster PDF, PNG, and JPEG files (including scans and photos of sheets) can be fed into the system. The goal of the actions embedded in the system is to obtain an image with stable visual properties, in which the topological relationships between lines, symbols, and text are not blurred.

To obtain results that are as close to reality as possible at the outset, the user defines information about the storeys by adding or subtracting records and entering the necessary data. For each storey to be analyzed, the ordinates of the base and top of the structural and partition walls should be assigned.

In the circulation of documentation, it often happens that projections are transmitted in the form of scans or photographs of printed sheets, which are captured in a perspective that distorts the drawing. Therefore, the first step is to straighten the image. This is done using a dedicated straightening tool, which, after the user has marked four corners of a known rectangle, gives the projection an appearance like the original drawing.

The next step in preparing the image is to improve its readability using ESRGAN [28], a machine-trained model for improving the quality of raster images. This stage involves

https://doi.org/10.3390/infrastructures11010006

correcting the sharpness of lines and characters, and for scans or photos with uneven lighting, brightness adjustment is used.

To later map to IFC, it is crucial to preserve information about units and scales, so the user first assigns a scale to the drawing by calibrating the design using dimensions in two perpendicular axes.

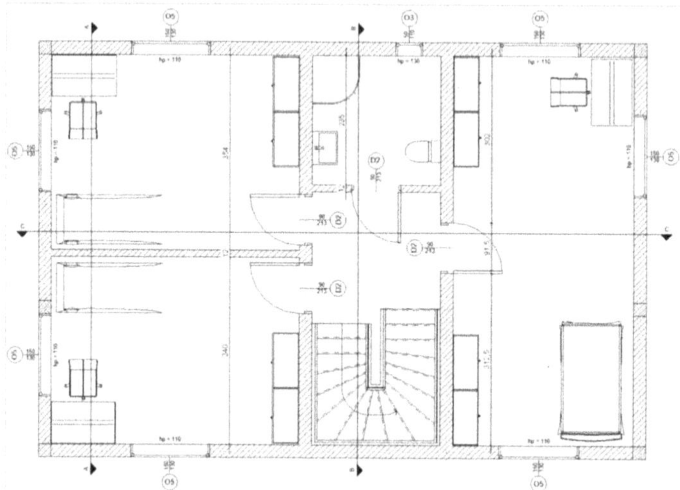

Figure 3. Image uploaded to the model at the input (source: own study).

This set of input data is standardized, described with metadata, and enriched with reliable annotations, thus providing CNN models and the OCR component with the conditions for stable detection of 2D elements and for subsequent deterministic transfer of information to 3D space in the BIM environment and IFC structures. The resulting representation can be mapped to an IFC-based parametric model used in further stages of the design process, following the Industry Foundation Classes (IFC) schema (ISO 16739-1:2018) [29] and related standards for BIM-based data exchange. As a result, a single 2D drawing becomes a fully fledged knowledge carrier that can be transformed into a coherent, parametric model used in further stages of the design process.

From a methodological perspective, the data used to train and evaluate Bimetria come from 250 real design projects, involving buildings with reinforced concrete, masonry, or mixed wall structures. A representative floor plan in raster format with a resolution of 96 to 300 DPI was selected from each project. A segmentation mask was created for each drawing using a consistent labeling protocol that distinguishes between classes of elements, e.g., masonry walls, reinforced concrete walls, doors, windows. The annotations serve as ground truth for training and validation of the image recognition model. In the next step, an extensive system of image augmentation and corresponding masks was prepared in terms of scale, rotation, deformation, and blurring. After the class labeling stage, this was the most crucial stage in terms of the final quality of the model being built.

The images prepared in this way, together with the corresponding masks, were divided into tiles of a size that allowed for efficient learning on limited hardware resources with optimal context coverage for the classes sought. After dividing into tiles, tiles with zero participation of classes other than the background were selected and removed. The final number of fragments of projections prepared in this way and used for model training exceeded 16,000 images. For the entire model, mIoU ~85% was obtained, and none of the classes achieved less than 72 IoU.

https://doi.org/10.3390/infrastructures11010006

3.2. Image Recognition Model

The image recognition process begins with dividing the image into square tiles, which optimize the model's performance and are used to calculate the number of times the application is used by the user. From this point on, the processes related to the recognition of elements in images take place on separate parts, thus minimizing the risk of failure of the operation due to the large size of the input data. The core of the system is an encoder–decoder convolutional network with skip connections, which predicts a semantic segmentation map of the drawing together with auxiliary masks needed for post-processing. The model is trained to classify each pixel into 5 classes corresponding to the object categories used in Bimetria (masonry walls, reinforced concrete walls, doors, windows) and background. The loss function combines a standard cross-entropy term with to handle class imbalance between large background regions and relatively small structural elements.

Bimetria is a two-stage image recognition model tailored to the specifics of technical drawings. The first part is segmentation based on a network that allows for the stable separation of three types of regions: walls with columns, openings, and the background constituting the rest of the drawing. The direct results of image recognition have imperfections resulting from the prediction process, which is why they are immediately subjected to automatic analysis and correction to generate output data in the form of rectangles with appropriate dimensions and angles while maintaining the relationships between elements (Figure 4).

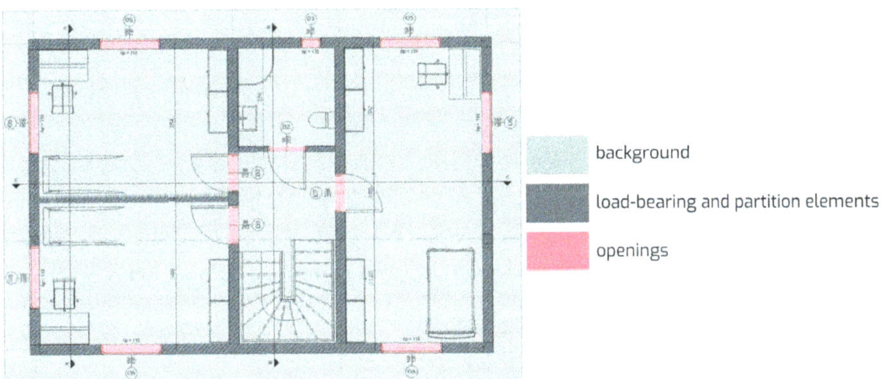

Figure 4. The first stage of image recognition (source: own work).

The model was trained in a multi-task scheme. The model's performance was designed for production use. The model uses quantization and agile convolution blocks, which enables the processing of large sheets in a time acceptable to the user using appropriate GPU (Graphics Processing Unit) cloud resources. Stability between drawing versions was achieved through input normalization and color calibration mechanisms. In the second stage of image recognition, the model separates masonry walls from reinforced concrete elements and distinguishes between windows and doors (Figure 5). The network is implemented in Python + deep-learning framework and trained on a standard GPU-equipped workstation. It should be noted that no formal, large-scale quantitative evaluation or systematic ablation study of the Bimetria model is reported in this paper. The case study is intended to illustrate a feasible integration of CNN-based perception with BIM/IFC workflows in an industrial setting, rather than to provide a comprehensive benchmark of model performance.

https://doi.org/10.3390/infrastructures11010006

The third stage of data recognition uses OCR, which analyses the image for text data. The model recognizes letters and numbers that carry information and then selects those that are relevant to the recognized elements. Information is filtered based on both content and geometric position. Ultimately, parameters such as the height and width of openings and the height of windowsills are selected.

Figure 5. The second stage of image recognition (source: own work).

After the data in the images have been recognized, the coordinates of the elements in the form of the beginning and end of the axes and the thickness of the elements are generated based on the generated raster data on walls, reinforced concrete elements, windows, and doors. The elements are assigned parameters read by OCR, and a JSON file is created to store all this information in a specific structure.

3.3. *Verification of Results*

After completing all stages related to the recognition of elements and information on the input data, the user has the option to verify the results. This is done based on vector data displayed on the input image. At this stage, the user has the option to add missing elements, remove redundant elements, and edit parameters. Once the user approves the design, the data in the JSON file is updated.

3.4. *Intermediate JSON Layer and IFC Mapping*

All recognition results are stored in an intermediate JSON representation that decouples perception outputs from the target BIM model. Each JSON record corresponds to a single building element and contains at least the following fields:

- 'id' (string): unique identifier of the element within a project;
- 'type' (string): semantic category, e.g., 'wall', 'opening', 'column', 'door', 'window';
- 'geom_2d' (object): 2D geometry in sheet coordinates, including the coordinates of endpoints or polygon vertices in sheet coordinates;
- 'level_id' (string): identifier of the storey/level the element belongs to;
- 'params' (object): dictionary of scalar parameters such as thickness, height, sill elevation, material code;
- 'text_refs' (array): references to OCR tokens that provided the parameter values;
- 'confidence' (float): model confidence score for the detection/segmentation of the element.

During IFC export, each JSON record is mapped to an IFC entity according to a set of deterministic rules. For example, elements with 'type = wall' and a non-zero thickness are converted to 'IfcWall' or 'IfcWallStandardCase'; openings in walls are mapped to

https://doi.org/10.3390/infrastructures11010006

'IfcOpeningElement'; doors and windows are represented as 'IfcDoor' and 'IfcWindow' inserted into the corresponding openings. Quantitative attributes such as length, area and volume are computed from the 2D geometry and level information and stored in quantity sets (e.g., 'Qto_WallBaseQuantities'), while descriptive attributes such as material and fire rating are stored in property sets (e.g., 'Pset_WallCommon'). The use of a JSON layer makes the pipeline robust to changes in perception models and facilitates exporting to different BIM authoring tools that consume IFC.

3.5. Quantity Takeoff

Based on the verification data, an IFC model containing all recognized elements is generated. The model is displayed in the application for better visual verification.

Quantities are calculated based on the approved data contained in the JSON file. The values that are calculated are areas, volumes, and numbers of elements. They are exported to tables in CSV/XLSX format, which can be downloaded and used for any calculations and analyses.

The application also generates a downloadable PDF report, allowing for quick printing and transfer of information if paper documentation is used in the process. It contains a 3D view of the BIM model and tables with quantities (Figure 6). Quantity surveying in Bimetria is the stage of converting image recognition results into measurement values used for cost estimation and planning purchases or construction works. The starting point is a structured graph of elements obtained from masks and object detection with neighborhood relations and links to OCR readings. On this basis, the system determines the base quantities (lengths, areas, volumes) in accordance with a set of measurement rules, then converts them to target units and combines them with the classification of elements.

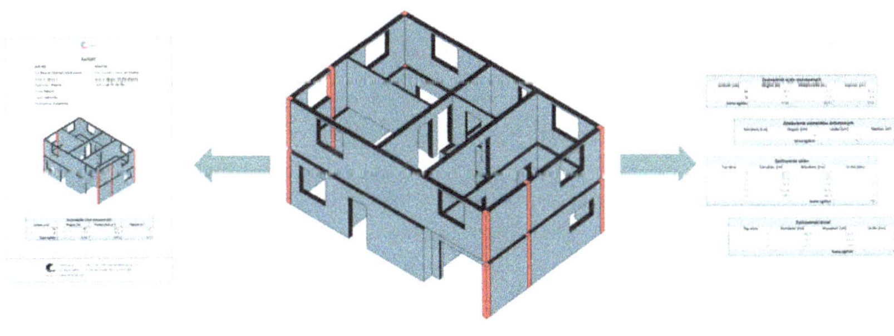

Figure 6. BIM model with bill of quantities values (source: own study, Bimetria, 2025).

The process begins with the normalization of units and versioning of the sheet: all dimensions are converted to SI metrics, and the sheet is given an identifier and scale signature, which guarantees the repeatability of calculations between successive revisions. Recording relationships allows for the distinction between gross and net values, which is crucial for most standard catalogs (subtracting openings, including/excluding recesses, treating columns in floors, etc.).

For masonry and reinforced concrete walls, the system calculates lengths from alignment and developed areas from thickness determined based on the legend or layer caption; if the thickness is not explicitly stated, inference rules based on symbols and line scale are used, and the result is marked with a certain degree of uncertainty. Openings are subtracted from the net areas according to a minimum area threshold to avoid interference from small marker recognition artifacts. The measurement values are accompanied by a margin of uncertainty resulting from error propagation: the system aggregates segmentation deviations, OCR reading quality, and scale calibration accuracy.

https://doi.org/10.3390/infrastructures11010006

The result of this stage is a bill of quantities organized according to the classification of elements. At the BIM level, the data is transferred to IFC structures. At the same time, CSV/XLSX reports and filter views are generated, enabling the mapping of items to cost estimate items, with the system able to use assignments to industry dictionaries and codes used in cost estimation. This integration allows for direct feeding into 5D processes, and thanks to input versioning and deterministic measurement rules, the lists are comparable between project revisions and suitable for tracking changes.

3.6. Scope and Limitations of the Case Study

It should be noted that no formal, large-scale quantitative evaluation or systematic ablation study of the Bimetria model is reported in this paper. The case study is intended to illustrate a feasible integration of CNN-based perception with BIM/IFC workflows in an industrial setting, rather than to provide a comprehensive benchmark of model performance.

4. Discussion

In interpreting the results, a distinction is made between observations that are directly grounded in the described pipeline and case study, and broader reflections informed by experience and the reviewed literature. Statements that refer to specific mechanisms or design choices in Bimetria should be understood as practice-oriented and illustrative, whereas more general claims about trends in BIM–AI are based on the narrative review rather than on new quantitative experiments.

The U-Net network was the starting point for the development of Bimetria and, in general, should be considered a breakthrough. In turn, thanks to the HR-Net network, i.e., its application to the analysis of documentation and details, much better results (accuracy) were obtained than in the case of U-Net. Due to its high resolution throughout the network, only HR-Net allows for the determination of, among other things, wall polygons in a manner acceptable from the point of view of quantity surveying.

The results of the literature review and the Bimetria case study allow us to look at the evolution of CNNs/RNNs in BIM from a more nuanced perspective than a simple exposure to "old" networks vs. "new" transformers. On the one hand, there is a clear shift in the focus of research towards attention-based architectures, graphs, and point clouds, which better capture 3D geometry and topological relationships in IFC models. On the other hand, practice shows that mature and well-understood components are still crucial for many critical tasks: convolutional networks segmenting 2D images, classic RNNs/LSTM for time series, and simple language/embedding models that can be relatively easily integrated into existing engineering processes. Bimetria fits into this logic of "reasonable conservatism": instead of chasing the latest benchmarks, it uses the proven HR-Net network and OCR modules to solve a specific, business-critical problem—the transition from 2D drawings to BIM models and bills of quantities.

An analysis of Bimetria's pipeline provides a partial answer to the question posed in the introduction about when classic CNNs/RNNs are "sufficient." For a highly structured domain such as technical drawings, with well-designed pre-processing (scale normalization, distortion correction, resolution improvement) and a limited but precise object dictionary, convolutional networks still provide stable segmentation and element detection. The key is not so much the architecture itself, but rather its integration with domain rules (e.g., inferring missing wall thickness based on context) and rigorous user-level quality control woven into the verification process. From this point of view, more complex graph models or transformers could improve some marginal cases (complex details, non-standard

https://doi.org/10.3390/infrastructures11010006

symbols), but they are not a prerequisite for achieving production usability in typical building designs.

An important conclusion that emerges from a comparison of literature and the Bimetria example is the central role of the intermediate layer between prediction and the BIM/IFC world. In the proposed solution, this function is performed by a structured description in JSON format, which stores both the coordinates and semantics of the recognized elements. It is this layer that enables the mapping of CNN/OCR results to IFC objects, their subsequent use in quantity surveying processes, and tracking changes between revisions. The literature often assumes that an "intelligent" model will find the right representation on its own, but from an AECOO practice perspective, explicit, auditable bridges between input data, prediction, and the BIM model are key. This approach also promotes explainability: the designer or user not only receives a numerical result, but also the ability to trace which parts of the drawing generated specific IFC objects and what the quantity survey values are.

The issue of balance between synthetic and real data and the problem of domain gap manifests itself in Bimetria in a slightly different way than in typical scientific research. Instead of building huge synthetic corpora, the system relies on so-called dirty documentation from real projects, and the diversity of drawing styles, formats, scanning artifacts, and human errors is treated as a natural problem. While this limits the possibility of precise comparison with other solutions (lack of public benchmarks), it better reflects the actual conditions of implementation. In the broader context of BIM–AI, this suggests that it may make sense to build smaller, "local" data ecosystems around specific applications and organizations, rather than striving for a single correct and universal reference set.

The final topic of discussion concerns explainability and trust in AI systems in BIM. In literature, XAI is often reduced to general techniques for visualizing attention or feature importance, while experience with Bimetria indicates that from the user's perspective, simple but well-integrated mechanisms are key: overlaying segmentation results on the original drawing, the ability to manually correct and immediately recalculate measurements, and versioning of input data and results. "Operational explainability" understood in this way may be less impressive than advanced XAI methods, but it better addresses the real needs of designers and cost estimators. It seems that in the coming years, it will be precisely such pragmatic solutions, combining mature CNN/RNN architecture with well-designed data semantics and user interfaces, that will determine whether AI becomes an integral part of everyday BIM workflows or remains mainly the domain of research experiments. This observation is consistent with experiential learning approaches in BIM education, where direct, hands-on interaction with BIM and AI-augmented tools is seen as a prerequisite for developing trust and deep understanding of digital workflows [30].

Given the current lack of public benchmarks dedicated to BIM-specific tasks, it seems useful to outline at least a minimal open protocol aligned with the three lenses used in this article. As a starting point, such a protocol could include a set of varied 2D sheets with ground-truth segmentation masks and quantities for key structural and architectural elements, a small number of annotated site videos capturing construction progress, and several time-series datasets (e.g., cost, schedule, energy) with standard train/validation/test splits. For each lens, a core set of baseline metrics (e.g., IoU/mAP for perception, MAE/RMSE for forecasting, and topology-aware measures for IFC integration) should be agreed upon. The typology and case study presented here are intended as a conceptual blueprint for designing such community benchmarks rather than as a complete specification.

5. Conclusions

The article provides a synthetic overview of the evolution of convolutional and recurrent neural networks in BIM applications against the backdrop of the broader development

https://doi.org/10.3390/infrastructures11010006

of deep learning methods, showing how the community has progressed from simple 2D perception tasks to models that operate directly on point clouds, IFC graphs, and event sequences in the object lifecycle. The authors organize the current state of knowledge into three complementary perspectives: computer visualization coupled with BIM models (detection, segmentation, and quality control in images, videos, and point clouds), sequence and time series modeling (cost, energy, work progress, and risk forecasts), and the integration of deep learning results with the semantics and topology of IFC models and CDE environments. Based on a review of the literature from the last two decades, the dominant architectures (including the U-Net and HR-Net family, LSTM/GRU, 3D and graph networks, transformers), typical pipelines, and main implementation barriers have been identified, among which the key ones are: the lack of standardized benchmarks dedicated to BIM tasks, the high cost of annotation in native environments, the domain gap between synthetic and real data, the limited interoperability of the IFC standard, and the insufficient explainability of AI models in cost- and schedule-critical applications. The literature review is supplemented by a case study of the Bimetria system, which uses a hybrid processing chain based on CNNs, OCR, domain rules, indirect representation in JSON format, and geometric reconstruction to strongly automate the transition from 2D drawings (scans, PDF) to an IFC model integrated with quantity takeoff and cost estimation processes, while illustrating that mature and well-understood CNN/RNN architectures can be "sufficient" in many tasks if embedded in a carefully designed workflow and user verification mechanisms. Conceptually, Bimetria can be seen as an example of "Active BIM". In such workflows, AI models are not limited to recognizing objects in drawings, but they actively construct and update a BIM/IFC model and its quantitative attributes. This means that the output of perception directly drives quantity takeoff, cost estimation and model management, instead of remaining a separate analytical layer detached from the BIM environment. The need to align advanced CNN/RNN-based workflows with the broader digital transformation of the AEC sector is also emphasized by Zawada et al. [31], who argue that the potential of BIM and 4.0 technologies, including AI, is still far from being fully exploited in everyday engineering practice [31]. Thus, the article proposes a framework typology of BIM-AI tasks and methods, formulates a set of fundamental questions concerning, among other things, the role of classical CNN/RNN models in relation to attention-based models, the balance between synthetic and real data, practically useful explainability, and minimum interoperability requirements, and indicates directions for further development aimed at building scalable, explainable, and comparable tools that can become a permanent part of everyday practice in the broadly understood construction industry—AECOO.

Author Contributions: Conceptualization, A.S.B., Ł.K. and K.R.; methodology, A.S.B., Ł.K. and K.R.; software, Ł.K. and K.R.; validation, A.S.B.; formal analysis, A.S.B., Ł.K. and K.R.; resources, Ł.K. and K.R.; data curation, Ł.K. and K.R.; writing—original draft preparation, A.S.B., Ł.K. and K.R.; writing—review and editing, A.S.B., Ł.K. and K.R.; visualization, A.S.B., Ł.K. and K.R.; supervision, A.S.B.; funding acquisition, A.S.B. All authors have read and agreed to the published version of the manuscript.

Funding: This research received no external funding.

Data Availability Statement: No new data were created or analyzed in this study.

Acknowledgments: The authors would like to thank the reviewers for their feedback, insightful comments, and assistance in improving the article.

Conflicts of Interest: Bimetria is a commercial software project developed and operated by Bimetria.com. Two of the authors are affiliated with this company and are directly involved in the design and implementation of the Bimetria system. Because Bimetria is an active business project, some implementation details (including selected aspects of the model architecture, training data curation and production infrastructure) are subject to trade-secret protection and therefore cannot be fully disclosed. The descriptions provided in this article focus on the conceptual pipeline, data flow and integration patterns with BIM/IFC, in a way that allows readers to understand and assess the approach without revealing proprietary business-critical details.

References

1. Criminale, A.; Langar, S. Challenges with BIM implementation: A review of literature. In Proceedings of the 53rd ASC Annual International Conference Proceedings, Seattle, WA, USA, 5–8 April 2017; pp. 329–335.
2. Meng, Q.; Zhang, Y.; Li, Z.; Shi, W.; Wang, J.; Sun, Y.; Xu, L.; Wang, X. A review of integrated applications of BIM and related technologies in whole building life cycle. *Eng. Constr. Archit. Manag.* **2020**, *27*, 1647–1677. [CrossRef]
3. Borkowski, A.S. Evolution of BIM: Epistemology, genesis and division into periods. *J. Inf. Technol. Constr.* **2023**, *28*, 646–661. [CrossRef]
4. Ding, S.; Li, H.; Su, C.; Yu, J.; Jin, F. Evolutionary artificial neural networks: A review. *Artif. Intell. Rev.* **2013**, *39*, 251–260. [CrossRef]
5. Xu, F.; Liu, Q. Building energy consumption optimization method based on convolutional neural network and BIM. *Alex. Eng. J.* **2023**, *77*, 407–417. [CrossRef]
6. Al-Sinan, M.A.; Bubshait, A.A.; Aljaroudi, Z. Generation of construction scheduling through machine learning and BIM: A blueprint. *Buildings* **2024**, *14*, 934. [CrossRef]
7. Jatobá, M.; Santos, J.; Gutierriz, I.; Moscon, D.; Fernandes, P.O.; Teixeira, J.P. Evolution of artificial intelligence research in human resources. *Procedia Comput. Sci.* **2019**, *164*, 137–142. [CrossRef]
8. Liu, Y.; Huang, H.; Gao, G.; Ke, Z.; Li, S.; Gu, M. Dataset and benchmark for as-built BIM reconstruction from real-world point cloud. *Autom. Constr.* **2025**, *173*, 106096. [CrossRef]
9. Ying, H.; Sacks, R.; Degani, A. Synthetic image data generation using BIM and computer graphics for building scene understanding. *Autom. Constr.* **2023**, *154*, 105016. [CrossRef]
10. Gerbino, S.; Cieri, L.; Rainieri, C.; Fabbrocino, G. On BIM interoperability via the IFC standard: An assessment from the structural engineering and design viewpoint. *Appl. Sci.* **2021**, *11*, 11430. [CrossRef]
11. Love, P.E.; Matthews, J.; Fang, W.; Porter, S.; Luo, H.; Ding, L. Explainable artificial intelligence in construction: The content, context, process, outcome evaluation framework. *arXiv* **2022**, arXiv:2211.06561. [CrossRef]
12. Borkowski, A.S.; Brożyna, J.; Litwin, J.; Rączka, W.; Szporanowicz, A. Use of the CDE environment in team collaboration in BIM. *Inform. Autom. Pomiary w Gospod. i Ochr. Sr.* **2023**, *13*, 93–98. [CrossRef]
13. LeCun, Y. LeNet-5, Convolutional Neural Networks. Available online: http://yann.lecun.com/exdb/lenet (accessed on 12 September 2025).
14. Krizhevsky, A.; Sutskever, I.; Hinton, G.E. Imagenet classification with deep convolutional neural networks. *Adv. Neural Inf. Process. Syst.* **2012**, *25*, 1097–1105. [CrossRef]
15. Mikolov, T.; Chen, K.; Corrado, G.; Dean, J. Efficient estimation of word representations in vector space. *arXiv* **2013**, arXiv:1301.3781. [CrossRef]
16. Sutskever, I.; Vinyals, O.; Le, Q.V. Sequence to sequence learning with neural networks. *Adv. Neural Inf. Process. Syst.* **2014**, *27*, 3104–3112.
17. Silver, D.; Huang, A.; Maddison, C.J.; Guez, A.; Sifre, L.; Van Den Driessche, G.; Schrittwieser, J.; Antonoglou, I.; Panneershelvam, V.; Lanctot, M.; et al. Mastering the game of Go with deep neural networks and tree search. *Nature* **2016**, *529*, 484–489. [CrossRef]
18. Vaswani, A.; Shazeer, N.; Parmar, N.; Uszkoreit, J.; Jones, L.; Gomez, A.N.; Kaiser, Ł.; Polosukhin, I. Attention is all you need. *Adv. Neural Inf. Process. Syst.* **2017**, *30*, 5998–6008.
19. Radford, A.; Narasimhan, K.; Salimans, T.; Sutskever, I. Improving language understanding by generative pre-training. *Comput. Sci. Linguist.* **2018**.
20. Available online: https://openai.com/index/better-language-models/?utm_source=chatgpt.com (accessed on 30 September 2025).
21. Ouyang, L.; Wu, J.; Jiang, X.; Almeida, D.; Wainwright, C.; Mishkin, P.; Zhang, C.; Agarwal, S.; Slama, K.; Ray, A.; et al. Training language models to follow instructions with human feedback. *Adv. Neural Inf. Process. Syst.* **2022**, *35*, 27730–27744.
22. Bubeck, S.; Chandrasekaran, V.; Eldan, R.; Gehrke, J.; Horvitz, E.; Kamar, E.; Lee, P.; Lee, Y.T.; Li, Y.; Lundberg, S.; et al. Sparks of artificial general intelligence: Early experiments with GPT-4 (2023). *arXiv* **2023**, arXiv:2303.12712.
23. Kahneman, D. *Thinking, Fast and Slow*; Farrar, Straus and Giroux: New York, NY, USA, 2011.
24. Sun, K.; Xiao, B.; Liu, D.; Wang, J. Deep high-resolution representation learning for human pose estimation. In Proceedings of the IEEE/CVF Conference on Computer Vision and Pattern Recognition, Long Beach, CA, USA, 15–20 June 2019; pp. 5693–5703.

https://doi.org/10.3390/infrastructures11010006

25. Begić, H.; Galić, M.; Klanšek, U. Active BIM system for optimized multi-project ready-mix-concrete delivery. *Eng. Constr. Archit. Manag.* **2024**, *31*, 5057–5084. [CrossRef]
26. Dasović, B.; Galić, M.; Klanšek, U. Active BIM approach to optimize work facilities and tower crane locations on construction sites with repetitive operations. *Buildings* **2019**, *9*, 21. [CrossRef]
27. Namini, S.S.B. Active BIM with Artifical Intelligence for Energy Optimisation in Buildings. Ph.D. Thesis, University of Technology Sydney, Sydney, Australia, 2017.
28. Wang, X.; Yu, K.; Wu, S.; Gu, J.; Liu, Y.; Dong, C.; Qiao, Y.; Change Loy, C. Esrgan: Enhanced super-resolution generative adversarial networks. In Proceedings of the European Conference on Computer Vision (ECCV) Workshops, Munich, Germany, 8–14 September 2018; pp. 1–16.
29. *ISO 16739-1:2018*; Industry Foundation Classes (IFC) for Data Sharing in the Construction and Facility Management Industries—Part 1: Data Schema. International Organization for Standardization: Geneva, Switzerland, 2018.
30. Borkowski, A.S. Experiential learning in the context of BIM. *STEM Educ.* **2023**, *3*, 190–204. [CrossRef]
31. Zawada, K.; Rybak-Niedziółka, K.; Donderewicz, M.; Starzyk, A. Digitization of AEC industries based on BIM and 4.0 technologies. *Buildings* **2024**, *14*, 1350. [CrossRef]

https://doi.org/10.3390/infrastructures11010006

infrastructures

MDPI

Article

Reframing BIM: Toward Epistemic Resilience in Existing-Building Representation

Ciera Hanson, Xiaotong Liu and Mike Christenson *

School of Architecture, University of Minnesota, Minneapolis, MN 55455, USA; hans7604@umn.edu (C.H.); liu02391@umn.edu (X.L.)
* Correspondence: mike001@umn.edu; Tel.: +1-612-624-7866

Abstract

Conventional uses of building information modeling (BIM) in existing-building representation tend to prioritize geometric consistency and efficiency, but often at the expense of interpretive depth. This paper challenges BIM's tendency to promote *epistemic closure* by proposing a method to foreground *relational ambiguity*, transforming view reconciliation from a default automated process into a generative act of critical inquiry. The method, implemented in Autodesk Revit, introduces a parametric reference frame within BIM sheets that foregrounds and manipulates reciprocal relationships between orthographic views (e.g., plans and sections) to promote interpretive ambiguity. Through a case study, the paper demonstrates how parameterized view relationships can resist oversimplification and encourage conflicting interpretations. By intentionally sacrificing efficiency for epistemic resilience, the method aims to expand BIM's role beyond documentation, positioning it as a tool for architectural knowledge production. The paper concludes with implications for software development, pedagogy, and future research at the intersection of critical representation and computational tools

Keywords: parametric; BIM; computational; representation; resilience

1. Introduction

The epistemological position of digital modeling software in representing existing buildings seems to differ fundamentally from its position in the design of new buildings. In the former case, a referent building exists, and software in combination with remote-sensing technology can be used to achieve a precise and efficient form of documentation [1]. In the latter case, there is a prevailing assumption that the act of architectural representation, particularly when executed iteratively, can be used as a designerly mechanism for generating new knowledge [2]. Thus, the act of digitally modeling existing buildings appears to prioritize knowledge display over the knowledge production that is more typical of new-building design. In a wider context, ref. [3] articulates the distinction between knowledge display and knowledge production. For [3], knowledge display is associated with a concealment of the decisions and processes on the basis of which it was produced. Similarly, ref. [4] introduce a distinction between "models of" and "models for," wherein the former are concerned with representations that map a presumably objective reality, and the latter are concerned with how models and their interrelationships "serve an epistemic purpose beyond depiction." Traditional applications of BIM for existing buildings typically function as devices for knowledge display, in that they present geometrically resolved models while concealing the interpretive decisions made during model construction. In contrast, the methodology proposed in this paper leverages BIM for knowledge production by directly exposing ambiguity within models.

Conventional uses of Building Information Modeling (BIM) software for representing existing buildings tend to treat architectural representation as depiction, with a heavy emphasis on achieving precision and geometric accuracy, resulting in "models of" buildings [5]. Nevertheless, digital models of existing buildings are not neutral containers of information but are instead the consequence of representational acts structured by context [6]. To paraphrase [7], architects do not make buildings; they make digital models of buildings, and interpreting those models is not the same as interpreting the buildings to which they refer [8]. It follows that digital models of existing buildings should not be viewed as transparent or neutral reflections of reality, but as situated constructs that actively shape what can be seen, how what can be seen is understood, and whether certain dimensions of knowledge are foregrounded, marginalized, or actively omitted. In the case of BIM, we suggest that its conventional use encourages a form of *epistemic closure*, or in other words, a false sense that the model both constitutes and represents a complete and finished reality.

Rather than reinforcing that conventional emphasis, this paper argues for BIM-based practices that encourage interpretive openness–where *relationships between views* become the source of architectural meaning, and where a BIM model's role expands its core function (namely, the efficient production of coordinated documentation) to encompass active and critical inquiry. In this way, existing-building BIM is positioned as a process that can approach the productive ambiguity of new-building design referred to above. The primary implication is that BIM's conventional emphasis on geometric consistency and epistemic closure can be productively disrupted to foreground *relational ambiguity*, which in this context means highlighting uncertainties between views. This has the effect of transforming BIM into a tool for *epistemic resilience*, which we define as a model's capacity to sustain multiple, conflicting interpretations without collapsing into false certainty.

Our approach differs from dominant trends in existing-building BIM research, which generally focuses on the problems of efficiency and accuracy in automated reconstruction [9,10]. Instead, this paper's emphasis is on knowledge effects as understood to derive from a plurality of representational modes. These effects may align with forms of sociopolitical and economic resilience, particularly in contexts where representational control is contested, as shown, for example, in the work of critical cartography [11,12]. More specifically, representational methods that foreground ambiguity can enable multiple narratives, which are critical in contexts where histories are contested [13]. Thus, the method introduced here offers a means for participatory and pluralistic architectural discourse. In this context, we see our proposed method as potentially valuable for investigative actors: heritage conservationists, architectural historians, and researchers generally. Unlike a professional BIM user in the new-building industry, these investigative actors do not necessarily measure the effectiveness of their methods by speed or efficiency: they can value the visualization of conflict and subjectivity as methodological ends in themselves.

While the method we describe prioritizes the need for critical interpretation over computational efficiency, the outcomes also carry practical implications for the creation of "digital twins" for existing structures. Specifically, the concept of epistemic resilience can be positioned practically as a strategy for managing uncertainty in contexts where documentation is often incomplete or conflicting. By foregrounding ambiguity rather than working to conceal it, our method can help to expose potentially costly assumptions while providing a more nuanced representation of a building's complex history.

https://doi.org/10.3390/infrastructures11020040

1.1. Reciprocal Relationships

Like other representational artifacts, digital models function as epistemic tools, mediating architectural knowledge by means of culturally specific representational practices [2]. Existing buildings participate in architectural discourse through being represented in artifacts like drawings and models. These drawings and models trigger and sustain an interpretive process of conceptual negotiation [14]. This interpretive process is neither simple nor immediate [15]. It involves, *inter alia*, the mental construction of spatial relationships that may not be explicitly depicted in drawings or models. An example may be found in the paired plan and section drawings of an existing building (Figure 1).

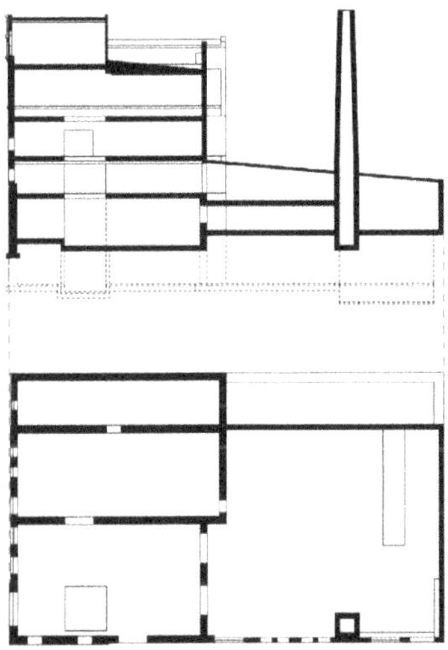

Figure 1. Paired plan and section as an example of a typical reciprocal relationship between orthographic views.

The two drawings (plan and section) reciprocally interpret each other in contributing to the mental formation of spatially coherent understandings. This complementary relationship is precisely what [16] identify as "reciprocity and ... [mutual] dependence." As [7] emphasizes, orthographic projections are "never exhaustive"–they always invite a kind of mental synthesis or what is referred to in this paper as *reconciliation*. Reconciliation is fundamentally generative, reflecting what [17] describe as the "hopeful possibility of plan, section and elevation being ... reversible." The transformative work involved in mentally relating drawings and the thoughts they initiate enables what [18] describe as "allowing a three-dimensional presence to form in the viewer's mind" as architectural understanding unfolds.

The idea of mentally reconciling projections (e.g., orthographic views) is linked to the capacity of representation to absorb and proliferate "ambiguities and inconsistencies" [7]. This assertion goes beyond any technical concern. It assumes that different architectural representations, such as orthographic views or perspectives, will be differently relevant to diverse stakeholders, and that processes of mentally relating representations into a coherent whole will operate in specific and culturally grounded contexts, perhaps even idiosyncratically. Reconciliation is also not something that can trivially be achieved by presenting photorealistic views of buildings or immersive walkthroughs, both of which are inseparable from perspectivism [17]. Instead, manual reconciliation is contingent on the idea that architecture and its representations operate on a level that "barely initiate[s] thought" [14]. For these reasons, architectural representation must be skeptical of any

https://doi.org/10.3390/infrastructures11020040

attempt to provide totalizing views, or to remove or sideline the importance of mental reconciliation as a process.

1.2. BIM's Effects of Knowledge Production for Existing Buildings

Choosing to engage BIM in existing-building representation implies acceptance of a way of organizing work, the knowledge-effects of which include increased specialization [19]. Moreover, consistency among representations–precisely, BIM's default position–is a long-recognized criterion for information-visualization validity [20,21]. Nevertheless, digitally modeling an existing building is an interpretive act [22,23], one that begins with a choice of approach, e.g., between geometric, parametric, or semantic approaches, or between parametric, geometrically accurate, or knowledge-generating ones [24–26]. Moreover, digital modeling inherently involves tacit judgment [6,27]. As such, digital modeling considered broadly has the capability to operate as an ambiguity-foregrounding method accommodating multiple knowledge domains.

BIM's most obvious and significant knowledge implication lies in the software's ability to provide an arbitrarily large number of consistent, coordinated representations of a geometrically complex reality, e.g., a building. This concept is referred to here as *automatic reconciliation* to emphasize its difference from the manual process of mental reconciliation described above. Automatic reconciliation takes place without any active engagement by the modelmaker; it is simply a consequence of BIM's central model being made visible by means of distinct projections (e.g., plan and section). It is in part due to this ability that BIM derives its obvious utility within architectural and engineering practice in support of new-building design. What in a pre-BIM context constituted a tedious process of manually coordinating two-dimensional drawings (e.g., plans with sections) becomes, within BIM, a banal consequence of the software's ontology. In the context of new-building design, the omission of this manual coordination process has had clear positive effects on production and efficiency [28].

So much is obvious within contemporary practice as it relates to the design of *new* buildings. However, BIM's relevance to representing *existing* buildings remains an active and contested area of research [29–34]. In large part, research in this area examines distinctions between BIM's ability to support objective documentation (e.g., through point-cloud scanning or photogrammetry) and semantic enrichment (i.e., the addition of meaningful relationships to a model). For example, ref. [35] introduce a distinction between "black box" and "white box" modeling approaches, in which the former treats the referent building as a coherent structure of scanned points modeled in 3d space, and the latter treats the referent building as a consequence of parametrically informed modeling decisions, with the potential to incorporate predefined element libraries. These two approaches result not only in two distinct kinds of models but in two different affordances for conceptually understanding the architecture. Similarly, ref. [36] raise the possibility of different model construction processes leading to differing interpretations, i.e., "different modes of critical re-reading." Model construction processes are further complicated in the case of demolished buildings, where a distinction arises between a "theoretical" model, that is, a model based on documentary sources, and an "as-built" model representing field-verified geometry [37]. That a building no longer physically exists means that any verification of its elements introduces a unique kind of ambiguity, i.e., among interpretations of an incomplete or internally contradictory documentary record [38,39]. Failure to observe the distinction between document-based and field-based models can result in reduced applicability to applications like structural health monitoring [37].

Fundamentally, BIM's hierarchical, component-based knowledge structure risks conflict with open-ended, interpretive approaches to representing and modeling built heritage.

In part, this is simply because BIM's ontology, to the extent that it depends on parametric relationships and predefined elements, can struggle to capture the formal and material idiosyncrasies of existing buildings [40]. The epistemological consequences arise in at least two ways. First, there are uncertainties regarding BIM standardization in heritage and conservation contexts [41]. Second, there is a risk to marginalizing or excluding culturally specific practices, e.g., construction and living patterns that defy BIM's expectations for clarity of organization and geometric consistency [42].

1.3. Cultural Assumptions and Epistemic Bias in BIM

Existing-building BIM aims for efficiency in producing complete, accurate, and comprehensive models. These aims have the effect of prioritizing dominant knowledge systems, i.e., the standards of the AEC industry [43]. Such standards are often positioned as epistemologically neutral, aimed at making interoperability between software platforms "seamless" [44]. For example, consider how an existing masonry building with sagging floors or out-of-plane walls could become "homogenized" in BIM. To maintain geometric consistency, a modelmaker might simplify observational data into the level planes and vertical walls encouraged by the software's internal logic [1,10]. Such an approach would result in an "interpretive homogenization" where the building's idiosyncrasies are flattened into a standard, industry-compliant representation. Even apparently objective standards such as IFC may carry specific Western, industry-centered logics that risk marginalizing nuanced or conflicting understandings of space [45,46]. For example, the IFC entity IfcWall refers to "a vertical construction that may bound or subdivide spaces" [47]. This apparently straightforward definition assumes dichotomies (e.g., container versus contained) and hierarchies (e.g., subdivisions) that arguably make more sense when applied to buildings like factories and offices rather than to nomadic shelters, reed mudhifs, or cave housing. Within a context informed by sharp dichotomies and clear hierarchies, BIM's major conceit is that ambiguity and conflict cannot exist, or rather that alignment and coordination between representations of the model are incontestable because all views refer to a single central model. This is the case for orthographic and perspective views as well as exploded or "displaced" views (Figure 2).

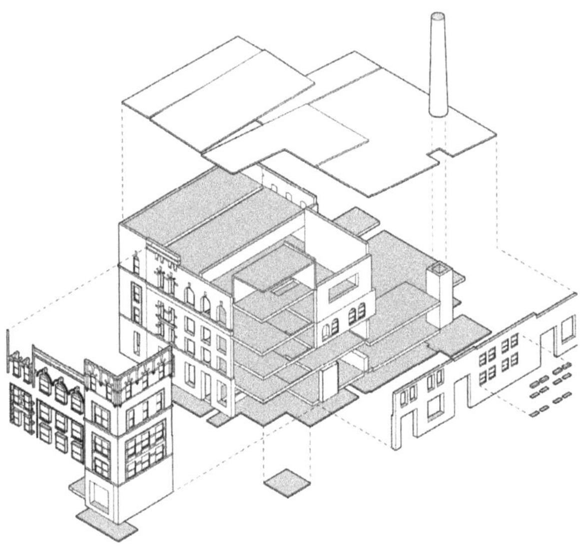

Figure 2. Displaced view in BIM, illustrating the possibility of producing views of BIM model components 'displaced' from their known positions.

https://doi.org/10.3390/infrastructures11020040

The central-model assumption guarantees alignment of BIM with Cartesian frameworks reflecting a Western privileging of geometric objectivity, marginalizing non-metric knowledge systems [17]. Yet even apparently neutral concepts such as geometric accuracy have different meanings, different degrees of relevance, and different extents of subjectivity within different cultural contexts [14,48].

1.4. Devices of Reciprocity in BIM

Apart from the software's ontological focus on automatic reconciliation, BIM relies on certain formal mechanisms to explicate reciprocal relationships. These mechanisms, referred to here as *devices of reciprocity*, may be categorized into model-based devices and sheet-based devices. The categories reflect important technical and conceptual differences between two modes of producing modeled or drawn information. Model-based devices include coordinate systems, reference planes, level lines, and parametric relationships that are explicitly coded to relate parameters to modeled form. Sheet-based devices identify or relate views to each other. While both model-based and sheet-based devices are designed to streamline automatic view coordination, they have the side effect of constraining the modelmaker's ability to engage a model's spatial ambiguity. The following section outlines a method for repurposing these mechanisms to prioritize manual reconciliation as a site of critical inquiry.

2. Materials and Methods

This section of the paper highlights a specific method for problematizing reciprocal relationships in a case-study BIM model. The method is explicitly not designed as an automated process but as a manual workflow intended to introduce subjective and critical ambiguity. In this way, the method is not positioned solely as a technical process but as a philosophical reorientation of how meaning is constructed through modeling.

The project relies on Autodesk Revit 2026 to particularize its argument. However, the underlying logic broadly reflects capabilities that are inherent to BIM software generally (e.g., ArchiCAD, Vectorworks): namely, the software's automated reconciliation of orthographic projections derived from a parametrically driven database. We consider a BIM model of a recently demolished building in Winona, Minnesota (USA), constructed by this paper's corresponding author (Figure 3). The case study is selected as an example of the kind of building susceptible to our method: that is, any one where some form of interpretive friction exists between as-built reality, traditional documentation, and digital model. The construction of such existing-building models implicates multiple sources of evidence [38]. In this case, the building's demolition introduces a specific kind of interpretive ambiguity, because it is no longer possible to conduct the kind of field verification or validation that an existing building would present.

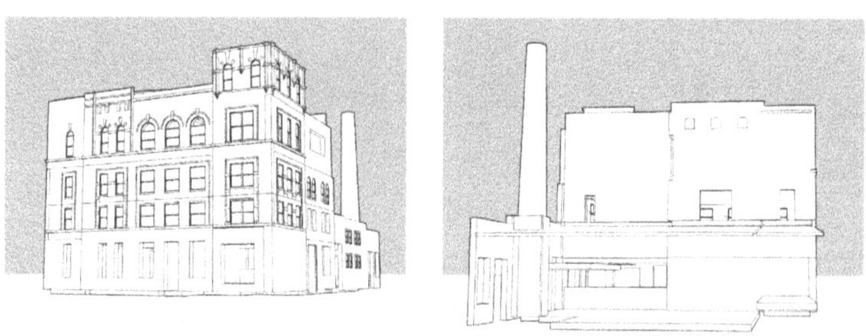

Figure 3. Two perspective views illustrating BIM model of recently demolished building, Winona, Minnesota (USA).

https://doi.org/10.3390/infrastructures11020040

2.1. Construction of Reference Frame

To counter BIM's default automation of view reconciliation, this method introduces an intentional layer of ambiguity and relationality by foregrounding reciprocal relationships between orthographic views. The ambiguous nature of orthographic views may be highlighted by deliberately juxtaposing and superimposing them, calling their presumed existence as independent registers of the model into question. Stated somewhat differently, reciprocal relationships between views can be *problematized:* that is, relations between views can be considered as an open question requiring active investigation, rather than as a fixed, automatic linkage. Our method establishes relationships graphically and then positions views relative to that drawn relationship. More specifically, we begin by identifying a possibly critical relationship, such as between two orthographic views. We then graphically deploy an armature or frame that provides a kind of intentional superimposition between those views. The resulting superimposition is then open to a process of manual reconciliation. To illustrate, suppose that a rectangular armature or frame *f* is produced in a Revit sheet, and that *f* is understood to signify a reciprocal relationship between two graphic elements positioned, respectively, at a and b (Figure 4).

Figure 4. A rectangular, parametric armature *f* placed on a Revit sheet can be used to define and regulate the reciprocal relationship between views positioned, respectively, at a and b. Lines a and b indicate the respective positions of two views. The dashed (vertical) lines indicate an "imaginal joint" between them.

f's rectangular frame and its use of dashed lines respond to the reciprocity mechanisms described above. The frame provides a fixed reference against which misalignment can be measured. Dashed lines, as non-material indicators of spatial relationships, build upon conventions of projection: they indicate what [49] calls an "imaginal joint between two realms," i.e., the graphic elements positioned at a and b. For example, paired plan and section views can be positioned at a and b, aligned positionally and at a consistent scale to emphasize formal and spatial relationships that exist between the two views. Figure 5 illustrates how *f* can emphasize a formal relationship between a space as it appears in plan and the same space as it appears in section.

The premise of Figure 5 is that the reference frame *f* initiates the possibility of a meaningful visual relationship between views. Once the frame is in place, the first view (plan or section) is placed relative to f, and then the second view (section or plan) is selected and positioned relative to the first, using *f* as a guide.

There is no reason that *f* needs to remain an inert or fixed element. Instead, like other elements in Revit–or in BIM software generally–*f* could incorporate parameters, enabling variation in visible features such as proportion, position, and size. More specifically, *f* may be implemented as a parametric family; its dimensions (length, height) and position on the sheet controlled by user-defined parameters. The crop regions of the plan and section views may be manually aligned and constrained to its edges. This technique effectively transforms the sheet into a dynamic modeling environment, allowing *f* to become a sheet-based device for explicating reciprocal relationships.

https://doi.org/10.3390/infrastructures11020040

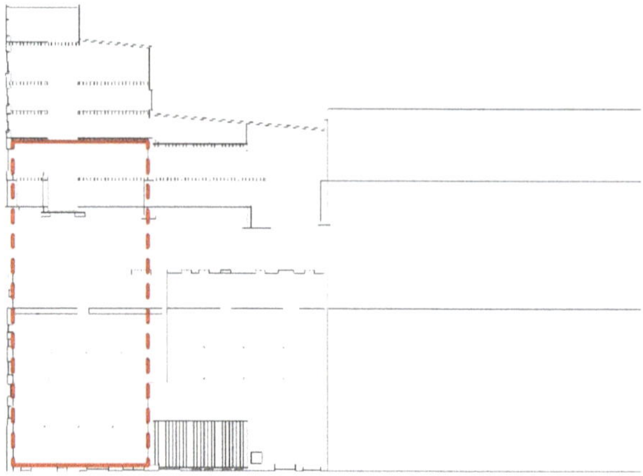

Figure 5. The parametric reference frame *f* (shown in red) is used to spatially organize a plan and a section view of the same architectural space within a Revit sheet. As in Figure 4, the solid red lines indicate the respective positions of the two views, and the dashed (vertical) lines indicate an "imaginal joint" between them. Aligning the views to a common visual framework highlights their formal and spatial reciprocity.

Adjusting the parameters enables iterative exploration of spatial ambiguity as the individual views become visually superimposed to a greater or lesser extent (Figure 6).

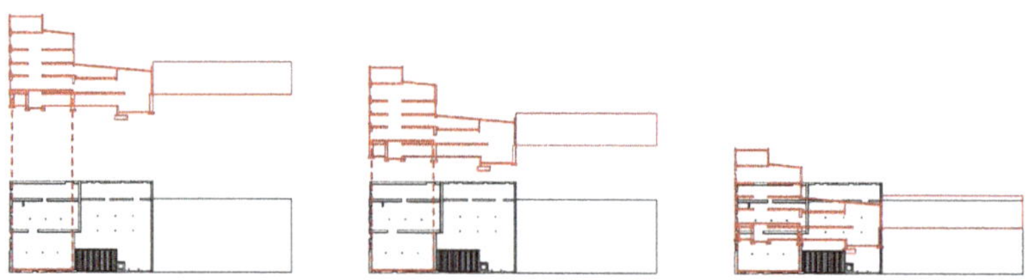

Figure 6. Parametric manipulation of *f*'s vertical dimension causes the plan view (shown in black) and the section view (shown in red) to increasingly overlap. Increased superimposition leads to visual ambiguity, demanding heightened interpretive engagement. This superimposition is deliberately created to provoke interpretive engagement with the spatial relationship between plan and section.

Traditional documentation relies, at least in part, on the visual juxtaposition of views–that is, placing plan and section side-by-side to encourage the mental construction of spatial relationships. In contrast, this method utilizes superimposition to force a collision of spatial information. By parametrically reducing *f*'s vertical dimension, the modelmaker collapses the orthogonal distance that typically separates these views. Critically, this transition is not merely graphical; it causes a shift from observation to active resolution of visual conflict. To state this idea somewhat more precisely, when the constituent views are separated and visually juxtaposed in the conventional manner, the modelmaker can rely on BIM's automatic reconciliation, implicitly trusting that the software has coordinated the model geometry to produce each view. However, as the modelmaker compresses *f* vertically, and the views increasingly overlap (superimposition), this implicit trust is disrupted by visual interference. Lines belonging to the floor plan overlap and intersect with lines in the section, leading to a visually dense mesh of information that cannot be passively read as reflecting certain knowledge. In turn, this requires that the modelmaker engage in manual reconciliation to mentally disentangle plan from section. In this way, superimposition acts as a heuristic device: by obscuring the easy readability of plan and section, it discloses latent logics that connect them to each other.

https://doi.org/10.3390/infrastructures11020040

As an information visualization strategy, visual superimposition is conventionally understood to have certain advantages, such as efficiency of comparison within a shared visual space, as well as apparent disadvantages, notably occlusion and clutter [50,51]. The iterative approach to overlapping in Figure 6 inverts those conventional assumptions. In this case, visual ambiguity is not presented as an obstacle to resolve but as a deliberate visual provocation. By parameterizing visual superimposition, the method makes new demands on interpretive labor, i.e., demanding active interpretation rather than passive consumption.

2.2. Expanding to Other Views

A similar approach can be followed using other views. Suppose that instead of addressing reciprocal relationships between plan and section, relationships are to be emphasized between a section and an oblique projection (Figure 7). A volumetric reference frame, *f* ′, establishes the possible relationship; views are placed in alignment with *f* ′. Parametric adjustments to *f* ′ allow for variable investigation of the reciprocal relationships between the section and the oblique.

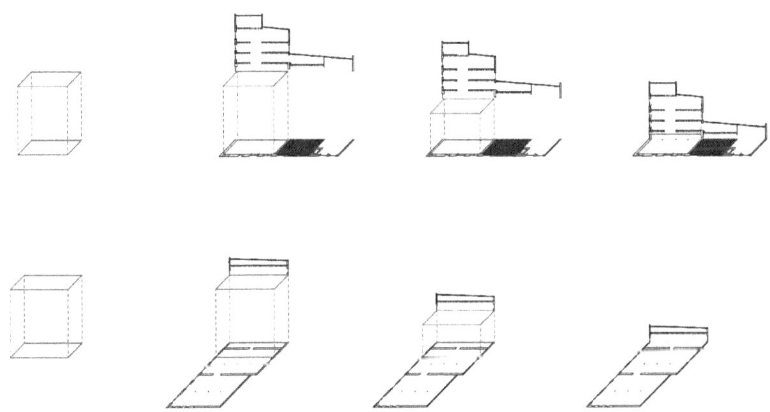

Figure 7. A volumetric variant of the reference frame, *f* ′, is used to position and relate a section view to an oblique projection. Here, the reference frame (shown in red) extends into a pseudo-3D spatial volume within the Revit sheet, emphasizing formal correspondence between the component views (shown in black). Parametric adjustment of *f* ′ enables iterative control over view relationships.

Introducing reference frame *f* has the effect of focusing modelmaker attention on a reciprocal relationship that exists between two views. This shift in prioritization adds an important function to BIM's representational capabilities. Rather than treating the central model as the primary source of knowledge, this approach recognizes the epistemological value of relationships between views. Consequently, understanding emerges from the active process of establishing important relationships using reference frame *f* and then attaching views, rather than simply deriving individual views from a central model.

3. Results

This method challenges BIM's default prioritization of geometric consistency by foregrounding dynamic relationships between views, such as plan-section reciprocity. A conventional set of drawings, with its focus on clearly distinct, separated views, implies a singular, coordinated reality. In contrast, variably superimposed views as discussed here lead to an ambiguity demanding interpretive labor. Reference frame *f* actively highlights relationships between different views rather than treating them as isolated derivatives of the central model. By parameterizing the reference frame, users can dynamically adjust the visual proximity and overlaps between views (as shown in Figure 6). Overlapping

https://doi.org/10.3390/infrastructures11020040

views can highlight contested spatial relationships. The visual ambiguity generated by the superimposition of views functions as a mechanism for knowledge production by resisting what we call epistemic closure.

Our approach acknowledges two fundamentally distinct aspects of BIM. First, BIM centers an underlying database containing information susceptible to meaningful (building-related) queries. Second, BIM provides a mechanism for producing meaningfully useful "views" of the information in that database. Conventionally, those "views" take the form of individual orthographic projections or tabular displays (e.g., room finish schedules). Simply, our method manipulates BIM's viewing mechanism to uniquely reveal the qualities of information in the database. Our manipulation, taking the form of visual superimposition, generates new knowledge in two ways. First, it demands the capacity of spatial synthesis from viewers. That is, visual overlap forces a simultaneous perception of section and plan, confounding the viewer's ability to reconcile the building planimetrically. Vertical readings (e.g., of height, daylight, and structure) must be directly tested against the floor plan. Second, in a related way, superimposition promotes a kind of relational visibility. By utilizing *f*'s parametric controls to overlap views, the modelmaker can identify and visually problematize spatially critical relationships. Conflicts and misalignments, along with unexpected but latent relationships (e.g., between the structural grid and the spatial enclosure), not obvious in isolated views, can become apparent in the form of visual interference.

This shift from static to dynamic representation creates new cognitive challenges, e.g., in the need to interpret visually overlapped drawings. Ambiguity is increased at the expense of clarity and objectivity, inviting contemplation. That is, the method intentionally sacrifices efficiency for deeper engagement in the form of active disambiguation, i.e., manual reconciliation. In its shift from defining individual views to establishing relationships, the workflow encourages new kinds of critical thinking about spatial meanings. Rather than assuming spatial ambiguity must always be resolved, the method encourages conflicting interpretations, thereby calling into question hidden frameworks structuring architectural knowledge (e.g., Cartesian systems). These trade-offs between efficiency and engagement challenge BIM's conventional assumptions of consistency, stability, and fidelity to referent geometry.

The method identifies the concrete problem of *source ambiguity* within existing-building model construction, namely, that such models implicate multiple sources of evidence, typically including drawings or photographs alongside some form of field verification [38]. As a recently demolished building, the Winona case study permits no field verification whatsoever, complicating the concrete issue of source ambiguity, and opening space for our method to play an important role. This attitude toward ambiguity engages critically with uncertainty visualization, a well-established concern in heritage studies [38,39]. Uncertainty visualization conventionally assumes that ambiguity is a consequence of incomplete data, and that better tools will result in increasingly accurate and comprehensive representations of a fixed referent. By contrast, the method proposed here rejects the possibility of a fully knowable referent building. Superimposition, with its inherent occlusion and clutter [50,51] demands active interpretation and departs from a search for technical certainty, opening new space for multiple voices and contested histories. Undoubtedly, similar methods could be implemented in other (non-BIM) software. However, the method described here is notable, insofar as by working within the BIM environment, it effectively subverts BIM's inherent drive for clarity from within. The method uses BIM's own tools against its default tendencies.

The method's interpretive flexibility aligns with a broader sociopolitical context. The flexibility that comes with ambiguity means that the method is open to shifting cultural contexts, and treating ambiguity as generative gives new support to developing pluralistic

https://doi.org/10.3390/infrastructures11020040

narratives. Expanding opportunities for different forms of knowledge production also implies an opportunity for improving a diversity of voices and constituents, i.e., foregrounding contested or community-reclaimed narratives. Relational ambiguity in BIM is thus able to disrupt conventional workflows; the implications extend beyond technique, with a potential to redefine BIM's cultural role. Nevertheless, while this method offers new opportunities for critical engagement, it will certainly be less efficient in practice, requiring a level of interpretive labor that challenges typical workflows.

Ultimately, the method demonstrates that BIM's value in existing-building representation extends beyond its conventional commitment to the efficient production of geometrically consistent and coordinated views. By parametrically enabling the collision and overlap of views, BIM can become a tool for epistemic resilience, where the difficulty of reading the drawing leads to a deeper engagement with the referent, i.e., the building itself.

Practical Implications

Renovation projects are complex, and risk management is essential to success [52]. Various methods exist to manage risk in renovation projects [53,54]. The method proposed here, precisely because it is idiosyncratically capable of highlighting ambiguity, provides a unique mechanism for anticipating expensive errors. For example, consider a hypothetical retrofit of an existing masonry structure with an out-of-plane wall. A standard BIM workflow would model the wall as a vertical element, consistent with software constraints. The resulting floor plan would be geometrically consistent with plan measurements, but epistemically false with respect to the existing as-built conditions. Consequently, precision interventions (i.e., renovations) made on the basis of the plan would inevitably conflict with on-site irregularities. Because the plan and section appear perfectly coordinated by the software, the modelmaker (and the retrofit contractor) may not realize the wall is out-of-plane. The software's drive for consistency would thereby "mask" spatial uncertainty, concealing a physical risk by a digital expediency. In contrast, our proposed method enables a wall's footprint to be visually superimposed, to varying degrees, with its sectional projection. The resulting visual interference–essentially, a misalignment of lines–would not offer a geometric solution but would instead signal a site for further inquiry. Such visual interference calls upon a modelmaker to manually reconcile plan and section, transforming a potential source of expensive error into an opportunity for critical attention.

The same issue can arise with the use of point-cloud scanning to produce "as-built" BIM models, as the conversion from a point cloud to a functioning BIM model relies largely on manual processes involving subjective judgment [9,10]. Similarly, ref. [55] remark that the process of "twinning" an existing building demands interpretative labor. Moreover, because contemporary renovation and adaptive reuse projects are characterized by the presence of diverse, often conflicting constituencies (e.g., architects, engineers, community members, historians, and clients), a robust and well-informed decision-making process is necessary to disclose, examine, and weigh comparative or conflicting knowledge claims. The method described here could inform this process alongside other established means [56].

While traditional BIM workflows in the new-building industry are characterized by design and construction phases, our method is firmly situated in the investigative and pre-design phases of existing-building projects (e.g., conservation or adaptive-reuse work). It aims to provide opportunities for investigative actors to examine conflicts before decisions are solidified into production-oriented BIM models. Similarly, in contrast to in-built mechanisms for interference checking [57], our method does not aim to automatically identify interfering or "clashing" elements within a model. Instead, it attempts to focus modelmaker

https://doi.org/10.3390/infrastructures11020040

attention on the inevitable frictions that arise when software's default assumptions are tested against built reality.

4. Discussion

This paper has identified a problem, namely, that BIM's native automated reconciliation in existing-building representation has the effect of prioritizing efficiency over critical interpretation. As a solution, the paper proposes a computational approach aimed at foregrounding relationships between views rather than relying on automatic reconciliation between independent views. The implication is that BIM's role will expand as a generative tool for architectural knowledge. The proposal is not to abandon BIM's obvious strengths in coordination or accuracy, but rather to suggest that interpretive ambiguity can be meaningfully introduced in a BIM workflow.

It is essential to distinguish this paper's approach from routine BIM tasks such as clash detection or cross-disciplinary coordination. These routine tasks necessarily seek to resolve conflicts in pursuit of a singular geometric truth. By contrast, the approach described here seeks to preserve and indeed to amplify conflict as a site of investigative work. To state this idea differently, BIM's value in routine practical contexts is largely a consequence of its efficiency in providing possibilities for reconciliation. But BIM-based reconciliation is always of a particular kind, in that it reinforces the idea of an unambiguous and universally understood representation, rather than opening space for the kinds of contradictions, conflicts, and gaps that prompt productive and critical interpretation. By calling BIM's knowledge structure into question, this paper acknowledges BIM's efficiency in the routine tasks of reconciliation while arguing that this efficiency may inadvertently minimize the knowledge-production effects inherent in reciprocal relationships. By problematizing those relationships–in fact, by leveraging BIM's capacity for representation–intentional ambiguity may be introduced, leading to epistemological gaps that invite nuanced interpretations. The intent is for ambiguity to support new forms of epistemic resilience, countering the possibility of historical oversimplification, and opening space for otherwise marginalized voices.

While much of the current literature in existing-building BIM focuses on automated reconstruction [9,10], we offer an original approach through an intentional subversion of automation. In our view, it is precisely the removal of the modelmaker's manual interpretive labor (as promised by automation) that constitutes a risk to both knowledge production and critical inquiry. Also, unlike uncertainty visualization methods [38,39] that often rely on color-coding or other forms of visual labeling, our approach is original in using parametrically driven superimposition to generate visual interference, in turn demanding manual reconciliation. Finally, by using Revit's sheets as active sites for investigation instead of as simple registers of a model, we use the software's own internal logic to critique its tendency toward epistemic closure.

The method leaves pragmatic questions unanswered. As suggested above, future research could proceed to develop plugins and addons aimed at expanding BIM's representational capacities. The implication is for a new class of plugins, not for automation and clash detection, but for narrative layering, annotation, and the visual management of interpretive ambiguity directly atop the coordinated model. Future research could also examine how relational workflows could be adapted to different BIM platforms (i.e., other than Revit) and to varying modelmaker competencies.

There are methodological implications for architectural and engineering pedagogy. Students can deepen their understanding of BIM's capabilities in critical modeling through exposure to ambiguity-aware workflows. Pedagogy should emphasize the importance of relationships and narrative multiplicity alongside practical needs for documentary clarity.

https://doi.org/10.3390/infrastructures11020040

Teaching strategies need not be aimed at selecting the best or most clearly explanatory individual views in a BIM project, as is typical, but rather at identifying critical relationships that can be visually problematized. Applications in history, conservation, and archaeology could similarly benefit from a workflow that actively engages ambiguity and conflict.

The method proposed here demonstrates that BIM's tendency toward epistemic closure can be resisted through deliberately ambiguous and relational view configurations. This adds value both practically and epistemically. In a practical sense, the method identifies spatial discrepancies and archival inconsistencies that would otherwise be obscured by BIM's internal logic. Epistemically, the method transforms a BIM model from its default position of automated certainty into a mechanism for active investigation. The method thereby "reframes" BIM from being a production-oriented document generator into a tool for critical inquiry. This is particularly important in contexts where multiple narratives may coexist. By sacrificing efficiency for investigative rigor, the method proposed here empowers conservationists, historians, and architects to leverage contemporary digital tools to disclose productive uncertainties of the existing built environment.

Author Contributions: Conceptualization, M.C.; methodology, M.C., C.H. and X.L.; formal analysis, M.C., C.H. and X.L.; investigation, M.C., C.H. and X.L.; writing, M.C.; supervision, M.C.; project administration, M.C.; funding acquisition, M.C. All authors have read and agreed to the published version of the manuscript.

Funding: This research was funded by the Digital Design Center at the University of Minnesota.

Data Availability Statement: The original contributions presented in this study are included in the article. Further inquiries can be directed to the corresponding author.

Conflicts of Interest: The authors declare no conflicts of interest.

References

1. Bolognesi, C.M.; Bassorizzi, D. Scan to BIM Processes: Heritage Accuracy versus BIM Semantic. *Int. Arch. Photogramm. Remote Sens. Spat. Inf. Sci.* **2023**, *48*, 267–272. [CrossRef]
2. Christenson, M. *Theories and Practices of Architectural Representation*; Routledge: London, UK, 2019.
3. Drucker, J. *Graphesis: Visual Forms of Knowledge Production*; Harvard University Press: Cambridge, MA, USA, 2014.
4. Gouvea, J.; Passmore, C. "Models of" versus "Models for": Toward an Agent-Based Conception of Modeling in the Science Classroom. *Sci. Educ.* **2017**, *26*, 49–63. [CrossRef]
5. Gimenez, L.; Hippolyte, J.-L.; Robert, S.; Suard, F.; Zreik, K. Review: Reconstruction of 3D Building Information Models from 2D Scanned Plans. *J. Build. Eng.* **2015**, *2*, 24–35. [CrossRef]
6. Huvila, I. The Subtle Difference Between Knowledge and 3D Knowledge. *Hambg. J. Für Kult.* **2017**, *7*, 99–111.
7. Evans, R. Architectural Projection. In *Architecture and Its Image: Four Centuries of Architectural Representation*; Blau, E., Kaufman, E., Eds.; Canadian Centre for Architecture: Montreal, QC, Canada, 1989; pp. 19–35.
8. Bianchini, C. Metamodellazione/Metamodelling. *Disegnare Idee Immagin.* **2023**, *66*, 80–95. [CrossRef]
9. Rocha, G.; Mateus, L.; Fernández, J.; Ferreira, V. A Scan-to-BIM Methodology Applied to Heritage Buildings. *Heritage* **2020**, *3*, 47–67. [CrossRef]
10. Tang, P.; Huber, D.; Akinci, B.; Lipman, R.; Lytle, A. Automatic Reconstruction of As-built Building Information Models from Laser-scanned Point Clouds: A Review of Related Techniques. *Autom. Constr.* **2010**, *19*, 829–843. [CrossRef]
11. Paez, R. *Operative Mapping: The Use of Maps as a Design Tool*; Actar Publishers: New York, NY, USA, 2019.
12. Counter Cartographies Collective; Dalton, C.; Mason-Deese, L. Counter (Mapping) Actions: Mapping as Militant Research. *ACME Int. J. Crit. Geogr.* **2015**, *11*, 439–466. [CrossRef]
13. Schreibman, S.; Papadopoulos, C. Textuality in 3D: Three-Dimensional (Re)constructions as Digital Scholarly Editions. *Int. J. Digit. Humanit.* **2019**, *1*, 221–233. [CrossRef]
14. Piotrowski, A. On the Practices of Representing and Knowing Architecture. In *The Discipline of Architecture*; Piotrowski, A., Robinson, J.W., Eds.; University of Minnesota Press: Minneapolis, MN, USA, 2001; pp. 40–60.
15. Fassi, F.; Achille, C.; Fregonese, L. Surveying and Modelling the Main Spire of Milan Cathedral Using Multiple Data Sources. *Photogramm. Rec.* **2011**, *26*, 462–487. [CrossRef]

https://doi.org/10.3390/infrastructures11020040

16. Clark, R.H.; Pause, M. *Precedents in Architecture: Analytic Diagrams, Formative Ideas, and Partis*; John Wiley & Sons: Hoboken, NJ, USA, 2005.
17. Pérez-Gómez, A.; Pelletier, L. *Architectural Representation and the Perspective Hinge*; MIT Press: Cambridge, MA, USA, 1997.
18. Fraser, I.; Henmi, R. *Envisioning Architecture: An Analysis of Drawing*; Van Nostrand Reinhold: New York, NY, USA, 1994.
19. Borkowski, A.S. Evolution of BIM: Epistemology, Genesis and Division into Periods. *J. Inf. Technol. Constr.* **2023**, *28*, 646–661. [CrossRef]
20. Tan, S.; Lai, C.; Zhang, X.; Yuan, X. An Empirical Guide for Visualization Consistency in Multiple Coordinated Views. In *Proceedings of the 2023 IEEE 16th Pacific Visualization Symposium (PacificVis), Seoul, Republic of Korea, 18–21 April 2023*; IEEE: Los Alamitos, CA, USA, 2023; pp. 31–40. [CrossRef]
21. Wang Baldonado, M.Q.; Woodruff, A.; Kuchinsky, A. Guidelines for Using Multiple Views in Information Visualization. In *Proceedings of the Working Conference on Advanced Visual Interfaces, Palermo, Italy, 23–26 May 2000*; ACM: New York, NY, USA, 2000; pp. 110–119. [CrossRef]
22. Dell'Unto, N. 3D Models and Knowledge Production. In *Archaeology and Archaeological Information in the Digital Society*; Huvila, I., Ed.; Routledge: London, UK, 2018; pp. 54–69. [CrossRef]
23. Perry, S. Crafting Knowledge with (Digital) Visual Media in Archaeology. In *Material Evidence: Learning from Archaeological Practice*; Chapman, R., Wylie, A., Eds.; Routledge: London, UK, 2014; pp. 189–210. [CrossRef]
24. Yang, X.; Grussenmeyer, P.; Koehl, M.; Macher, H.; Murtiyoso, A.; Landes, T. Review of Built Heritage Modelling: Integration of HBIM and Other Information Techniques. *J. Cult. Herit.* **2020**, *46*, 350–360. [CrossRef]
25. Castellano-Román, M.; Pinto-Puerto, F. Dimensions and Levels of Knowledge in Heritage Building Information Modelling, HBIM: The Model of the Charterhouse of Jerez (Cádiz, Spain). *Digit. Appl. Archaeol. Cult. Herit.* **2019**, *14*, e00110. [CrossRef]
26. Christenson, M. Parametric Analysis of Existing-Building Geometry: Generative, Dissective, and Morphological Approaches. *Int. J. Archit. Comput.* **2024**. [CrossRef]
27. Wylie, A. How Archaeological Evidence Bites Back: Strategies for Putting Old Data to Work in New Ways. *Sci. Technol. Hum. Values* **2017**, *42*, 203–225. [CrossRef]
28. Parsamehr, M.; Perera, U.S.; Dodanwala, T.C.; Perera, P.; Ruparathna, R. A Review of Construction Management Challenges and BIM-Based Solutions. *Asian J. Civ. Eng.* **2023**, *24*, 353–389. [CrossRef]
29. Dauda, J.A.; Bower, N.; Adebisi, W.A.; Saad, A.; Saka, A.B.; Kazemi, H.; Ajayi, S.O. Exploring the Application of Heritage Building Information Modelling (HBIM) for Heritage Conservation: Insights from Industry Practitioners. *J. Archit. Conserv.* **2025**, *31*, 99–112. [CrossRef]
30. Parente, M.; Bruno, N.; Ottoni, F. HBIM and Information Management for Knowledge and Conservation of Architectural Heritage: A Review. *Heritage* **2025**, *8*, 306. [CrossRef]
31. Patil, J.; Kalantari, M. Automatic Scan-to-BIM—The Impact of Semantic Segmentation Accuracy. *Buildings* **2025**, *15*, 1126. [CrossRef]
32. Mirarchi, C.; Gholamzadehmir, M.; Daniotti, B.; Pavan, A. Semantic Enrichment of BIM: The Role of Machine Learning-Based Image Recognition. *Buildings* **2024**, *14*, 1122. [CrossRef]
33. Yang, S.; Hou, M.; Li, S. Three-Dimensional Point Cloud Semantic Segmentation for Cultural Heritage: A Comprehensive Review. *Remote Sens.* **2023**, *15*, 548. [CrossRef]
34. Bastem, S.S.; Cekmis, A. Development of Historic Building Information Modelling: A Systematic Literature Review. *Build. Res. Inf.* **2021**, *50*, 527–558. [CrossRef]
35. Saygi, G.; Remondino, F. Management of Architectural Heritage Information in BIM and GIS: State-of-the-Art and Future Perspectives. *Int. J. Herit. Digit. Era* **2013**, *2*, 695–713. [CrossRef]
36. Attenni, M.; Bianchini, C.; Griffo, M.; Senatore, L.J. HBIM Meta-Modelling: 50 (and More) Shades of Grey. *ISPRS Int. J. Geo-Inf.* **2022**, *11*, 468. [CrossRef]
37. Nieto-Julián, J.E.; Antón, D.; Moyano, J.J. Implementation and Management of Structural Deformations into Historic Building Information Models. *Int. J. Archit. Herit.* **2019**, *14*, 1384–1397. [CrossRef]
38. Kensek, K.M.; Dodd, L.S.; Cipolla, N. Fantastic Reconstructions or Reconstructions of the Fantastic? Tracking and Presenting Ambiguity, Alternatives, and Documentation in Virtual Worlds. *Autom. Constr.* **2004**, *13*, 175–186. [CrossRef]
39. Schäfer, U.U. Uncertainty Visualization and Digital 3D Modeling in Archaeology: A Brief Introduction. *Int. J. Digit. Art Hist.* **2019**, *3*, 86–106. [CrossRef]
40. Brusaporci, S.; Maiezza, P.; Tata, A. A Framework for Architectural Heritage HBIM Semantization and Development. *Int. Arch. Photogramm. Remote Sens. Spat. Inf. Sci.* **2018**, *42*, 179–184. [CrossRef]
41. Lovell, L.J.; Davies, R.J.; Hunt, D.V.L. The Application of Historic Building Information Modelling (HBIM) to Cultural Heritage: A Review. *Heritage* **2023**, *6*, 6691–6717. [CrossRef]
42. Prizeman, O.; Barazzetti, L. Low-Cost 3D Acquisition of Geometric Data for Living Heritage: Attempting to Record the Pudhu Mandapam, Madurai. *Int. Arch. Photogramm. Remote Sens. Spat. Inf. Sci.* **2021**, *46*, 555–562. [CrossRef]

https://doi.org/10.3390/infrastructures11020040

43. Diara, F.; Rinaudo, F. IFC Classification for FOSS HBIM: Open Issues and a Schema Proposal for Cultural Heritage Assets. *Appl. Sci.* **2020**, *10*, 8320. [CrossRef]
44. Yu, Y.; Kim, S.; Jeon, H.; Koo, B. A Systematic Review of the Trends and Advances in IFC Schema Extensions for BIM Interoperability. *Appl. Sci.* **2023**, *13*, 12560. [CrossRef]
45. Lara, F.L. Spatial Abstraction as a Colonizing Tool. In *The Routledge Companion to Decolonizing Art History*; Flores, T., San Martín, F., Black, C.V., Eds.; Routledge: London, UK, 2023; pp. 331–341.
46. Cardoso Llach, D.; Burbano, A. Other Computations: Digital Technologies for Architecture from the Global South. *Dearq* **2020**, *27*, 6–21. [CrossRef]
47. BuildingSMART International. IfcWall. Available online: https://standards.buildingsmart.org/IFC/DEV/IFC4_2/FINAL/HTML/schema/ifcsharedbldgelements/lexical/ifcwall.htm (accessed on 12 January 2026).
48. Kitayama, S.; Duffy, S.; Kawamura, T.; Larsen, J.T. Perceiving an Object and Its Context in Different Cultures: A Cultural Look at New Look. *Psychol. Sci.* **2003**, *14*, 201–206. [CrossRef] [PubMed]
49. Emmons, P. The Means and Meanings of Dashed Lines. In *Proceedings of the 89th ACSA Annual Meeting, Baltimore, MD, USA, 16–20 March 2001*; Association of Collegiate Schools of Architecture: Washington, DC, USA, 2001; pp. 458–463.
50. Javed, W.; Elmqvist, N. Exploring the Design Space of Composite Visualization. In *Proceedings of the 2012 IEEE Pacific Visualization Symposium, Songdo, Republic of Korea, 28 February–2 March 2012*; Hauser, H., Kobourov, S., Qu, H., Eds.; IEEE: Piscataway, NJ, USA, 2012. [CrossRef]
51. Gleicher, M.; Albers, D.; Walker, R.; Jusufi, I.; Hansen, C.D.; Roberts, J.C. Visual Comparison for Information Visualization. *Inf. Vis.* **2011**, *10*, 289–309. [CrossRef]
52. Garrido Martins, C.; Bogus, S.M.; Valentin, V. Quantitative Risk Assessment Model and Optimization in Infrastructure Fast-Track Construction Projects. *Infrastructures* **2023**, *8*, 78. [CrossRef]
53. Macek, D.; Vitásek, S. Risk Analysis in Building Renovations: Strategies for Investors. *Buildings* **2024**, *14*, 2219. [CrossRef]
54. Numan, M. BIM and Risk Management: A Review of Strategies for Identifying, Analysing and Mitigating Project Risks. *J. Eng. Res. Sci.* **2024**, *3*, 20–26. [CrossRef]
55. Lu, R.; Rausch, C.; Bolpagni, M.; Brilakis, I.; Haas, C.T. Geometric Accuracy of Digital Twins for Structural Health Monitoring. In *Structural Integrity and Failure*; Oyguc, R., Tahmasebinia, F., Eds.; IntechOpen: London, UK, 2021. [CrossRef]
56. Della Spina, L. Cultural Heritage: A Hybrid Framework for Ranking Adaptive Reuse Strategies. *Buildings* **2021**, *11*, 132. [CrossRef]
57. Wang, L.; Qian, B.; Cui, X.; Chao, L. Application of BIM Technology in Interference Check in Construction Work. In *Proceedings of the 2018 7th International Conference on Energy and Environmental Protection (ICEEP 2018), Shenzhen, China, 14–15 July 2018*; Kim, Y.-H., Ed.; Atlantis Press: Dordrecht, The Netherlands, 2018; pp. 116–123. [CrossRef]

https://doi.org/10.3390/infrastructures11020040

infrastructures

MDPI

Article

RACI–AHP–BIM Methodology in Projects with High Functional Complexity and Conservation Constraints

Urszula Kwast-Kotlarek [1] and Mariusz Szóstak [2,*]

[1] Central Administration Unit, Department of Investments and Renovations, University of Wroclaw, plac Uniwersytecki 1, 50-137 Wroclaw, Poland; urszula.kwast-kotlarek@uwr.edu.pl

[2] Department of Building Engineering, Faculty of Civil Engineering, Wroclaw University of Science and Technology, Wybrzeże Stanisława Wyspiańskiego 27, 50-370 Wroclaw, Poland

* Correspondence: mariusz.szostak@pwr.edu.pl

Abstract

The article presents an integrated RACI–AHP–BIM methodology that supports responsibility management, decision-making, and information management in complex construction projects delivered under the design–build model, with particular emphasis on conservation-orientated investments. The approach combines three complementary components: the RACI responsibility matrix, the analytic hierarchy process (AHP), and building information modeling (BIM). The methodology is validated on a higher-education conservation project using a BIM execution plan (BEP), scan-to-BIM procedures, and structured decision-making. The integration of RACI with BIM reduced accountability gaps and improved stakeholder coordination, while linking AHP with BIM data enabled data-driven design decisions using the BOCR model. The findings demonstrate measurable benefits, including clearer responsibility allocation, improved interdisciplinary coordination, and more transparent decision-making. The application of laser scanning and scan-to-BIM supported the creation of a digital model of historic elements for both design and future facility management. The main contribution is a holistic integration of RACI, AHP, and BIM into a unified methodology for conservation-orientated projects with high functional complexity, providing a reference framework for public-sector investment management.

Keywords: RACI matrix; analytic hierarchy process; building information modeling; design–build; conservation projects; decision-making

1. Introduction

1.1. Context and Problem Statement

Digital transformation in public construction projects requires the combined use of advanced management methods and information technologies to support decision-making [1,2]. Contemporary investment projects are carried out in environments that involve not only core project participants (client, designer, site manager, supervising inspector) but also a wide range of additional stakeholders [3,4]. In such settings, a clear allocation of responsibilities, the transparency of decision-making processes, and the efficient flow and integration of project information are critical.

Conservation projects in historic buildings require a simultaneous consideration of technical, functional, economic, and cultural constraints and involve a broader group of stakeholders, including heritage authorities [5,6]. This complexity necessitates the use of tools enabling the integration of responsibility management, decision-making processes, and the project information environment.

 https://doi.org/10.3390/infrastructures11030105

Building information model/modeling/management (BIM) has become a key component of the construction project delivery, supporting multidisciplinary coordination and data management [7,8]. However, BIM alone does not provide built-in mechanisms for responsibility management or structured decision-making, which are still handled through dispersed organizational and analytical tools [9,10].

In parallel, project management practice employs responsibility assignment tools such as the RACI matrix (Responsible, Accountable, Consulted, Informed), which structure relationships among project participants and increase organizational transparency [11]. Multi-criteria decision-support methods, particularly the analytic hierarchy process (AHP), are used to evaluate design alternatives and prioritize investment actions [12]. However, these tools typically operate independently of the BIM environment and formal information management structures [13].

The lack of integration between responsibility allocation (RACI), multi-criteria decision-making mechanisms (AHP), and the BIM information environment leads to fragmentation of the decision-making process [14]. This issue is particularly significant in conservation-orientated projects, where design and construction decisions have long-term technical and cultural consequences and require strict coordination among multiple stakeholders.

Previous studies on construction project management models have focused mainly on individual aspects: the RACI matrix has been analyzed in organizational management contexts, AHP has been applied to the selection of alternative construction investment, and BIM, according to ISO 19650 [15], has been implemented primarily as an information management system. The literature still lacks a holistic approach that integrates these three methodologies into a single coherent decision-support model, particularly in relation to projects implemented in heritage-protected buildings [16].

1.2. Aim and Scope of the Article

The purpose of this article is to develop an RACI–AHP–BIM model that integrates responsibility management, multi-criteria decision-making, and the BIM environment into a coherent methodology for conservation-orientated construction projects. The methodology was designed as a tool that supports the allocation of responsibility, the structuring of variant selection processes, and the integration of decisions with the digital information environment of the project.

The scope of the study includes the following:

- development of a conceptual model integrating RACI, AHP, and BIM into a unified project management framework;
- analysis of the structure of project roles and assigned responsibilities within the investment delivery process;
- identification of information requirements and risk mitigation mechanisms through the implementation of a BIM execution plan (BEP) according to ISO 19650;
- verification of the proposed approach based on a real case study of a design–build project implemented in a heritage-protected facility.

The reconstruction of the greenhouse complex at the Botanical Garden of the University of Wrocław serves as a validation case, illustrating the application of the proposed model in a project that combines conservation, functional, and technological requirements.

The reconstruction of the greenhouse complex at the Botanical Garden of the University of Wrocław, located on Sienkiewicza Street 23, represents a unique challenge that combines conservation requirements with modern project management methods. The facility was entered in the heritage register under No. A-2374/209/4 (5 February 1974) and is located in the Grunwaldzki Square area in Wrocław. The former greenhouse complex (Palm House–Victoria–Australia), established in the nineteenth century, deteriorated technically and was

https://doi.org/10.3390/infrastructures11030105

dismantled in 2013 by administrative decision due to its poor structural condition and lack of funding for conservation.

Only selected elements were preserved: fragments of the steel structure of the Australia greenhouse, including wrought-iron columns and parts of the steel roof skylight structure, as well as the stone basin of the Victoria greenhouse with four corner basins and masonry walls finished with stone copings. The reconstruction implemented in the design–build procurement model required the application of an integrated methodological approach that combined:

- the RACI matrix—to ensure clear allocation of responsibilities among four key stakeholders: the Client, the Contractor, the Designer, and the Users;
- the analytic hierarchy process (AHP) method—for multi-criteria evaluation of implementation alternatives considering conservation requirements, functional needs (culturing of tropical plants), and economic and organizational factors;
- Building information model/modeling/management (BIM) methodology according to ISO 19650—for effective project information management, multidisciplinary coordination, and integration with preserved historic structural elements retained in situ.

2. State of the Art

2.1. RACI Matrix in Construction Project Management

The RACI matrix (Responsible, Accountable, Consulted, Informed) is a fundamental tool for assigning responsibilities in construction projects, including those implemented in BIM environments characterized by complex stakeholder structures [17,18]. Its implementation in BIM projects enables the reduction in accountability gaps by 32% compared with traditional project management methods, increases communication transparency, and helps avoid redundant task assignments [19]. In the context of design–build projects, the RACI matrix gains particular importance due to the integrated responsibility of the contractor for both design and construction, which requires the precise definition of roles across project phases [20].

In conservation projects, the RACI matrix must incorporate additional stakeholders, such as the heritage conservation authority, acting as the key decision-maker (Accountable) regarding interventions in historic fabric, and specialist users (Consulted), e.g., botanists and collection curators in the case of a botanical garden, serving as consultants on functional requirements related to plant cultivation [21]. In academic projects, it is also necessary to take into account the multi-level administrative structure of universities, where investment decisions require approval at multiple administrative levels [22].

The RACI matrix constitutes a structured framework that assigns four levels of involvement for a combination of tasks and roles [23]. A project-specific RACI matrix enables:

- elimination of accountability gaps;
- avoidance of task redundancy (multiple or overlapping assignments);
- communication transparency within multifunctional teams.

In multidisciplinary BIM projects, the RACI matrix integrates traditional roles (designer, contractor, client) with specialized roles (BIM Manager, BIM Coordinator, Data Exchange Process Manager), ensuring the consistency of decision-making and coordination processes.

2.2. Analytic Hierarchy Process (AHP) in Investment Decision-Making

The analytic hierarchy process (AHP) structures complex decisions as hierarchies of objectives, criteria, and alternatives and supports the multi-criteria evaluation of design and technological options in construction projects [24]. In the conceptual stage, it helps align client requirements with technical feasibility and reduce interdisciplinary conflicts through explicit criteria and priorities [25].

https://doi.org/10.3390/infrastructures11030105

Research, including [22], has demonstrated the effectiveness of AHP in evaluating linear infrastructure investments in environmentally protected areas, confirming the usefulness of the method in projects requiring multi-criteria analyses of alternatives and consideration of environmental, technical, and economic factors.

The application of AHP in conservation projects requires the inclusion of criteria specific to historic buildings, such as the degree of preservation of historic fabric and architectural form [26], reversibility of conservation interventions [27], authenticity of applied materials and technologies [28], and potential for future conservation research [29].

In Polish investment practice, the BOCR model (Benefits, Opportunities, Costs, Risks) constitutes an extension of the classical AHP structure, allowing for a complete evaluation of investment alternatives considering both positive factors (benefits, opportunities) and negative factors (costs, risks) within a single analytical framework [17,19,30,31].

2.3. Building Information Model/Modeling/Management (BIM)

The BIM methodology, implemented according to the ISO 19650 standard, integrates digital modeling of the built asset with information management processes at all stages of the project lifecycle (concept, design, construction, operation) [32–34]. The ISO 19650 standard defines organizational frameworks, including organizational information requirements (OIR) [35], project information requirements (PIR) [36], asset information requirements (AIR) [37], and employer's information requirements (EIR) [38].

Employer's information requirements (EIR), particularly those orientated toward facility management (FM), are of particular importance. In specialized projects such as greenhouses for tropical plant cultivation, the BIM methodology enables precise management of microclimate parameters, irrigation systems, lighting that support photosynthesis, and maintenance schedules for technical equipment [26,27]. In conservation projects, BIM also integrates data on heritage values and the condition of historical fabric [39–41].

In the context of conservation projects, the application of BIM requires, among others, the integration of point clouds obtained from 3D scanning preserved historic elements with the design model [39]; parameterization of historic elements considering their condition, material, manufacturing technique, and conservation interventions [40]; definition of the levels of detail (LOD) appropriate to the level of knowledge about historic fabric [41], and management of heritage-value information within the IFC data structure [42].

In the Polish public sector, particularly in public procurement aligned with the PL BIM Standard (2020) [26], the BIM methodology includes elements such as multidisciplinary geometric modeling in IFC 4.0 [27], CDE (Common Data Environment) platforms for data exchange [38], clash detection procedures, and information quality control mechanisms.

The BIM execution plan (BEP) is a contractor's implementation document in response to the employer's information requirements (EIR), defining the strategy and procedures to achieve the BIM objectives within the project [43]. According to ISO 19650:2018 [15], BEP should include the following: the definition of BIM objectives and their impact on schedule and budget, modeling standards, rules for working within the CDE, quality control mechanisms, roles and responsibility matrices (RACI), data exchange procedures between stakeholders, as well as legal aspects, data security, training, etc.

When properly developed, the BEP constitutes a key document, as it defines the CDE platform for data exchange, data exchange formats (native and IFC), the levels of detail LOD/LOIN for individual project phases [30], interdisciplinary coordination and clash detection procedures, and the BIM team structure together with the responsibility matrix.

https://doi.org/10.3390/infrastructures11030105

2.4. Scan-to-BIM and Historic Building Information Modeling (HBIM) in Architectural Heritage Conservation

Scan-to-BIM and Historic BIM (HBIM) enable the precise digitalization of existing historic buildings [44].

Scan-to-BIM converts terrestrial laser scanning (TLS) and photogrammetric data into parametric BIM models, providing precise digital representations of existing assets. In conservation projects, it supports the accurate documentation of historic fabric through scanning, point cloud registration, parametric modeling, and preparation of as-built documentation [45].

Historic BIM (historic building information modeling) extends BIM to historic buildings by modeling characteristic architectural elements, allowing adaptations to measured dimensions and geometric irregularities, integrating historical data with 3D models, and supporting the planning of conservation intervention [46]. HBIM enables not only the representation of geometry and materials but also the preservation of information about heritage values, previous interventions, and historical phasing, thus supporting conservation planning [47].

The benefits of scan-to-BIM/HBIM include high accuracy, non-invasive data acquisition, rapid data collection, and access to hard-to-reach areas, while limitations are associated with equipment costs, the need for specialized expertise, and large data file size [48].

2.5. Research Gap and Scientific Contribution of the Article

The literature review indicates that previous studies have focused primarily on:

- the application of the RACI matrix in BIM projects for newly designed buildings, overlooking the specificity of conservation projects requiring inclusion of additional stakeholders (heritage conservation authorities, specialist users);
- the use of AHP in design–build projects without incorporating the BOCR model, which enables comprehensive evaluation of benefits, opportunities, costs, and risks within a single hierarchical structure;
- the implementation of BIM and BEP in conservation projects to a limited extent, the aspect of which requires the development of specific information requirements (EIR) that incorporate the management of heritage-value information, which remains an unexplored research area.

Therefore, it should be stated that there is a lack of studies that integrate the three proposed methodologies (RACI, AHP, BIM) into a comprehensive model for the implementation of construction projects that combine conservation, functional, and technological requirements. This article fills the identified research gap through the following:

- development of a RACI matrix for 21 processes in the construction, commissioning, and operation phases, taking into account the specificity of a design–build project implemented in a conservation-protected facility;
- application of the AHP–BOCR model for the multi-criteria evaluation of implementation alternatives, including conservation, functional, economic, and organizational criteria;
- development of the employer's information requirements (EIR) and a BIM execution plan (BEP) according to ISO 19650 for a conservation project, integrating point clouds with the design model;
- validation of the proposed model based on a real case study involving the reconstruction of a historic greenhouse complex in the Wroclaw Botanical Garden.

The proposed RACI–AHP–BIM model constitutes an innovative tool that supports decision-making and information management in conservation projects, with potential for adaptation to other complex projects characterized by a high degree of functional and organizational complexity.

https://doi.org/10.3390/infrastructures11030105

3. Research Methodology: RACI-AHP-BIM Methodology

3.1. Assumptions of the RACI-AHP-BIM Methodology

The proposed RACI–AHP–BIM methodology was developed as an integrated model to manage projects characterized by a high level of complexity, particularly conservation projects implemented in the design–build delivery model. Its fundamental premise is the integration of three complementary dimensions of project management (methodological domains):

- a management system represented by the RACI matrix;
- a decision-making process based on the AHP method;
- information management implemented within the BIM environment according to ISO 19650.

The developed methodology was intentionally designed as a methodological framework prior to project implementation and was not the result of ex post analysis. The case study serves a validation function, providing evidence of the applicability of the model in project practice.

The integration of the methods is systemic in nature: the RACI matrix structures process responsibilities, the AHP method supports the selection of alternatives and prioritization of actions, while the BIM methodology provides an information environment enabling the operationalization of decisions and monitoring of project implementation.

The RACI–AHP–BIM methodology is based on a three-layer functional structure comprising the organizational, decision-making, and informational layers.

1. Organizational layer (RACI matrix): Within the organizational layer, project roles are identified, responsibilities are assigned, and decision-making relationships among participants in the investment process are defined. The RACI matrix performs the function of the project's management and decision-making structure and forms the basis for mapping project processes.
2. Decision-making layer (AHP–BOCR method): In this layer, the processes for selecting design, technological, and organizational alternatives are modeled using the AHP structure in the BOCR framework, enabling a balanced evaluation of alternatives in the context of conservation, functional, technological, and economic requirements.
3. Informational layer (BIM–ISO 19650): BIM constitutes an environment that integrates project data, geometric models, information requirements, and management documentation. The BIM execution plan (BEP), the employer's information requirements (EIR), and the CDE ensure the flow of information among project participants and data quality control. The interconnection of these three layers is iterative and interdependent.

The responsibility structure defined in the RACI matrix influences the generation and approval of project information within BIM; decisions made using AHP modify information requirements and the scope of project analyses; and data generated within the BIM environment provide a basis for updating both responsibility assignments and decision-making priorities. As a result, the methodology operates as a project management system based on continuous information flow between the organizational structure, the decision-making process, and the digital environment.

3.2. Research Steps

The study consisted of four steps.

Step I—Identification of processes and roles (RACI matrix), including:

- analysis of source documents, including EIR, BEP, and design documentation (architectural and construction design, technical design, site development design);

- identification of project processes across the phases of construction, commissioning, and operation;
- definition of the stakeholder structure and assignment of R/A/C/I responsibilities;
- mapping of BIM processes onto the responsibility structure.

Step II—Development of the decision problem structure (AHP/BOCR method):

- decomposition of project problems into a hierarchical structure;
- identification of criteria and sub-criteria within the BOCR model;
- pairwise comparisons according to the Saaty scale (1–9);
- determining priority weights for implementation alternatives.

Step III—Integration with the BIM methodology:

- analysis of information requirements (EIR) and the BIM execution plan (BEP);
- identification of key data delivery points (PDD);
- analysis of levels of detail (LOD/LOIN);
- integration of design models and scanning data.

Step IV—Risk analysis:

- identification of categories of project risks;
- mapping of risks into RACI processes and BIM information requirements (EIR);
- use of AHP results to prioritize mitigation actions.

3.3. Research Object

The study adopted a case study approach using the triangulation of data sources. The analysis included project documentation, EIR and BEP documentation, the development of a RACI matrix covering 21 processes and 80 responsibility assignments, the construction of an AHP–BOCR decision model, and the performance of risk and BIM process analyses. This approach enabled a simultaneous capture of the organizational, decision-making, and informational aspects of the project and evaluation of their interdependencies.

The case study concerns the reconstruction project of the Palmiarnia–Wiktoria–Australia greenhouse complex in the Botanical Garden of the University of Wrocław, implemented under conditions of complex conservation, environmental, and technological constraints. The Botanical Garden area was entered into the register of historic monuments under no. A-2374/209/4 dated 5 February 1974 and is located within a conservation protection zone and an archeological heritage protection zone, according to Local Development Plan No. 362 (Resolution No. XXXVII/855/13 of the Wrocław City Council of 17 January 2013).

The 19th-century greenhouse complex consisted of four parts: the Palm House, Australia, Wiktoria, and a connector. Due to poor technical conditions and a lack of funding, it was solved in 2013. The preserved elements included wrought-iron columns and roof fragments of the Australia section, the stone basin of the Wiktoria section, and the concrete slab over the underground heating node.

In the implemented project, conservation requirements include the following:

- preservation of historic wrought-iron columns of the Australia section as decorative elements;
- preservation of stone cappings of the Wiktoria walls and their conservation prior to reinstallation;
- reconstruction of the historic architectural form: Palm House (cascading roof), Australia (triple-pitch skylights), Wiktoria (central pool basin);
- visual differentiation of historic elements from the new structural components;
- conducting rescue archeological investigations.

Environmental protection requirements include the following:

- protection of two natural monuments: ginkgo trees;

- protection of valuable species: Atlas cedar and American beech.

The project specificity includes the necessity to:

- preserve and incorporate historic wrought-iron columns as decorative elements into the new steel structure;
- reconstruct the historic architectural form of three greenhouses (Palm House—single-pitch cascading roof, Australia—triple-pitch roof skylights, Wiktoria—central pool with aquatic vegetation),
- meet technical requirements for tropical plant cultivation (minimum temperature 23 °C, humidity 70–80%, double-glazed Optiwhite glazing with Ug = 0.6 W/m^{2}K);
- ensure compliance with the requirements of the Voivodeship Heritage Conservation Officer and the Department of Environment (protection of two natural monuments—ginkgo trees no. 16A and 16B, and protection of old-growth trees—Atlas cedar and American beech);
- implement the project within an operating Botanical Garden.

The complexity of these conditions makes the project an appropriate environment for verifying the integrated RACI–AHP–BIM approach.

The study has typical limitations for a single case. It focuses on one facility at the executive design stage, so data from construction and operation were not available. Access to detailed financial data was limited, and AHP assessments relied on expert judgment, introducing subjectivity. Despite these limitations, the study provides an empirical basis for evaluating the functioning of the proposed methodology and its potential adaptation in other complex construction and conservation projects.

4. Research Results

4.1. Structure of the Project Team–RACI Matrix

As a result of applying the RACI matrix, 80 responsibility assignments related to 21 project processes were identified. Table 1 presents the developed RACI matrix. The following roles and types of responsibility were identified:

- Responsible (R): 23 assignments;
- Accountable (A): 18 assignments;
- Consulted (C): 22 assignments;
- Informed (I): 17 assignments.

Total number of assignments of responsibility: 80.
Role abbreviations:

- UWr—Investor (University of Wrocław)
- Contractor—Design–build contractor
- Designer—Design team (part of the contractor's team)
- BG—End user (Botanical Garden of the University of Wrocław)
- External stakeholders, such as the heritage conservation authority (Provincial Heritage Conservation Officer-WKZ), were assigned to specific processes (e.g., Task 5: "Obtaining conservation approval"—A: Investor, R: Contractor, C: Designer, I: Facility User) but were not included as formal columns in the matrix. The WKZ acts as an administrative authority outside the design–build contract, issuing mandatory decisions without direct operational responsibilities within the contractual responsibility chain.

https://doi.org/10.3390/infrastructures11030105

Table 1. RACI Matrix—Project Roles and Responsibility Assignments.

No.	Process/Task	Investor (UWr)	Contractor	Designer	End User (BG)
	Implementation phase (7 processes)				
1	Definition of functional requirements of the facility	**A**	C	C	**R**
2	Development of the architectural concept	C	**A**	**R**	C
3	Preparation of multidisciplinary design documentation	I	**A**	**R**	C
4	Analysis of technical alternatives (AHP method)	**A**	**R**	**R**	C
5	Obtaining conservation approval (Heritage Officer)	**A**	C	**R**	I
6	Preparation of the BIM Execution Plan (BEP)	C	**A**	**R**	I
7	3D scanning of historic elements (scan-to-BIM)	I	**A**	**R**	C
	Commissioning phase (4 processes)				
8	Verification of compliance between construction and design	**A**	**R**	C	C
9	Quality control of the BIM model (clash detection)	I	**A**	**R**	I
10	Technical acceptance of the facility	**A**	**R**	C	**R**
11	Handover of as-built documentation	**A**	**R**	**R**	I
	Operation phase (10 processes)				
12	Preparation of HVAC system operation manuals	C	**A**	**R**	C
13	Training of BG personnel in the operation of technical systems	I	**R**	C	**A**
14	Management of microclimate parameters (temp. 23 °C, humidity 70–80%)	I	C	I	**A**
15	Monitoring of the technical condition of historic elements	C	C	I	**A**
16	Maintenance of irrigation systems for tropical plants	I	C	I	**A**
17	Planning of technical inspections according to BEP	**A**	**R**	C	**R**
18	Updating the BIM model (LOD 500) for facility management	C	**R**	**R**	C
19	Reporting failures and corrective interventions	I	**R**	C	**A**
20	Management of plant collections (cultivation requirements)	I	I	I	**A**
21	Archiving operational data in the CDE	**A**	**R**	C	**R**

The RACI matrix underwent a two-stage internal consistency validation to ensure organizational rigor:

- Completeness check: Each of the 21 processes was verified to ensure the presence of at least one Responsible role (R) responsible for task execution and one Accountable role (A) acting as the decision owner. This procedure eliminated potential accountability gaps. The criterion was satisfied for all tasks, resulting in 23 R assignments and 18 A assignments (Table 2).
- Redundancy minimization: For each task, the number of Responsible roles was limited in order to avoid overlapping responsibilities. As a standard rule, a single R role

https://doi.org/10.3390/infrastructures11030105

was assigned. Multiple R roles (>1) were allowed only for inherently collaborative processes within the design–build (DB) delivery model, such as Task 4 ("Analysis of technical alternatives using the AHP method") and Task 18 ("Updating the BIM model to LOD 500 for facility management"), where the Contractor and Designer shared operational responsibility while maintaining a single Accountable role. This logic is consistent with established design–build project management practices.

Table 2. Statistical summary.

Role	R	A	C	I	Total	Engagement Profile
UWr (Investor)	1	9	2	9	21	DECISION-MAKER (A: 40%, I: 43%)
Contractor	10	6	4	1	21	IMPLEMENTER (R: 48%, A: 29%)
Designer	8	1	9	3	21	TECHNICAL ADVISOR (C: 43%, R: 38%)
BG (End user)	4	2	7	4	17	INFORMATION RECIPIENT (C: 41%, I: 24%)
TOTAL	**23**	**18**	**22**	**17**	**80**	—

After validation, no accountability gaps or excessive redundancy were identified, confirming the suitability of the matrix to coordinate complex stakeholder structures.

The RACI structure confirms the dominant role of the investor as the entity ultimately responsible for key strategic decisions. The high number of Accountable (A) assignments for the University of Wrocław shows that functional requirements, organizational frameworks, and design solutions remain under its authority, which is typical for public clients. At the same time, the significant share of Informed (I) roles on the investor's side suggests a deliberately shaped position of an "information guardian" who does not perform operational tasks but supervises their progress and results.

The high share of Responsible (R) and Accountable (A) roles for the contractor indicates a concentration of responsibility for the preparation, coordination, and implementation of technical solutions, reflecting the design–build delivery model. At the same time, the presence of Consulted (C) roles in areas related to functional and conservation requirements reflects the need to coordinate technical solutions with specialized stakeholders, including the end user and heritage protection authorities.

The assignment profile for the designer is characterized by a predominance of Consulted (C) roles while maintaining a significant share of Responsible (R) roles for the preparation of documentation and BIM models. This configuration indicates that the designer performs the function of a specialized technical advisor, whose expert knowledge is used both at the stage of formulating design solutions and in assessing their consequences within multi-criteria analyses (AHP–BOCR). The limited number of Accountable (A) roles on the designer's side confirms that the ultimate responsibility for decisions lies with the investor and the contractor, while the designer co-shapes these decisions by providing structured technical information.

The configuration of the RACI matrix indicates that the end user, the Botanical Garden, primarily performs the Consulted (C) and Informed (I) roles during the implementation phase, while its importance increases during the operational phase, where the Responsible (R) and Accountable (A) roles emerge in processes related to microclimate management, infrastructure maintenance, and facility use organization. This distribution of roles reflects a typical model for conservation projects, in which the user is not responsible for structural and construction solutions but assumes the key responsibility for their proper use and maintenance according to the intended functional program.

The balanced distribution of R, A, C, and I roles and the absence of tasks without Accountable (A) assignments indicate effective mitigation of so-called accountability gaps,

https://doi.org/10.3390/infrastructures11030105

typical for complex projects involving multiple stakeholders. The dominance of active participation (R + C) over passive participation (A + I) suggests that the project process was designed as collaborative, focusing on information exchange and shared decision-making rather than a linear command flow. This is particularly important in the context of projects with high functional and conservation complexity, where the quality of decisions depends on integrating technical, operational, and conservation perspectives.

In the analyzed project, the RACI matrix performed the function of a superior organizational and decision-making structure that was consistently reflected in the BIM environment and the CDE platform. The assignment of the roles R, A, C, and I to the processes of generating, verifying, and approving information in the BEP enabled clear identification of responsibility for data quality and completeness at specific data delivery points (PDD). Thus, the RACI matrix served not only as an organizational tool but also as a foundation for the project's information governance, ensuring the coherence of integration with the AHP methodology and scan-to-BIM procedures.

The largest share in the RACI matrix consisted of Responsible (R) and Consulted (C) assignments, accounting for 23 (29%) and 22 (28%) entries, respectively. Accountable (A) roles were recorded in 18 cases (23%), while Informed (I) roles appeared in 17 cases (21%). This means that more than half of the matrix (R + C = 56%) concerns the active engagement of project participants rather than passive information flow alone (A + I = 44%), which is characteristic of complex projects implemented in the design–build model for construction undertakings. The distribution of responsibilities is presented in Figure 1; Figure 2 shows the percentage share of engagement types; and Figure 3 presents the distribution of RACI roles by stakeholder.

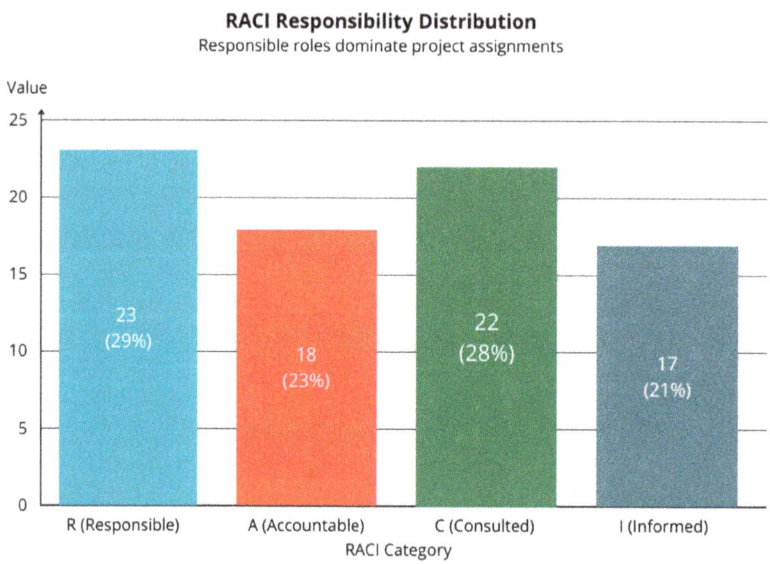

Figure 1. Distribution of responsibilities in the RACI matrix by engagement type.

https://doi.org/10.3390/infrastructures11030105

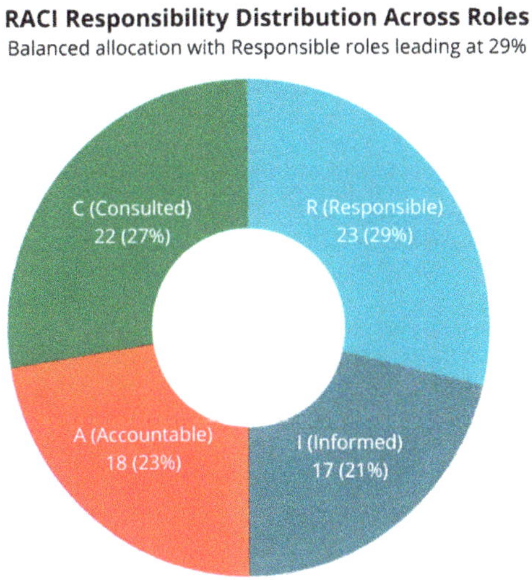

Figure 2. RACI matrix: Percentage share of engagement types.

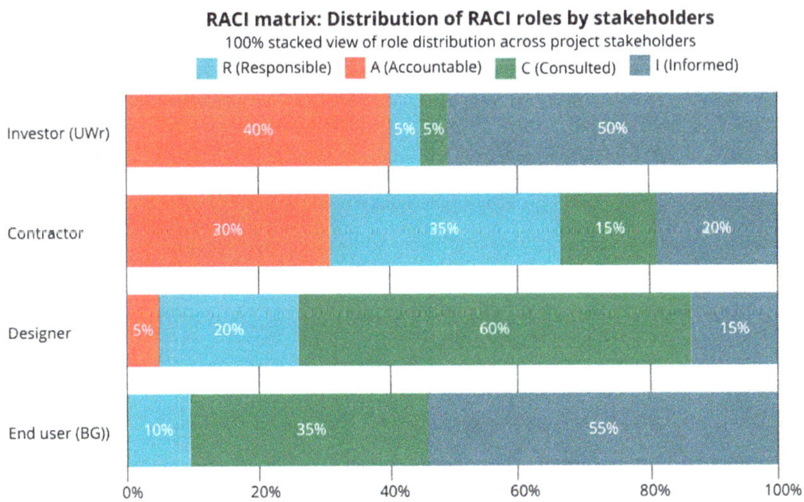

Figure 3. RACI matrix: Distribution of RACI roles by stakeholders.

Distribution of all RACI roles:

- I (Informed): 17 (21%)—dominance of information flow
- C (Consulted): 22 (28%)—consultative structure
- A (Accountable): 18 (23%)—decision-making structure
- R (Responsible): 23 (29%)—execution structure

The distribution of RACI roles by stakeholders shows differentiated engagement profiles: the investor performs a decision-making function (dominance of A and I roles), the contractor an implementation function (predominance of R and A roles), the designer an advisory function (dominance of C roles), and the end user acts primarily as an information recipient and co-creator of functional requirements (predominance of I and C roles).

The distributed responsibilities obtained (presented in Figure 1) indicates a collaborative character of the investment process with active participation of all parties—from

https://doi.org/10.3390/infrastructures11030105

the investor, through the heritage conservation authority, to the specialists of the Botanical Garden—typical for complex projects implemented in the design–build model. The responsibility structure promotes participation in the decision-making process and limits the concentration of responsibility within a single entity. Figure 4 presents a comparison of active versus passive engagement.

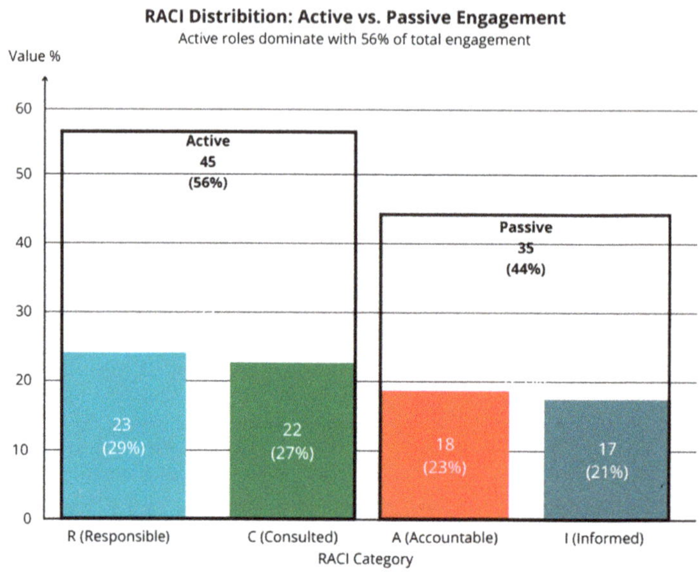

Figure 4. RACI matrix: Division into active vs. passive engagement of project participants.

Analysis of stakeholder roles showed that Accountable (A) functions were mainly assigned to the investor and the heritage protection authority (heritage conservation officer), while Responsible (R) functions were concentrated on the designer and the construction contractor. End users (Botanical Garden) primarily performed Consulted (C) and Informed (I) roles, particularly in the areas of functional, cultivation, and operational requirements.

Based on the analyses conducted, the identified structure indicates:

- an even distribution of responsibilities (no dominance of any single group);
- clear assignment of decision-making roles (elimination of accountability gaps—each task has both R and A);
- structured communication relationships.

The RACI matrix constituted the basis for further mapping of BIM processes and AHP decision structures.

4.2. Information Management Structure—BIM Execution Plan (BEP)

The results of the BIM environment analysis showed that information management was organized according to the employer's requirements (EIR) and the ISO 19650 standard. The BIM execution plan (BEP) includes seven main information management domains that define:

- the structure of project data;
- procedures for their generation and validation;
- data delivery points;
- collaboration rules within the CDE.

Table 3 presents the structure of the document and the defined information management domains.

https://doi.org/10.3390/infrastructures11030105

Table 3. BEP structure and information management domains.

BEP Area	Content	Document Form	Responsible
1. BIM Objectives	Increase management efficiency, reduce design errors (clash detection), improve interdisciplinary collaboration, prepare for FM LOD 6.	Text description + objectives table	BIM Manager
2. Modeling Standards	IFC 4.0, LOIN (LOD 3 → 4 → 6 according to EN 17412), GIS catalogs, element attributes, parametric relationships.	Technical specification + schedules	BIM Modeler + Lead Designer
3. CDE Work Rules (DALUX)	Workflow: WIP → Shared → Published → Archived, naming convention, backup (10 years), access control, version management.	Operational procedures	CDE Coordinator
4. Quality Control Procedures (MQC)	Model verification after each update, clash detection at least once per month, data audit, non-conformance reports.	Quality Control Plan (QCP)	Clash Detection Specialist
5. Roles and Responsibilities	RACI matrix 21 × 80, escalation processes, interdisciplinary communication, coordination meetings every 2 weeks.	RACI matrix + CP/RM procedures	BIM Coordinator
6. Data Security	PN-EN ISO/IEC 27001 [49], IP confidentiality, archiving, encryption, role-based access control.	ISMS policy	IT/CDE Manager
7. Training and Competencies	4 DALUX training sessions, BIM certification (minimum 40 h), tool updates.	Training plan (CDP)	HR + BIM Manager

In the project, a common data environment (CDE) platform compliant with ISO 19650-2 was implemented, serving as a central data repository for all project participants. The CDE platform was divided into four zones:

- WIP (Work in Progress)—working area for design teams;
- SHARED—area for consultation and verification;
- PUBLISHED—approved documentation area;
- ARCHIVE—historical version archiving area.

The implementation of the CDE contributed to the following:

- reduction in document search time by 65%;
- elimination of work on outdated versions (100% version consistency);
- reduction in document approval time by 40%;
- increased transparency of the decision-making process (complete history of changes);
- statistical data on the use of the CDE platform are presented in Table 4.

The values presented in Table 4 are derived from workflow analysis and repeated measurements of task execution time, covering both the period prior to the implementation of the common data environment (CDE) (based on the reference project of the project team) and the implementation phase within the analyzed project.

https://doi.org/10.3390/infrastructures11030105

Table 4. Statistical data on the use of the CDE platform in the project.

Category	Indicator	Value	Data Type	Comment
CDE structure	Number of zones in CDE	4	Measured [1]	WIP, SHARED, PUBLISHED, ARCHIVE—according to ISO 19650 and BEP
Access to information	Average time to locate a required document	45 → 5 min (89%)	Measured (CDE)	Reduction of approx. 45 min in the traditional model (approx. 89% time savings)
Approval process	Reduction in documentation approval time	approx. 40%	Measured (CDE)	Compared with projects without formalized CDE/BEP (based on team experience)
Interdisciplinary coordination	Number of clashes detected in the BIM model (pre-construction)	61 (8 hard, 4 soft, 49 clearance)	Measured (Navisworks reports)	8 hard, 4 soft, 49 clearance—identified in the CDE/BIM environment
Interdisciplinary coordination	Actual cost of resolving clashes (in the project)	approx. 12,000 PLN	Measured (order records)	Corrective work performed at the design stage in the CDE
Interdisciplinary coordination	Estimated cost savings	approx. 168,000 PLN	Expert estimate [2]	Difference between the traditional scenario and the implemented RACI–AHP–BIM + CDE methodology; PLN 2500/collision on site
Error reduction	Estimated	60–80%	Literature + experience	Literature, projects ref.

[1] Measured = system actions/reports; Expert estimate = benchmarks + team experience. [2] Scenario: costs of resolving conflicts on the construction site (team reference projects).

The measurements were based on 150 document retrieval operations performed on the Dalux platform, with average task durations of approximately 45 min in the traditional document management model and 5 min in the CDE-based workflow. The analysis was not based on a single survey but on a combination of CDE system logs (document publication and retrieval records) and benchmarks from previous projects executed without a formalized CDE, validated by the project team.

The reported indicators (e.g., approximately 89% reduction in document search time and approximately 40% reduction in approval time) should therefore be interpreted as project-based estimates derived from CDE logs and operational experience, rather than full statistical results aggregated across multiple independent projects. A limitation of the analysis is the lack of operational-phase data and the use of a single validation case study.

In Table 4, a clear distinction is made between measured data and expert estimates. Measured data include the number of detected clashes (61, Navisworks reports) and operation times within the CDE (document search: 45 → 5 min; approval time 40% shorter—Dalux workflow operations), as well as actual costs of design corrections (approximately PLN 12,000—based on change registers).

Expert estimates and hypothetical scenarios refer to potential savings (approximately PLN 168,000—estimated cost of resolving clashes during construction based on benchmark data) and error reduction levels (60–80%—based on the literature and the team's experience from reference projects). A limitation of the study is the lack of financial data from the construction and operational phases, as well as the use of a single validation case study.

https://doi.org/10.3390/infrastructures11030105

Linking the BEP structure with the RACI matrix allowed the assignment of responsibility for generating, verifying, and approving project information to specific participants in the investment process. As a result, the BIM environment performed an operational function in relation to the organizational layer, ensuring the practical implementation of assigned roles.

4.3. Implementation of the Analytic Hierarchy Process (AHP): Selection of Technical Alternatives

The application of the AHP method in the analyzed project aimed to structure the process of selecting technical alternatives for the glass façade of the greenhouse complex, taking into account simultaneous conservation, functional, economic, and organizational requirements. Decision problems were represented as a hierarchy comprising four levels:

1. main objective—selection of the façade alternative;
2. main criteria groups within the BOCR model (Benefits, Opportunities, Costs, Risks);
3. detailed sub-criteria;
4. analyzed technical façade alternatives.

Three technical alternatives were adopted for the evaluation:

- Variant A—stainless steel frame + tempered safety glass;
- Variant B—painted carbon steel frame + laminated glass;
- Variant C—prefabricated modular system.

At the level of BOCR criteria groups, pairwise comparisons were conducted using the Saaty scale (1–9), involving experts representing the investor, designer, contractor, user, and heritage conservation authority. The global weights obtained for the BOCR groups were as follows:

$$w_B = 0.35 \text{ (Benefits)};$$

$$w_O = 0.25 \text{ (Opportunities)};$$

$$w_C = 0.25 \text{ (Costs)};$$

$$w_R = 0.15 \text{ (Risks)},$$

which reflects the priority assigned to conservation and functional aspects while simultaneously considering costs and implementation risks.

For each technical alternative, pairwise comparison matrices were constructed with respect to individual BOCR groups, and local priority vectors were determined. In the next step, these values were aggregated into a BOCR-type decision index calculated according to the formula:

$$\text{Score} = w_B \cdot B + w_O \cdot O - w_C \cdot C - w_R \cdot R, \quad (1)$$

where B, O, C, R denote normalized evaluation values of alternatives in the groups of benefits, opportunities, costs, and risks.

In the baseline analysis, with the above set of weights, Variant B achieved the highest aggregated result (Score = 0.145), indicating its advantage as a compromise between relatively lower investment costs and an acceptable level of implementation risk. Variant A and Variant C obtained score values close to zero (approx. 0.002 and −0.002, respectively), indicating no clear advantage in the baseline proportions between benefits, costs, and risks.

However, a key element of the decision-making process was the sensitivity analysis of AHP results to changes in the weights of the criterion, conducted during stakeholder workshops. Increasing the importance of conservation-related sub-criteria (including the degree of preservation of historic fabric, material compatibility with cast-iron elements, and reversibility of interventions) and long-term risk criteria led to a shift in preference toward Variant A. Variant B remained economically more advantageous, whereas Variant

https://doi.org/10.3390/infrastructures11030105

A, despite higher initial costs, ensured better protection of heritage values and greater durability with a lower risk of future repair interventions.

Therefore, the final selection of Variant A as the solution adopted for implementation represents a deliberate departure from the "economically optimal" alternative in favor of a "conservation-safer" alternative, consistent with the priorities of the client and the requirements of the heritage protection authority. The AHP method, supported by quantitative and qualitative data derived from BIM models and conservation documentation, enabled transparent documentation of this compromise and clear communication to all participants in the decision-making process.

For the purpose of synthetic presentation of the results in the article, Table 5 presents the aggregated AHP priority values for the alternatives analyzed and their decision status (preferred, alternative, rejected), Table 6 presents the threshold for weight changes in the AHP-BOCR sensitivity analysis, while Figures 5 and 6 present, respectively, the distribution of alternative priorities and the score values for the BOCR model.

Table 5. Results of the AHP analysis—selection of technical alternatives for the glass façade of the greenhouse complex. AHP priorities for variants A, B, and C after workshops and sensitivity analysis.

Technical Alternative	Weight (AHP Priority)	Status and Justification
Variant A: Stainless steel frame + tempered safety glass	0.42 (42%)	PREFERRED—Highest conservation compatibility, durability > 50 years, compliance with heritage authority requirements
Variant B: Painted carbon steel frame + laminated glass	0.35 (35%)	ALTERNATIVE—Lower costs (~20%), lower durability (30–40 years), requires maintenance every 5 years
Variant C: SCHÜCO modular prefabricated system	0.23 (23%)	REJECTED—Lack of compatibility with historic cast-iron elements, unacceptable to the heritage authority

Table 6. Threshold for weight changes in AHP-BOCR sensitivity analysis.

Scenario	Changed Key Weights	Variant A	Variant B	Variant C	Rank
Baseline (economic)	B = 0.35; O = 0.25; C = 0.25; R = 0.15	0.002	0.145	–0.002	B > A > C
Conservation (±15% on B/R maintenance)	B = 0.42; O = 0.22; C = 0.20; R = 0.16	0.152	0.140	–0.010	A > B > C
Cost (±15% on C/O)	B = 0.30; O = 0.28; C = 0.30; R = 0.12	–0.015	0.132	0.008	B > C > A
Reversal threshold (conservation > 55–60%)	Maintenance criterion+risk > 0.55	>0.145	<0.145	–	A > B

Note: Reversal of rank (A > B) occurs when the combined weight of conservation and risk criteria (B/R) exceeds approximately 55–60%, while the weight assigned to cost criteria decreases below approximately 20–22%. Under these conditions, Variant A remains stable across a relatively wide range of weight variations (approximately ±15%).

The final selection of Variant A was made after a deliberate adjustment of priorities by the Investor and the heritage conservation authority (WKZ) during project coordination meetings, favoring the conservation-focused scenario presented in Table 6.

This decision does not contradict the AHP methodology. In contrast, the sensitivity analysis confirms that AHP functions as a decision-support tool that flexibly responds to changes in weighting assumptions aligned with the strategic objectives of the project.

https://doi.org/10.3390/infrastructures11030105

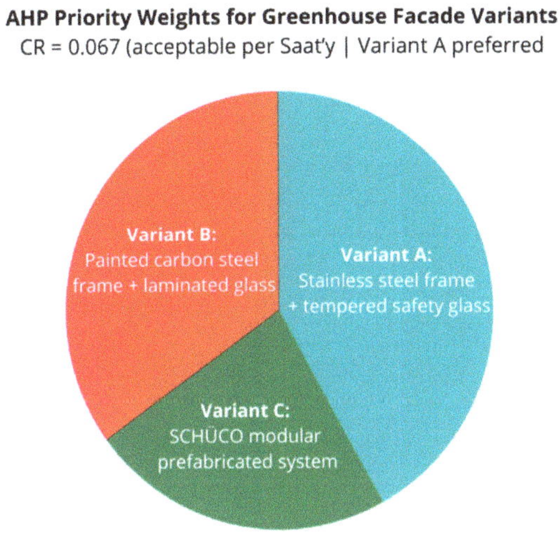

Figure 5. Results of the AHP analysis—priorities of technical alternatives for the glass façade (CR = 0.067).

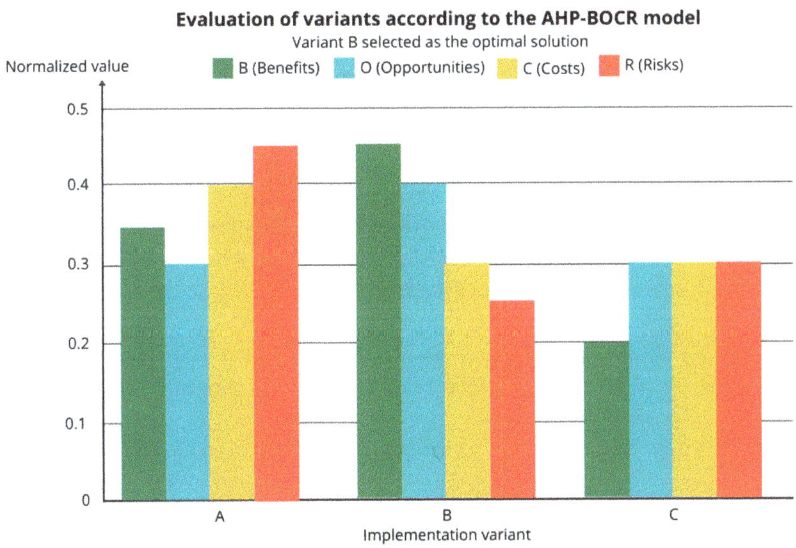

Figure 6. Evaluation of AHP–BOCR.

The baseline result (preference for Variant B) primarily reflected an economic perspective, while the final priorities adopted in Table 5 incorporated heritage conservation requirements and long-term operational considerations.

Values represent the geometric mean of the judgments of from five experts. All aggregated matrices satisfy the CR < 0.10 consistency requirement, while two individual expert matrices exceeding CR > 0.12 were recalibrated during a second evaluation round.

Table 7 presents example pairwise comparison matrices used in the AHP analysis. The procedure for aggregation and verification of expert judgment involved five experts representing the Investor, Designer, Contractor, User of the Facility, and the Heritage/Environmental Authority (WKZ).

https://doi.org/10.3390/infrastructures11030105

Table 7. Example pairwise comparison matrices in AHP (Saaty 1–9 scale).

(a) BOCR Level Matrix (5 Experts—Geometric Mean Aggregation)					
BOCR	**Benefits (B)**	**Opportunities (O)**	**Costs (C)**	**Risks (R)**	**Local Priority**
Benefits (B)	1	3	5	7	0.642
Opportunities (O)	1/3	1	3	5	0.248
Costs (C)	1/5	1/3	1	3	0.088
Risks (R)	1/7	1/5	1/3	1	0.022
Column sum	1.67	4.53	9.33	16.00	CR = 0.045
(b) Sub-Criterion Matrix: "Compatibility with Historic Fabric" (Benefits Group)					
Sub-criterion B	Durability	Authenticity	Reversibility	Priority	
Durability	1	1/3	5	0.385	
Authenticity	3	1	7	0.579	
Reversibility	1/5	1/7	1	0.036	
CR = 0.062				1.000	

Expert judgments were aggregated using the geometric mean of individual pairwise comparisons, which represents the standard aggregation method of decision-making of the AHP group according to Saaty. Each expert independently completed pairwise comparison matrices using the Saaty scale 1–9 for the four-level hierarchy (BOCR → sub-criteria → alternatives).

Individual consistency was verified using the criterion CR < 0.12 for each expert matrix. Two individual assessments that exceeded this threshold (CR = 0.15 and CR = 0.14) were rejected and recalibrated through a second evaluation round after clarification of the inconsistencies.

For the aggregated group matrices, the consistency was verified using the stricter threshold CR < 0.10 at each hierarchy level, generating the following results:

- BOCR level: CR = 0.045
- Sub-criteria matrices: average CR = 0.062
- Alternative comparison matrices: CR = 0.067

The fourth stage of the procedure involved the calculation of synthetic priorities by multiplying local priorities through the hierarchy, according to Equation (1). The overall consistency ratio of the model (CR = 0.067) confirms an acceptable level of consistency in the final decision model.

The AHP model supported the process as an analytical tool, responding to a deliberate shift in priorities introduced by stakeholders. Figure 7 presents the scheme of the decision-making process that was applied.

As shown in Figure 7, the decision path was iterative, starting from the baseline BOCR result (preference for Variant B under economically oriented weights), followed by sensitivity analysis, stakeholder workshops, and finally leading to the adjusted AHP priorities presented in Table 5. The final selection of Variant A resulted from a conscious shift in priorities by the Investor and the Provincial Heritage Conservation Officer (WKZ) toward a "conservation-focused" scenario. This outcome is consistent with the role of AHP as a decision-support tool, rather than an automatic ranking mechanism.

https://doi.org/10.3390/infrastructures11030105

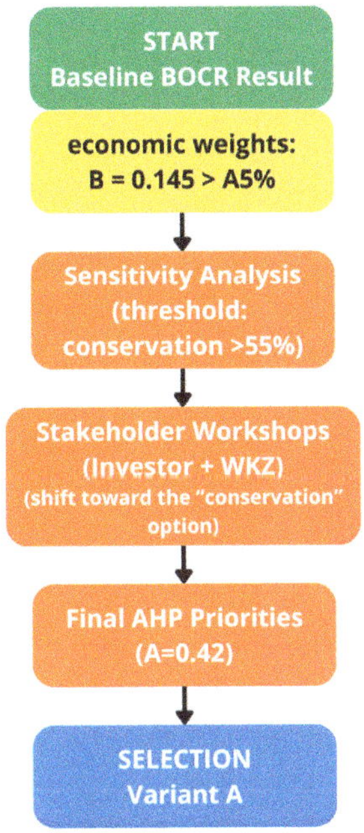

Figure 7. Decision path from the BOCR baseline result to the selection of option A.

4.4. Integration of Spatial Data: Scan-to-BIM

As part of the preparation of historical documentation for the preservation of heritage elements of the Australian system (cast-iron columns and beams), terrestrial laser scanning technology (TLS) (Leica ScanStation P50) was applied, enabling the acquisition of high-resolution geometric data of preserved structural fragments with an accuracy of ±3–5 mm (in according to ISO 17123 [50]). The obtained point clouds (eight scans, 360° coverage, resolution of 5 million points) were subjected to registration, filtering, and modeling processes. Point cloud data was stored in the E57 format (ASTM E2807), an open standard for laser scanning data exchange. The BIM model was shared in the IFC 4.0 format, compliant with buildingSMART standards [51], enabling software-independent information exchange among project participants.

The scan-to-BIM process included the following:

- point cloud processing (registration, noise filtering, cleaning);
- geometry extraction (semi-automatic modeling in CloudCompare);
- enrichment of the model with material attributes and conservation information;
- integration with the coordination model (IFC 4.0).

A level of detail (LOD) of 4 was achieved for historic elements (geometric accuracy + full attribute documentation), and a data structure was prepared that allowed the development of the model to an LOD of 6 (as-built) for facility management and conservation purposes.

During the geometric extraction process, differential tolerances appropriate for the project phase were applied. Table 8 presents the objective quality indicators for reconstruction. Manual segmentation in CloudCompare used a tolerance of approximately ±2 mm for precise edges (e.g., profiles of cast-iron columns), while semi-automatic extraction procedures (Segment/Mesh operations) used tolerances of approximately ±6 mm for irregular surfaces such as deformations or welded joints.

https://doi.org/10.3390/infrastructures11030105

Table 8. Objective quality indicators of the scanning-to-BIM reconstruction.

Quality Indicator	Value	Measurement Method	Accepted Tolerance	Comment
Point cloud registration error (RMS)	3–5 mm	CloudCompare (ICP registration)	<6 mm (ISO 17123 [50])	8 scans, full 360° coverage
BIM model deviation vs. point cloud	82% (0–5 mm) 15% (5–10 mm) 3% (>10 mm)	CloudCompare Distance analysis	±10 mm for LOD 4	Main structural elements and heritage components
Completeness of historic elements	95%	Comparison with heritage inventory	>90%	Inventory catalog: 42 heritage elements
Extraction uncertainty (semi-automatic)	±4 mm	CloudCompare Segment/Mesh analysis	Manual: ±2 mm Semi-automatic: ±6 mm	Irregular edges and complex geometry

The general uncertainty of the modeling of approximately ±8–12 mm does not limit the applicability of the model to facility management (FM) purposes, such as planning conservation interventions or monitoring environmental conditions.

To ensure transparency for future facility managers, an additional parameter heritage_uncertainty = ±10 mm was included in the IFC Property Sets, indicating the expected geometric uncertainty of the reconstructed heritage elements.

In the case of the Australia Hall cast-iron columns, which represent key heritage components, the BIM geometry was intentionally modeled with a +5 mm dimensional margin relative to the measured geometry and subsequently validated in situ through physical verification, ensuring that the model remained safe for operational use despite minor scanning uncertainties.

4.5. Relationships Between the Layers of the RACI–AHP–BIM Methodology

The research results indicate strong interconnections between the three layers of the methodology:

- the responsibility structure (RACI matrix) determined the way information was generated and approved within the BIM environment;
- the BIM environment constituted the data source for AHP decision analyses;
- decisions made within the AHP model influenced the scope of project analyses and information requirements.

These relationships were iterative in nature, with changes in one layer triggering modifications in the others. As a result, the methodology functioned as a coherent project management system based on information flow between the organizational, decision-making, and informational layers.

5. Discussion

Previous research on BIM has focused primarily on technological and information-related aspects, less frequently integrating them with organizational and analytical management methods [52]. Similarly, the literature on AHP most often treats this method as a decision-support tool, without integration with project information systems [53,54]. In turn, the RACI matrix is generally applied as an organizational tool independent of digital environments.

The presented study indicates that only the integration of these three approaches creates a coherent model for managing projects of high functional complexity and conservation constraints. The results confirm the following:

- the BIM environment can function as a platform integrating decision-making processes;
- the AHP method can be fed with data derived from digital models;

- the RACI matrix constitutes a mechanism institutionalizing decision-making and communication processes.

Such an approach aligns with the growing research trend on the digital transformation of project management, in which the integration of organizational, informational and analytical processes is recognized as a key direction of development.

5.1. Synergistic Effects of Integrating RACI–AHP–BIM

The research conducted indicates that the integration of three approaches—organizational structure (RACI matrix), AHP decision analysis, and the BIM information environment—generates synergistic effects that go beyond the sum of benefits resulting from their separate application. Unlike fragmented approaches, the proposed methodology functions as a system of interconnected layers: organizational, decision-making and informational.

Table 9 presents the specific linkages showing how the clash detection data, the geometric parameters, and the TLS results fed the decision variables in the AHP-BOCR model. The data flow is operational and fully documented.

Table 9. Mapping of BIM/CDE with AHP–BOCR decision variables.

BIM/CDE Data Source	Specific Value	BOCR Criterion	Sub-Criterion	Influence on Variant Evaluation
Clash detection results	61 clashes (8 hard, 4 soft, 49 clearance)	Risks (R)	Implementation risk	Variant C: −15% (highest number of MEP conflicts)
Design correction costs	approx. 12,000 PLN	Costs (C)	Cost of design modifications	All variants: real correction costs observed during the design phase
Façade geometry parameters	Glazing area: 1250 m^2 Ug = 0.6 W/m^{2}K	Costs (C)	Material costs	Variant B: +12% advantage due to cheaper profiles
Durability parameters	Estimated service life: 25 years	Benefits (B)	Structural durability	Variant A: +18% advantage (Optiwhite glazing system)
TLS scan-to-BIM data	Column accuracy ±5 mm Completeness: 95%	Benefits (B)	Compatibility with historic fabric	Variant A: +25% (preservation of original cast-iron columns)
TLS reconstruction parameters	Oversize modeling margin +5 mm	Risks (R)	Reversibility of intervention	Variants A/B: low intervention risk

The first identified mechanism of synergy is the link between the RACI matrix and the BIM processes. In traditional BIM projects, there is often the problem of ambiguous assignment of responsibility for actions within the model environment, including clash management and data validation. Incorporating the RACI matrix into the BIM execution plan (BEP) structure enables the assignment of responsibility for generating, verifying and approving project information to specific roles, which reduces decision ambiguity and structures communication [55,56]. For example, in a traditional design process (without a BIM-based methodology), the following question may arise: who is responsible for resolving a clash detected in the federated model between the HVAC installation (MEP designer) and the structural system (structural designer)? By linking the RACI matrix with BIM, it becomes possible to identify the person responsible on the contractor's side for the detection of a clash, who initiates the resolution process and moderates the discussion, while the discipline designers (C) present possible alternatives, and the BIM Manager (A) grants the final approval.

The second area of synergy is the integration of AHP with BIM-derived data. Multi-criteria analysis based on data extracted directly from the digital model enables decision-making grounded in actual technical, cost and spatial parameters [25]. Automatic infor-

https://doi.org/10.3390/infrastructures11030105

mation extraction (e.g., cost data) from the IFC model and the possibility of simulating design variants directly from BIM geometry reduce subjectivity in assessments and increase the transparency of the decision-making process [57]. In the analysis of the investment project, the use of model-derived data enabled the identification of installation clashes (61: hard—8, soft—4, clearance—49), allowing the precise determination of collision risks in the installation interfaces and their inclusion in the variant analysis, which is of direct importance to reduce implementation risks.

The third element of synergy is the link between the RACI matrix and the AHP analysis. The responsibility matrix indicates which stakeholders should participate in evaluating particular decision criteria, thereby structuring the expert process and reducing arbitrariness in assigning weights. As a result, the AHP model becomes not only a mathematical tool but also a component of the project governance structure.

The results obtained confirm that the value of the methodology stems not from individual tools but from their systemic interconnection.

5.2. Importance of the BEP Structure in Reducing Project Risks

The analysis of results indicates that the BIM execution plan functions as a key mechanism for managing informational and process-related risks [58]. Identified risks related to unclear information requirements, inconsistencies in data standards and the lack of procedures for information exchange can be mitigated through process formalization within the BEP framework. Table 10 presents potential risks and mitigation mechanisms, while Figure 8 illustrates the effects of BEP risk reduction through mitigation mechanisms proposed by buildingSMART [51].

Table 10. Effects of the RACI–AHP–BIM methodology according to the literature and professional experience.

No.	Risk Category	Symptom	Mitigation Mechanism in the Project	Effect
1	Unknown roles and responsibilities	Information gaps, errors, delays	RACI matrix 21 × 80, coordination meetings every 2 weeks, escalation procedures	Elimination of accountability gaps (32% → 0%)
2	Inconsistent standards and procedures	Difficulties in interdisciplinary coordination, incompatible models	IFC 4.0, naming convention [Code][Discipline][Type][Revision][Date], LOIN EN 17412	Clash detection: 3 hard + 7 soft, resolved before construction
3	Inadequate LOIN level	Too high = excessive costs, too low = limited BIM usefulness	Precise definition: LOD 3 (design) → LOD 4 (construction) → LOD 6 (as-built + FM)	PLN 500k–800k saved
4	Inefficient data exchange process	Working on outdated versions, conflicts	CDE Dalux: WIP → Shared → Published → Archived, 10-year backup, naming convention	100% model currency, zero version conflicts
5	Lack of quality control procedures	Errors transferred to the construction stage	MQC, detection of clash at least once per month, Data Audit, QCP (Quality Control Plan)	Reduction in construction errors by 60–80%
6	No link between BEP and project objectives	BEP treated as a "paper document"	BEP linked to 21 RACI roles, BIM objectives (scan-to-BIM, clash detection, FM LOD 6), and the schedule	100% team engagement in BIM processes

https://doi.org/10.3390/infrastructures11030105

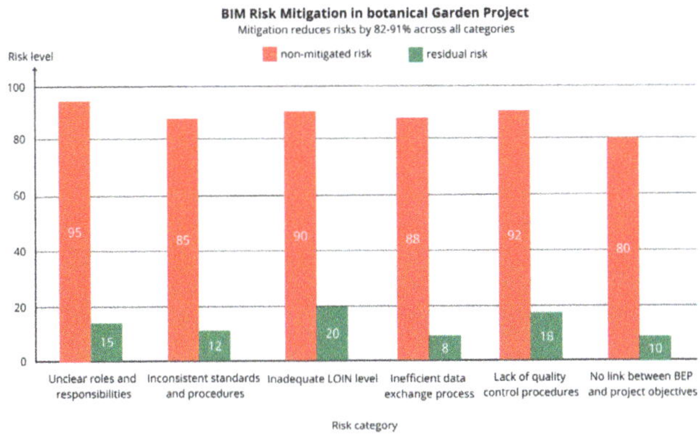

Figure 8. Effects of BEP risk reduction through mitigation mechanisms proposed by buildingSMART [51].

The research results indicate that integrating BEP with the RACI matrix and AHP analysis leads to risk reduction in three areas: informational, process-related and organizational. The BIM literature emphasizes that the lack of a coherent information management structure is one of the main causes of inefficiencies in digital projects; the results obtained confirm this relationship in a project of high functional complexity and conservation constraints.

5.3. Role of BEP in Systemic Risk Management

In the context of project management, BIM documents (EIR and BEP) play a key role in transferring the client's requirements to the contractor and in harmonizing the design and construction processes [59]. This document, updated across successive phases of the project lifecycle, enables the maintenance of information consistency between the design, construction and operation stages.

The research results indicate that, in terms of risk management, the application of BEP enables the reduction in:

- informational risks—through the precise definition of requirements and levels of information detail (formal EIR + BEP structure);
- process-related risks—through the formalization of cooperation and communication procedures;
- organizational risks—through linking the competencies of the participants with the assigned roles.

In the literature on BIM project management, BEP is increasingly perceived as a strategic management tool rather than merely an operational document. The findings of this study confirm this interpretation.

The experience of the project team indicates the following expected range of effects for projects of similar scale (public infrastructure, design–build delivery, heritage constraints):

- Reduction in accountability gaps from approximately 32% to 0% through the implementation of RACI in a BIM environment.
- Reduction in construction errors by approximately 60–80% due to clash detection and the use of a Common Data Environment (CDE).
- Potential cost savings of approximately PLN 500,000–800,000 resulting from optimized definition of LOIN/LOD (internal benchmark data).

These values should be interpreted as hypothetical benchmark estimates, which cannot yet be verified using empirical data from the analyzed investment because the construction and operational phases have not been completed. Full validation of these effects will only be possible after the completion of the project (planned after 2027).

https://doi.org/10.3390/infrastructures11030105

5.4. Role of the RACI Matrix in Limiting Project Conflicts

The results obtained confirm that the application of the RACI matrix contributes to structuring decision-making and communication relationships within the project. Clear assignment of Responsible and Accountable roles reduces the risk of interdisciplinary conflicts, which often arise from competency ambiguities, through:

- elimination of "black holes," where each task has at least one Responsible (R) and one Accountable (A) role,
- reduction in redundancy by preventing multiple assignments of the same task to two roles,
- communication transparency by clearly indicating who should be informed (I) versus consulted (C).

In the analysis project, a significant outcome of applying the RACI matrix was the extension of the project team structure with new roles related to facility operation management, indicating a shift in perspective from the design phase to the entire investment lifecycle.

6. Summary

The study conducted confirmed that the integration of the RACI matrix, the AHP method, and the BIM methodology in accordance with ISO 19650 constitutes an effective and comprehensive approach to the management of complex construction projects implemented in the design–build delivery model, particularly in the context of conservation-oriented and academic construction projects.

The application of the integrated methodology enabled the structuring of organizational relationships, decision-making processes, and information management within a single coherent project management system. The developed RACI matrix (21 processes × 80 responsibility assignments) allowed the clear allocation of executive and decision-making roles in all key tasks, reducing the appearance of accountability gaps and shortened decision-making processes in critical stages of the project.

The application of the AHP method within the BOCR model enabled a multi-criteria evaluation of design alternatives and the selection of a solution that accounted not only for economic criteria but also for conservation, functional, and risk-related considerations. The results of the analysis indicated a preference for the structural variant with the highest compatibility with the historic fabric of the facility, confirming the suitability of the decision-making model for projects subject to specific cultural and environmental constraints.

The implementation of BIM, supported by the BIM execution plan (BEP) and the common data environment (CDE), enabled the integration of project data, automation of quality control processes, and reduction in interdisciplinary clashes prior to the commencement of construction works. The use of laser scanning technology and the scan-to-BIM procedure made it possible to develop a digital model of preserved historic elements and to prepare operational documentation based on facility management (FM) data.

The findings indicate that the value of the proposed methodology is primarily the results of the integration of the organizational, decision-making, and information layers, rather than the separate application of individual tools.

This study fills a significant gap in the scientific literature:

- It provided a holistic integration of three approaches into a single RACI-AHP-BIM methodology. Previous research have analyzed RACI, AHP, and BIM as independent methodologies. This article represents one of the first comprehensive studies that integrates these three approaches into a coherent methodological framework. Integration is systemic in nature and demonstrates the mutual reinforcement of methods: RACI structures the organizational framework, AHP supports the decision-making process, and BIM constitutes the operational environment for information management;

https://doi.org/10.3390/infrastructures11030105

- It extended the application of the RACI matrix to conservation projects, in which heritage protection institutions and future facility users played a significant role. The introduction of roles related to conservation supervision and operational management indicates the possibility of applying RACI throughout the lifecycle of the facility. The proposed adaptation is also of particular relevance for the Polish public sector, where conservation projects constitute a substantial share of construction investments undertaken by higher-education institutions;
- It provided empirical verification of the application of the BOCR model within the AHP method in design–build projects implemented under conservation and environmental constraints. Although the AHP method has been widely recognized for decades, its application with the BOCR model in design–build projects carried out in protected heritage contexts has not been systematically examined to date. The article offers empirical confirmation of the suitability of the BOCR model for the simultaneous evaluation of conservation benefits, development opportunities, construction costs, and technical risks within a single integrated decision hierarchy;
- It presented an operational implementation of the ISO 19650 standard in a public project, indicating practical mechanisms for integrating BEP, RACI, and the CDE; documenting the application of scan-to-BIM technology in a conservation project, including technical parameters, data integration procedures; and the potential use of the model during the operational phase (HBIM/digital twin).

The article represents one of the few examples of the comprehensive integration of RACI, AHP, and BIM in the specific context of academic facilities delivered under the design–build system. The proposed RACI-AHP-BIM methodology may serve as a tool supporting the management of construction projects implemented in the public sector, particularly in cases involving a large number of stakeholders and complex formal and conservation constraints. Integration of RACI, AHP, and BIM increases transparency in responsibility allocation and communication, supports data-driven decision-making, reduces informational and organizational risks, and enables a life-cycle-oriented approach to project management. The methodology may be used as a reference framework in the development of BIM execution plans in construction and infrastructure projects, particularly under public procurement conditions.

The results of the study indicate the need for further research in several areas. First, it is necessary to verify the effectiveness of the methodology during the construction phase and the operational period of the facility, including the analysis of cost, time, and operational data. Second, it is justified to assess the potential adaptation of the RACI–AHP–BIM methodology in other types of projects, including infrastructure, linear, and public utility projects. Third, future research should focus on the integration of BIM models with facility management systems and on the development of the digital twin concept in conservation projects. Comparative studies involving a larger number of public projects are also recommended to evaluate the scalability and repeatability of the proposed methodology.

Author Contributions: Conceptualization, U.K.-K.; methodology, U.K.-K.; software, U.K.-K.; validation, U.K.-K.; formal analysis, U.K.-K.; investigation, U.K.-K.; resources, U.K.-K.; data curation, U.K.-K.; writing—original draft preparation, U.K.-K.; writing—review and editing, M.S.; visualization, M.S.; supervision, M.S. All authors have read and agreed to the published version of the manuscript.

Funding: This research received no external funding.

Data Availability Statement: Dataset available on request from the authors.

Conflicts of Interest: The authors declare no conflicts of interest.

https://doi.org/10.3390/infrastructures11030105

References

1. Rinchen, S.; Banihashemi, S.; Alkilani, S. Driving Digital Transformation in Construction: Strategic Insights into Building Information Modelling Adoption in Developing Countries. *Proj. Leadersh. Soc.* **2024**, *5*, 100138. [CrossRef]
2. Zhao, S.; Guan, Y.; Zhou, H.; Hu, F. Accelerating Digital Transformation in Construction: Best Practices for the Successful Implementation of Integrated Digital Delivery. *Autom. Constr.* **2025**, *178*, 106417. [CrossRef]
3. Meng, J.; Ye, Z.; Wang, Y. Financing and Investing in Sustainable Infrastructure: A Review and Research Agenda. *Sustain. Futures* **2024**, *8*, 100312. [CrossRef]
4. Hadfield, P.; Coenen, L. Contemporary Financial Capitalism and Sustainability Transitions in Urban Built Environments. *Environ. Innov. Soc. Transit.* **2022**, *42*, 285–300. [CrossRef]
5. Ripp, M.; Gustafsson, C.; Jones, Z.; Majumdar, S.; Ginzarly, M. Historic Urban Landscapes and Heritage Systems as the Basis for Sustainable Urban Development. *Land* **2025**, *14*, 1783. [CrossRef]
6. Lanciotti, A. Preservation of Historical Buildings through the Lens of International Law. *Real Estate* **2024**, *1*, 198–211. [CrossRef]
7. Wang, T.; Chen, H.M. Integration of Building Information Modeling and Project Management in Construction Project Life Cycle. *Autom. Constr.* **2023**, *150*, 104832. [CrossRef]
8. Kaur, R.; Mwambegele, B.J.; Abraham, A.G.; Basheer, S.A.; Garia, S. A Comprehensive Review on Building Information Modelling (BIM), Its Implementations and Applications. *Discov. Civ. Eng.* **2025**, *2*, 177. [CrossRef]
9. Morin, X.; Romero-Torres, A. How Does Building Information Modeling Influence Decision-Making Process in the Project Design? An Input, Process and Output Analysis. *Proj. Leadersh. Soc.* **2024**, *5*, 100160. [CrossRef]
10. Abanda, F.H.; Oti, A.H.; Tah, J.H.M. Sustainability and Building Information Modelling: Integration, Research Gaps, and Future Directions. *Autom. Constr.* **2024**, *163*, 105420. [CrossRef]
11. Lee, W.Y.; Lee, S.H.; Jin, C.; Hyun, C.T. Development of the RACI Model for Processes of the Closure Phase in Construction Programs. *Sustainability* **2021**, *13*, 1806. [CrossRef]
12. Cinelli, M.; Kadziński, M.; Miebs, G.; Gonzalez, M.; Słowiński, R. Recommending Multiple Criteria Decision Analysis Methods with a New Taxonomy-Based Decision Support System. *Eur. J. Oper. Res.* **2022**, *302*, 633–651. [CrossRef]
13. Madanchian, M.; Taherdoost, H. Applications of Multi-Criteria Decision Making in Information Systems for Strategic and Operational Decisions. *Computers* **2025**, *14*, 208. [CrossRef]
14. Costa, A.P.d.A.; Choren, R.; Pereira, D.A.d.M.; Terra, A.V.; Costa, I.P.d.A.; Junior, C.d.S.R.; Santos, M.d.; Gomes, C.F.S.; Moreira, M.Â.L. Integrating Multicriteria Decision Making and Principal Component Analysis: A Systematic Literature Review. *Cogent Eng.* **2024**, *11*, 2374944. [CrossRef]
15. *ISO 19650-1:2018*; Organization and Digitization of Information About Buildings and Civil Engineering Works, Including Building Information Modelling (BIM)—Information Management Using Building Information Modelling—Part 1: Concepts and Principles. ISO: Geneva, Switzerland, 2018.
16. Blavier, C.L.S.; Huerto-Cardenas, H.E.; Aste, N.; Del Pero, C.; Leonforte, F.; Della Torre, S. Adaptive Measures for Preserving Heritage Buildings in the Face of Climate Change: A Review. *Build. Environ.* **2023**, *245*, 110832. [CrossRef]
17. Dziadosz, A. Estimation and Selection of Building Investment Using the AHP. *Czas. Techniczne. Bud.* **2008**, *105*, 41–51.
18. Pan, N.F. Fuzzy AHP Approach for Selecting the Suitable Bridge Construction Method. *Autom. Constr.* **2008**, *17*, 958–965. [CrossRef]
19. El-Abbasy, M.S.; Zayed, T.; Ahmed, M.; Alzraiee, H.; Abouhamad, M. Contractor Selection Model for Highway Projects Using Integrated Simulation and Analytic Network Process. *J. Constr. Eng. Manag.* **2013**, *139*, 755–767. [CrossRef]
20. Dikmen, I.; Birgonul, M.T.; Han, S. Using Fuzzy Risk Assessment to Rate Cost Overrun Risk in International Construction Projects. *Int. J. Proj. Manag.* **2007**, *25*, 494–505. [CrossRef]
21. Saaty, T.L.; Ozdemir, M.S. Why the Magic Number Seven plus or Minus Two. *Math. Comput. Model.* **2003**, *38*, 233–244. [CrossRef]
22. Kwast-Kotlarek, U.; Hełdak, M. Evaluation of the Construction and Investment Process of a High-Pressure Gas Pipeline with Use of the Trenchless Method and Open Excavation Method. Analytic Hierarchy Process (AHP). *Sustainability* **2019**, *11*, 2438. [CrossRef]
23. Cheung, S.O.; Suen, H.C.H. A Multi-Attribute Utility Model for Dispute Resolution Strategy Selection. *Constr. Manag. Econ.* **2002**, *20*, 557–568. [CrossRef]
24. Saaty, R.W. The Analytic Hierarchy Process—What It Is and How It Is Used. *Math. Model.* **1987**, *9*, 161–176. [CrossRef]
25. Aladayleh, K.J.; Aladaileh, M.J. Applying Analytical Hierarchy Process (AHP) to BIM-Based Risk Management for Optimal Performance in Construction Projects. *Buildings* **2024**, *14*, 3632. [CrossRef]
26. BIM Standard PL—Urząd Zamówień Publicznych—Portal Gov.Pl. Available online: https://www.gov.pl/web/uzp/bim-standard-pl (accessed on 18 February 2026).
27. Kacprzyk, Z. Polskie Normy BIM—Norma IDM. *Bud. Praw.* **2017**, *20*, 1–243.
28. Project BIM Execution Planning (BEP) Standard—National Institute of Building Sciences. Available online: https://nibs.org/nbims/v4/bep/ (accessed on 18 February 2026).
29. Borkowski, A.S. A Literature Review of BIM Definitions: Narrow and Broad Views. *Technologies* **2023**, *11*, 176. [CrossRef]

30. Borrmann, A.; König, M.; Koch, C.; Beetz, J. *Building Information Modeling: Technology Foundations and Industry Practice*; Springer Nature: Berlin/Heidelberg, Germany, 2018; pp. 1–584. [CrossRef]
31. Borkowski, A.; Michalak, M. Błędy Poznawczei Nowa Emfatyczna Definicja BIM. *BUILDER* **2024**, *322*, 54–60. [CrossRef]
32. Osello, A.; Lucibello, G.; Morgagni, F. HBIM and Virtual Tools: A New Chance to Preserve Architectural Heritage. *Buildings* **2018**, *8*, 12. [CrossRef]
33. Sacks, R.; Eastman, C.; Lee, G.; Teicholz, P. *BIM Handbook*; Wiley: Hoboken, NJ, USA, 2018. [CrossRef]
34. Shawky, K.A.; Abdelalim, A.M.; Sherif, A.G. Standardization of BIM Execution Plans (BEP's) for Mega Construction Projects; A Comparative and Scientometric Study. *Trans. Mach. Learn. Artif. Intell.* **2024**, *12*, 103–129. [CrossRef]
35. Construction to Operations Building Information Exchange (COBie) V3—National Institute of Building Sciences. Available online: https://nibs.org/nbims/v3/cobie (accessed on 18 February 2026).
36. Mapa Drogowa Dla Wdrożenia Metodyki BIM w Zamówieniach Publicznych—Ministerstwo Rozwoju i Technologii—Portal Gov.Pl. Available online: https://www.gov.pl/web/rozwoj-technologia/mapa-drogowa-dla-wdrozenia-metodyki-bim-w-zamowieniach-publicznych (accessed on 18 February 2026).
37. Kensek, K.M.; Noble, D. *Building Information Modeling: BIM in Current and Future Practice*; Wiley: Hoboken, NJ, USA, 2015; pp. 1–415.
38. Godager, B.; Mohn, K.; Merschbrock, C.; Klakegg, O.J.; Huang, L. Towards an Improved Framework for Enterprise BIM: The Role of ISO 19650. *J. Inf. Technol. Constr.* **2022**, *27*, 1075–1103. [CrossRef]
39. Al Khalil, M.I. Selecting the Appropriate Project Delivery Method Using AHP. *Int. J. Proj. Manag.* **2002**, *20*, 469–474. [CrossRef]
40. Al-Harbi, K.M.A.S. Application of the AHP in Project Management. *Int. J. Proj. Manag.* **2001**, *19*, 19–27. [CrossRef]
41. Barzilai, J. Deriving Weights from Pairwise Comparison Matrices. *J. Oper. Res. Soc.* **1997**, *48*, 1226. [CrossRef]
42. Barzilai, J. On the Decomposition of Value Functions. *Oper. Res. Lett.* **1998**, *22*, 159–170. [CrossRef]
43. Knitter-Piątkowska, A.; Kotlarek, U. Znaczenie i Struktura Dokumentu BIM Execution Plan (BEP) w Procesie Inwestycyjno-Budowlanym. *BUILDER* **2025**, *342*, 20–25. [CrossRef]
44. Salah, R.; Károlyfi, K.A.; Szép, J.; Géczy, N. A Structured Framework for HBIM Standardization: Integrating Scan-to-BIM Methodologies and Heritage Conservation Standards. *Digit. Appl. Archaeol. Cult. Herit.* **2025**, *37*, e00420. [CrossRef]
45. Rocha, G.; Mateus, L.; Fernández, J.; Ferreira, V. A Scan-to-BIM Methodology Applied to Heritage Buildings. *Heritage* **2020**, *3*, 47–67. [CrossRef]
46. Lovell, L.J.; Davies, R.J.; Hunt, D.V.L. The Application of Historic Building Information Modelling (HBIM) to Cultural Heritage: A Review. *Heritage* **2023**, *6*, 6691–6717. [CrossRef]
47. Javier Sánchez Aparicio, L.; Luis Del Blanco García, F.; Martín Talaverano, R.; Klapa, P.; Pepe, M. 3D Heritage Reconstruction Through HBIM and Multi-Source Data Fusion: Geometric Change Analysis Across Decades. *Appl. Sci.* **2025**, *15*, 8929. [CrossRef]
48. Abd ElWahab, H.A.; Bakr, A.F.; Raslan, R.A. Heritage Building Information Modeling (HBIM) for Heritage Conservation: Framework of Challenges, Gaps, and Existing Limitations of HBIM. *Digit. Appl. Archaeol. Cult. Herit.* **2024**, *35*, e00366. [CrossRef]
49. *PN-EN ISO/IEC 27001:2023-08*; Bezpieczeństwo Informacji, Cyberbezpieczeństwo i Ochrona Prywatności—Systemy Zarządzania Bezpieczeństwem Informacji—Wymagania. Polski Komitet Normalizacyjny: Warsaw, Poland, 2024. (In Polish)
50. *PN-ISO 17123-3:2005*; Optyka i Instrumenty Optyczne—Terenowe Procedury Testowania Instrumentów Geodezyjnych i Pomiarowych. Polski Komitet Normalizacyjny: Warsaw, Poland, 2005. (In Polish)
51. buildingSMART Solutions and Standards. Available online: https://www.buildingsmart.org/standards/ (accessed on 20 February 2026).
52. Turk, Ž. Reflections on Three Decades of Building Information Modeling. *Buildings* **2025**, *15*, 231. [CrossRef]
53. Madzík, P.; Falát, L. State-of-the-Art on Analytic Hierarchy Process in the Last 40 Years: Literature Review Based on Latent Dirichlet Allocation Topic Modelling. *PLoS ONE* **2022**, *17*, e0268777. [CrossRef]
54. Vaidya, O.S.; Kumar, S. Analytic Hierarchy Process: An Overview of Applications. *Eur. J. Oper. Res.* **2006**, *169*, 1–29. [CrossRef]
55. Farouk, A.M.; Zulhisham, A.Z.; Lee, Y.S.; Rajabi, M.S.; Rahman, R.A. Factors, Challenges and Strategies of Trust in BIM-Based Construction Projects: A Case Study in Malaysia. *Infrastructures* **2023**, *8*, 13. [CrossRef]
56. Da Silva, T.F.L.; Vieira, D.R.; De Carvalho, M.M. Exploring the Challenges in Building Information Modeling (BIM) During the Design Phase: Evidence from Cross-Country Studies. *IEEE Trans. Eng. Manag.* **2024**, *71*, 14846–14860. [CrossRef]
57. Wong Chong, O.; Zhang, J. Automated Extraction of Locational Information from IFC-Based Building Information Models. In Proceedings of the Construction Research Congress 2024, Des Moines, IA, USA, 20–23 March 2024; pp. 148–156. [CrossRef]
58. Panagiotidou, N.; Pitt, M.; Lu, Q. Building Information Modelling Execution Plans: A Global Review. *Proc. Inst. Civ. Eng. Smart Infrastruct. Constr.* **2023**, *176*, 126–147. [CrossRef]
59. Chapter 4—BIM Project Implementation. OpenBIM KnowledgeBase. Available online: https://openbim-knowledgebase.org/en/docs/bimcert-manual-2024/chapter-4-bim-project-implementation/ (accessed on 20 February 2026).

https://doi.org/10.3390/infrastructures11030105

infrastructures

MDPI

Review

Smart Surveillance of Structural Health: A Systematic Review of Deep Learning-Based Visual Inspection of Concrete Bridges Using 2D Images

Nasrin Lotfi Karkan [1], Eghbal Shakeri [1,*], Naimeh Sadeghi [2] and Saeed Banihashemi [3,*]

[1] Department of Civil and Environmental Engineering, Construction Engineering Management, Amirkabir University of Technology, Tehran 15916-34311, Iran; nasrinlotfi@aut.ac.ir
[2] Faculty of Civil Engineering, K.N. Toosi University of Technology, Tehran 19967-15433, Iran; sadeghi@kntu.ac.ir
[3] School of Built Environment, University of Technology Sydney, Sydney, NSW 2007, Australia
* Correspondence: eshakeri@aut.ac.ir (E.S.); saeed.banihashemi@uts.edu.au (S.B.)

Abstract: Timely and accurate inspection of concrete bridges is critical to ensuring structural integrity and public safety. Traditional visual inspections conducted by human inspectors are labour-intensive, inconsistent, and often limited in their ability to access all structural components, particularly in hazardous or inaccessible areas. Image-based inspection techniques have emerged as a safer and more efficient alternative, and recent advancements in deep learning have significantly enhanced their diagnostic capabilities. This systematic review critically evaluates 77 studies that applied deep learning approaches to the detection and classification of surface defects in concrete bridges using 2D images. Relevant publications were retrieved from major scientific databases, screened for eligibility, and analyzed in terms of model type, training strategies, and evaluation metrics. The reviewed works encompass a wide spectrum of algorithms—spanning classification, object detection, and image segmentation models—highlighting their architectural features, strengths, and trade-offs in terms of accuracy, computational complexity, and real-time applicability. Key findings reveal that transfer learning, data augmentation, and careful dataset composition are pivotal in improving model performance. Moreover, the review identifies emerging research trajectories, such as integrating deep learning with Building Information Modeling (BIM), leveraging edge computing for real-time monitoring, and developing rich annotated datasets to enhance model generalizability. By mapping the current state of knowledge and outlining future research directions, this study provides a foundational reference for researchers and practitioners aiming to deploy deep learning technologies in bridge inspection and infrastructure monitoring.

Keywords: bridge inspection; defect detection; machine learning; deep learning; computer vision; image classification; image segmentation; object detection

1. Introduction

Bridges play a crucial role in our society by connecting different regions and serving as key points in the urban transportation network. Hence, substantial financial resources are invested in inspecting and identifying defects in these critical infrastructures [1]. Concrete bridges are more common than steel bridges due to their versatility, resilience, durability, low maintenance, and quick construction [2]. Hence, maintaining bridges in good condition is important because environmental factors, aging, and heavy loads can cause damage. This can decrease their lifespan and increase the risk of collapse, causing economic and

 https://doi.org/10.3390/infrastructures10120338

physical harm. For instance, a bridge collapse in Minneapolis (MN, USA) in 2007 led to the loss of 13 lives and 145 injuries [3].

Inspection serves as the initial phase of maintenance with this respect. Traditional bridge inspection is performed by human inspectors; this process can be time-consuming and may not be very precise due to human judgment's inconsistency and uncertainty. It also requires inspector training and requires some equipment, like man lifts, which can be dangerous for both the inspector and the public [4,5]. Using artificial intelligence (AI) for bridge inspections can make the process more efficient and accurate.

Deep learning, a branch of AI, has become increasingly important for bridge defect detection and measurement. Utilizing sensor data and images, these algorithms can be trained to identify and quantify defects in bridges, resulting in quicker inspections and reduced human errors. While acceleration, strain, and displacement sensors are commonly used for bridge inspections, they cover only a limited part of the bridge. Furthermore, they are challenging to maintain and access after installation. Consequently, there has been a shift towards using images to identify defects and inspect bridges. This has led to a significant emphasis on utilizing images for bridge inspection and defect identification [6].

Despite the significant importance of bridge defect detection, systematic reviews of computer vision and deep learning approaches specifically targeting concrete defects remain limited. Most existing literature predominantly focuses on crack detection [1,7]. Recently, two review studies have addressed broader applications of computer vision techniques in bridge damage detection. Luo, et al. [8] explored three key aspects—surface defect detection, vibration measurement, and vehicle parameter identification—while Zhang and Yuen [9] reviewed various AI-driven techniques, including image-based methods, point clouds, infrared thermography (IRT), ground penetrating radar (GPR), and vibration responses. However, these reviews do not focus specifically on deep learning architectures for concrete bridge surface inspection using standard RGB images, nor do they provide a task-structured comparative synthesis of model families in terms of both performance metrics and computational efficiency.

This study extends the existing body of knowledge in three main ways. First, it focuses explicitly on concrete bridges and standard 2D image data, excluding non-visual sensing modalities, over the period 2018–2023, during which deep learning became the dominant approach for image-based bridge inspection. Second, it follows PRISMA-based systematic literature review procedures, with clearly defined search, screening, and risk-of-bias assessment steps, resulting in a curated set of 77 deep learning studies on concrete bridge surface defects. Third, it provides a task-structured comparative synthesis of major deep learning model families—CNN classifiers, one- and two-stage object detectors, and segmentation networks—organized by task (classification, detection, segmentation) and summarized in terms of accuracy-related metrics (e.g., F1-score, mAP, mIoU), inference time, and architectural characteristics (e.g., deep vs. lightweight backbones, one-stage vs. two-stage detectors). This analysis clarifies speed–accuracy trade-offs and helps relate model structures to practical deployment contexts such as UAV-based surveys, robotic inspection, and detailed defect mapping for SHM and maintenance decisions.

The research questions that this paper aims to address are:

1. How have deep learning-based visual inspection methods for concrete bridge surface defects evolved since 2018 in terms of: (a) defect types, and (b) computer-vision tasks (classification, object detection, segmentation, and hybrid methods)?
2. For each main task (classification, object detection, segmentation), which families of deep learning models are most commonly used for detecting concrete bridge defects, and what trade-offs do they show between performance (e.g., accuracy, F1-score, mAP, mIoU) and computational efficiency (e.g., inference time)?

3. What are the main characteristics and limitations of the datasets used to train and validate deep learning models for concrete bridge visual inspection (e.g., dataset size, defect classes, annotation level)?

A systematic literature review was conducted following the PRISMA guidelines to address these questions. Databases such as Web of Science, Scopus, and Google Scholar were searched for papers published between 2018 and 2023 that used deep learning for defect detection and visual inspection of concrete bridges using standard 2D images. Predefined inclusion and exclusion criteria regarding publication type, language, bridge material, imaging modality, and methodological content were applied as described in Section 3.

The paper is organized into six sections. In Section 2, deep learning and its subsets are briefly described. Section 3 details our search strategy for identifying relevant studies. Section 4 categorizes the deep learning methods utilized in bridge damage detection and provides an analysis. Afterward, Section 5 compares the performance of different deep learning algorithms in the context of bridge damage detection. Finally, Section 6 presents the conclusion and future study recommendations.

2. Background on Deep Learning

Computer vision has different key subsets; this research focuses on image classification, segmentation, and object detection. Image classification determines whether an image contains a specific object or not. In object detection, in addition to that, the object location is defined using a bounding box, and finally, in image segmentation, the object is precisely defined at the pixel level. The emergence of deep learning algorithms has transformed the field of computer vision. Deep learning refers to multi-layer neural networks, which have recently been increasingly used for computer vision tasks. Various algorithms, such as Convolutional Neural Networks (CNNs), AlexNet, VGG, DeepLab, U-Net, Region-based CNN (R-CNN), and You Only Look Once (YOLO) were highly effective in this revolution, each offering distinct advantages [10,11]. Moreover, improvements in deep learning and computer vision techniques have enhanced traditional algorithms functions. This section provides the basic concepts and algorithms of deep learning relevant to this review and highlights their strengths compared to earlier models.

2.1. Deep Learning-Based Image Classification Algorithms

Most algorithms for computer vision tasks are based on CNNs. A CNN has multiple layers: convolutional, pooling, and fully connected. In the convolutional layer, kernels (or filters) convolve with the input images to extract feature maps. These filters have weights that are adjusted during training. An activation function, such as sigmoid, ReLU, or tanh, is then applied to these feature maps to introduce non-linearity and determine which features should be passed to the next layer. The pooling layer reduces the dimensions of the convolved features, typically through operations like max pooling or average pooling, to decrease computational load and control overfitting. Finally, the last convolutional layer is connected to the fully connected layer, which integrates the extracted features to make the final classification or prediction [12].

An important component in the training process of CNNs is the loss function, which evaluates their accuracy in predicting the given labels. Backpropagation minimizes the loss function by calculating the gradient of the loss with respect to each weight in the network, moving backward from the output layer. This gradient is then used to adjust the weights through an optimization algorithm, aiming to minimize the loss and improve the network's performance [13].

Key deep learning models include AlexNet and VGG-16.AlexNet, introduced by Krizhevsky et al. [14]. It is composed of eight weight layers, which include five convolutional layers and three fully connected layers with ReLU activations. This was followed by the VGG-16 model in 2014, developed by Simonyan and Zisserman [15], which enhanced depth by employing 16 layers, including thirteen convolutional layers with a 3 × 3 filter size and three fully connected layers. However, deeper networks, such as ResNet [16], address the vanishing gradient problem by using shortcut connections, enabling networks to have more layers without degradation in performance. DenseNet [17] builds on ResNet by connecting every layer to every subsequent one, improving information flow across the network and its architecture is displayed in Figure 1.

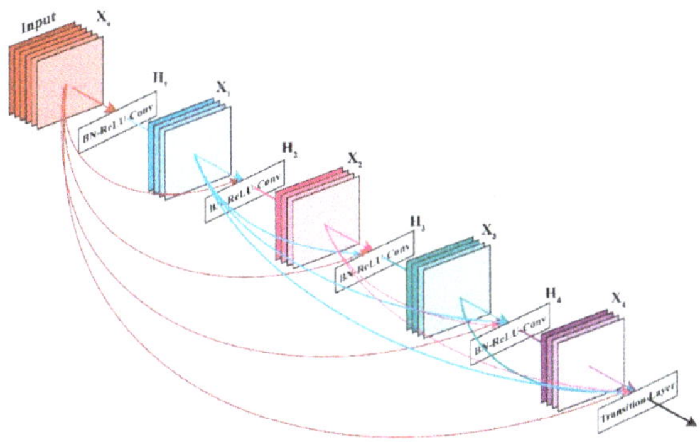

Figure 1. A five-layer DenseNet structure [18].

To address the computational challenges, models like MobileNet [18] and Xception [19] used depthwise separable convolutions, reducing the number of parameters and improving efficiency, making them ideal for real-time inspection tasks such as those performed by UAVs during bridge assessments.

2.2. Deep Learning-Based Image Segmentation Algorithms

Several approaches exist for image segmentation. Two important approaches are encoder–decoder structures and fully convolutional networks [20]. Fully convolutional network (FCN), developed by Long et al. [21], and U-Net [22] are two major models used in segmentation tasks. U-Net has a U-shaped architecture consisting of a contracting path (encoder) and an expansive path (decoder). The contracting path is similar to a standard convolutional network, utilizing convolutional and pooling layers for downsampling. The expansive path enhances the resolution and merges it with the information from the contracting path to create accurate segmentations.

Another important model is DeepLab [23], which uses dilated convolutions to expand the receptive field and capture more context without reducing spatial resolution. This feature is important for accurately identifying defects in concrete surfaces. Additionally, Mask R-CNN [24] extends Faster R-CNN by adding a mask branch, allowing for pixel-wise segmentation of detected objects. DeepLabv3+ further improves on the segmentation accuracy by using atrous spatial pyramid pooling (ASPP), which allows the model to capture features at various scales without losing spatial resolution, making it ideal for identifying defects of different sizes in concrete surfaces.

2.3. Deep Learning-Based Object Detection Algorithms

Object detection algorithms like R-CNN [25], Fast R-CNN [26], Faster R-CNN [27], and YOLO [28] have greatly improved the ability to detect and localize defects in concrete surfaces. The YOLO (You Only Look Once) family is especially useful for real-time defect detection due to its ability to process entire images in a single pass. YOLOv3 and YOLOv5 have improved accuracy and speed, making them ideal for use in UAV-based inspections. YOLOv5 introduced attention mechanisms to improve the detection of small and overlapping defects, which is a common challenge in concrete bridge inspection.

Deep learning algorithms improved the weaknesses of their previous algorithms by modifying their structure and adding new features. This continuous improvement made them more accurate, faster, and more reliable.

2.4. Evaluation Metrics

Deep learning performance is measured using different evaluation metrics. Some common evaluation metrics are described here.

- Accuracy: Classification algorithms don't always predict correctly, and their performance is measured according to their true and false predictions. True positives (TP) are when the model accurately predicts the positive class, while true negatives (TN) are when it accurately predicts the negative class. False positives (FP) happen when the model wrongly predicts the negative class as positive, and false negatives (FN) occur when the model incorrectly predicts the positive class as negative. Accuracy is the ratio of correct predictions to the total predictions, and mean accuracy is the average of accuracy across multiple model runs.

$$Accuracy = \frac{TP + TN}{Total\ predictions} \tag{1}$$

$$Mean\ Accuracy = \frac{1}{n}\sum_{i=1}^{n} Accuracy_i \tag{2}$$

where n is the number of runs and $Accuracy_i$ is the accuracy for each run.

Mean accuracy is obtained by averaging accuracy across multiple runs or folds. However, in highly imbalanced datasets—such as images where one class (e.g., cracks) occupies only a small portion of the pixels or where the "background" class dominates—overall accuracy can be misleading, because a classifier that almost always predicts the majority class may still achieve a high accuracy. Therefore, many studies also report precision, recall, and F1-score to better characterize model performance.

- Precision:

$$Precision = \frac{TP}{TP + FP} \tag{3}$$

- Recall (Sensitivity):

$$Recall = \frac{TP}{TP + FN} \tag{4}$$

- F1 Score:

$$F1\ Score = 2 \times \frac{Precision \times Recall}{Precision + Recall} \tag{5}$$

Precision reflects how many predicted defects are correct, recall reflects how many true defects are actually detected, and F1 balances these two aspects. In imbalanced settings, per-class precision, recall, and F1 (e.g., for crack vs. spallation vs. efflorescence) are more informative than a single global accuracy value.

For object detection, performance is usually summarized using Average Precision (AP) and Mean Average Precision (mAP). AP is the area under the precision–recall curve for a given class and Intersection-over-Union (IoU) threshold, whereas mAP is the mean of AP values over all classes. Many works report mAP at a fixed IoU threshold (e.g., mAP@0.5) or averaged over multiple thresholds (e.g., mAP@[0.5:0.95]) to simultaneously capture localization and classification quality.

For semantic segmentation, the dominant metric is Intersection-over-Union (IoU), defined for a given class as:

$$IoU = \frac{Area\ of\ overlap\ between\ prediction\ and\ ground\ truth}{Area\ of\ union\ between\ prediction\ and\ ground\ truth} \quad (6)$$

The mean IoU (mIoU) is the average IoU over all classes.

In addition to these accuracy-related metrics, several studies also report computational efficiency, typically in terms of inference time per image, frames per second, or model size (number of parameters).

3. Research Methodology

Following the PRISMA guidelines, this study conducted a Systematic Literature Review (SLR) to investigate the use of deep learning in concrete bridge automated defect detection. The literature review was conducted in three phases of data acquisition, screening, and in-depth analysis as illustrated in Figure 2.

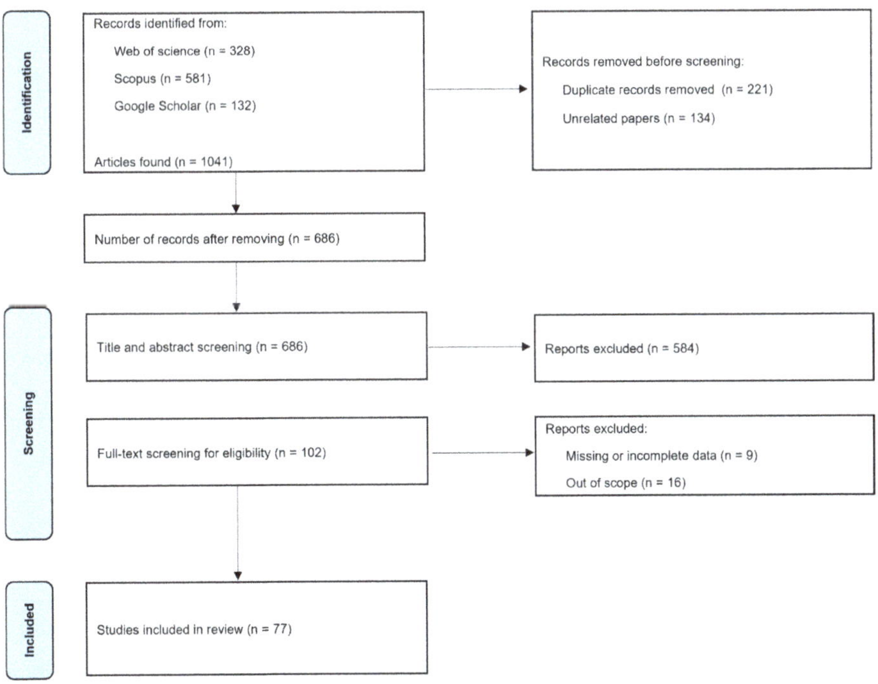

Figure 2. Flowchart of systematic literature review (*n* = number of papers).

The main goal of data acquisition was identifying the most relevant academic articles that could provide comprehensive data for our analysis. Scopus, Web of Science (WoS), and Google Scholar were used as our research engines. The keywords in Table 1 were utilized to set a comprehensive search query. The search was restricted to peer-reviewed journal articles and conference papers written in English that applied deep learning to visual inspection of concrete bridges using standard 2D images.

Table 1. Search query.

String
("Viaducts" OR "Bridge") AND ("Artificial intelligence" OR "Image processing" OR "Machine learning" OR "Deep learning" OR "CNN" OR "convolutional neural network") AND ("Defect detection" OR "Crack detection" OR "Damage detection" OR "spalling" OR "vegetation" OR "scaling" OR "Delamination" OR "Efflorescence" OR "bulging" OR "Pop-outs" OR "Honeycombs" OR "Reinforcing steel corrosion" OR "Exposed bar" OR "Exposed rebar" OR "corrosion stain " OR "Water leak" OR "Water leakage" OR "material loss" OR "Loss of section" OR "displacement" OR "deformation" OR "degradation" OR "damage quantification" OR "damage measurement" OR "damage evaluation" OR "damage severity") AND ("image" OR "video")

In the screening phase, the keywords were used to search the selected databases. We chose to restrict the time frame of our study to include only papers published after 2018 and up to October 2023. Searches were conducted based on article titles, abstracts, and keywords within the defined scope, which resulted in the finding of 1041 papers.

While conducting the literature review, we diligently ensured no duplicate entries and eliminated them accordingly. After conducting an initial database screening, it was found to be necessary to narrow down the number of papers only to include those that met the research criteria (Table 2). This was accomplished by carefully examining each document's title and abstract. Papers with unclear titles were excluded, and only those related to bridge defect detection were retained.

Table 2. Filtration criteria.

Phases	Criteria
Title and Abstract Screening	Review papers were excluded. Titles lacking clarity were excluded. Only articles related to visual defect detection on concrete bridges were included. Articles with abstracts lacking sufficient methodological detail to determine whether deep learning was applied to concrete bridge surface defect detection using standard 2D images were excluded. Non-English language articles and non-peer-reviewed documents (e.g., theses, technical reports) were excluded.
Full-text Screening	Articles focusing on hardware-only aspects (e.g., camera hardware, UAV trajectory planning, or sensor placement) without proposing or evaluating a deep learning model for defect detection were excluded. Articles focusing on other bridge defects like cable defects instead of concrete damages were excluded.

Articles whose abstracts did not provide sufficient information to determine whether deep learning was used for concrete bridge surface defect detection with standard 2D images were also excluded.

In the full-text screening phase, we further checked that the bridges were primarily concrete, that the input data consisted of standard 2D images or video (excluding thermal, 3D, radar, and other non-visual modalities). Studies that focused only on hardware configuration (e.g., camera placement, UAV trajectory planning) without proposing or evaluating a deep learning defect detection model were excluded. Studies with missing or incomplete data, which could not be reliably included in the synthesis, were also excluded. This selection process resulted in 77 papers being selected for analysis.

Finally, in-depth analysis was used to investigate the use of deep learning algorithms to detect defects in concrete bridges. Content analysis was used to identify patterns in concrete bridge defect detection using AI, while descriptive analysis was employed to address trends and focuses in deep learning for bridge defect detection, as described in Section 4. One reviewer extracted data from each study using a piloted extraction form in Excel, and a second reviewer verified the data. The form collected information on authors, dataset details, defect types, deep-learning algorithms, and performance metrics.

Studies were grouped according to the type of detection task: classification, segmentation, object detection, and hybrid methods. The study's results were organized into summary tables. These tables were used to facilitate comparisons across studies. To explore potential causes of heterogeneity in the study results, we examined patterns in the summary tables, which included details on study characteristics like input image size, dataset size, and defect types.

We prespecified task-appropriate outcomes and consistently extracted them across studies. For semantic segmentation, we collected mean intersection over union (mIoU), precision, and recall when available. For object detection, we gathered precision, recall, and mean average precision (mAP) when available. For classification, we collected accuracy. Results were taken from the test set when available; otherwise, they were obtained from the validation set. When multiple results were reported within an outcome domain (e.g., alternative training procedures or different input sizes), we retained the highest value according to the authors' stated evaluation protocol. The risk of bias in the studies included in this review was evaluated through a qualitative approach by two independent reviewers. The assessment concentrated on important factors such as the study design, the handling of data, and the transparency of reporting. Any differences in the reviewers' assessments were addressed through discussion and consensus (Appendix A).

4. Analysis and Results

4.1. Descriptive Analysis

After carefully reviewing the selected articles, it was found that deep learning algorithms were first used in 2018 to detect defects in bridges. Before this, conventional image processing techniques, such as various filters and the Otsu method [29], were utilized. However, these methods had limitations due to the impact of lighting conditions, image resolution, and the level of noise present in the images [1]. Creating an image processing algorithm that could handle these scenarios was a significant challenge. Deep learning algorithms then emerged as a promising solution to overcome these limitations. Figure 3 demonstrates the trend of deep learning usage in bridge damage detection.

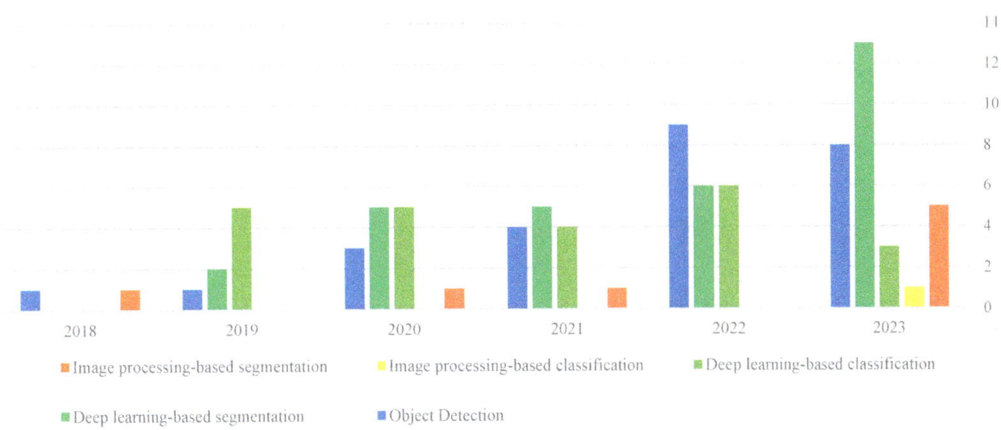

Figure 3. The trend of using deep learning methods.

Additionally, it is evident from Figure 3 that segmentation algorithms have become increasingly prominent in recent years. This may be because of their ability to identify damage at the pixel level and assess their severity. It is worth noting that only 28% of the researchers focused on measuring bridge defects. Among these, a significant majority (61%) specifically concentrated on measuring cracks, as illustrated in Figure 4.

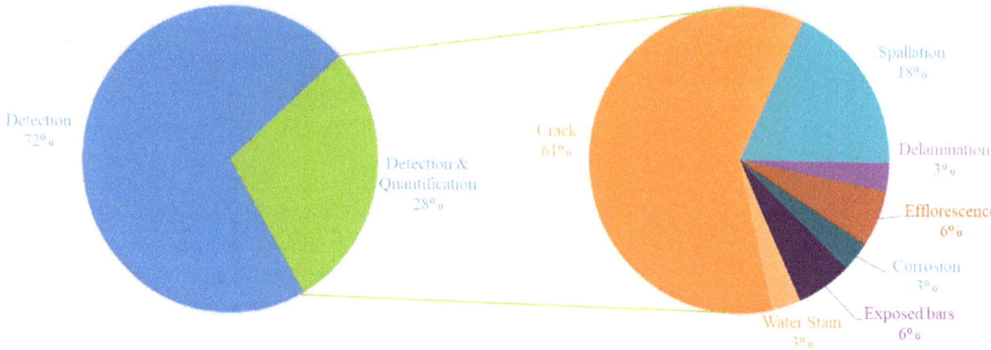

Figure 4. Frequency of damage detection and quantification methods.

In the same vein, Figure 5 emphasizes the critical role of crack detection in maintaining the structural integrity of concrete bridges. Cracks are the most common type of damage because they can lead to more severe issues if not managed on time. The figure highlights various defects requiring attention, including spallation and exposed bars, which can compromise the bridge's safety and longevity.

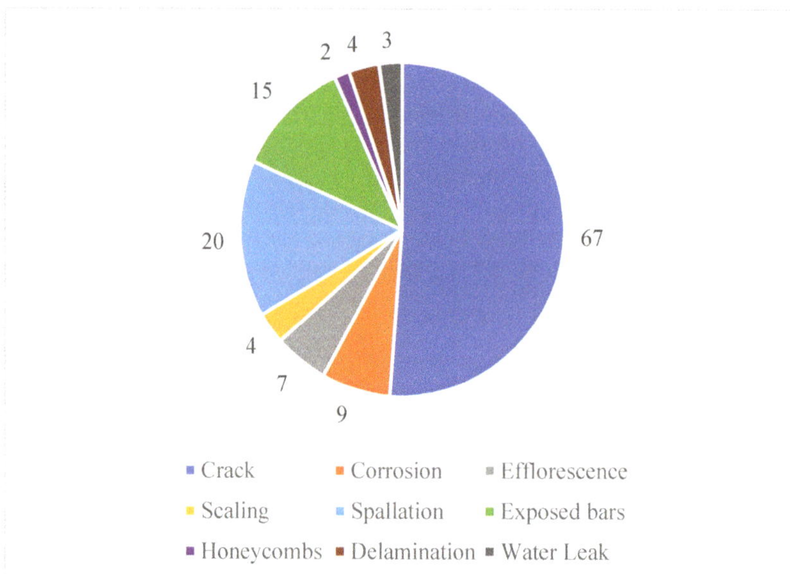

Figure 5. The frequency of detection methods used for each damage type.

4.2. Content Analysis

The traditional method of inspecting concrete bridges involves manual inspection. An inspector or a team of inspectors visits the bridge site to assess the structure visually, identifying any defects and determining their severity [3]. This information is then recorded in bridge inspection reports and transferred into BMS systems. However, this transfer process can result in data loss and is also time-consuming [30–32].

Though, new technologies are recently utilized as a substitute for this method. These methods can rapidly access bridge conditions, enhance data accuracy, and gain information about inaccessible parts of bridges. Among these technologies is the use of sensors, which record data continuously. However, sensors usually cover limited parts of bridges and are inaccessible once installed, resulting in poor sensor system maintenance. To overcome these limitations, digital images and computer vision techniques prove to be useful [6].

Some image-processing methods relied on filters, morphological analysis, and edge detection techniques without model training. However, these methods were insufficient due to the impact of lighting conditions, image resolution, and the level of noise present in the images. Developing an algorithm that could handle all the unpredictable situations in the real world was not easy [1,33]. Therefore, some researchers used machine learning algorithms such as support vector machines to overcome these limitations. For instance, Prasanna et al. [34] developed a histogram-based classification algorithm to detect cracks in concrete bridge decks, utilizing SVM. However, machine learning methods typically use a limited number of feature extraction levels, making it challenging to address the complexities of real-world image data [35].

As a result, deep learning algorithms emerged in bridge defect detection (Figure 6). Deep learning algorithms can automatically learn features from the data and handle large and complex datasets. Researchers using deep learning to detect bridge defects using images typically classify their approaches into image classification, image segmentation, object detection, and hybrid methods combining multiple algorithms. Figure 7 Illustrates the process of detecting bridge defects using deep learning techniques.

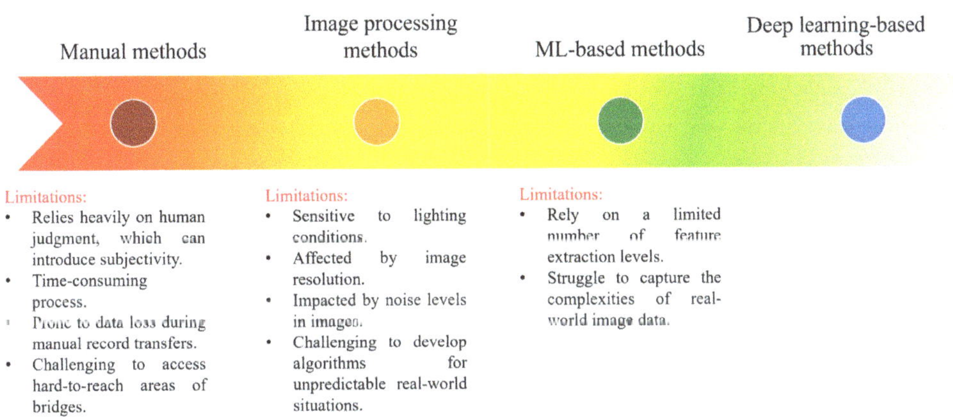

Figure 6. The transition from Traditional to AI-Powered bridge inspections.

4.2.1. Image Classification

Classification algorithms can assign a bridge defect label to each image. CNN is a pioneering image classification algorithm with this respect. Xu et al. [36] have successfully utilized end-to-end CNN, combined with atrous convolution, Atrous Spatial Pyramid Pooling (ASPP), and depthwise separable convolution, to develop a model for detecting concrete bridge cracks. Using atrous convolution allows for acquiring a wider receptive field without any reduction in resolution. ASPP module enables the network to extract multi-scale data, and depthwise separable convolution reduces the computational time. CNN is a powerful tool for feature extraction, but feature fusion is more critical, and CNN relies on fully connected layers to achieve this. Although these layers perform well in data fusion, they are insufficient for high-level information. To address this limitation, Zhang et al. [37] employed Convolutional Neural Networks and long short-term memory (LSTM) to detect concrete bridge deck cracks. LSTM was utilized as a feature fusion layer. This approach enabled faster crack detection compared to other CNN algorithms, making it particularly advantageous for real-time applications.

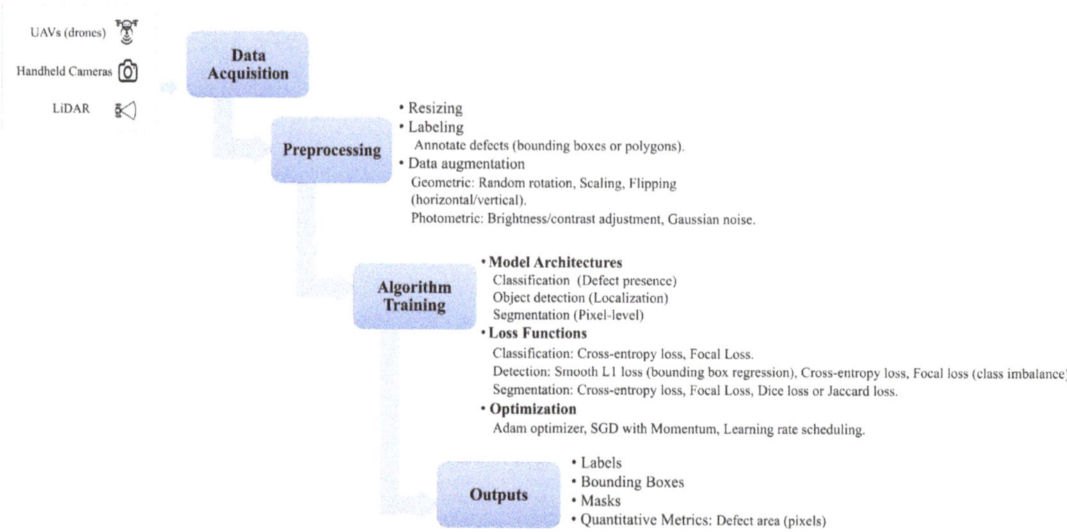

Figure 7. Workflow for bridge defect detection using deep learning.

As mentioned in Section 2.1, the VGG algorithm emphasizes expanding the network depth to 16–19 weight layers. VGG input images are fixed-size. Chen [38] trained a variant of the VGG structure over 1200 images captured by a UAV and a handheld DSLR camera to detect cracks. They utilized a migration learning technique to overcome the challenge of insufficient data.

DenseNet with dense connections allows for higher accuracy in smaller epochs, making it more suitable for small datasets. Alfaz et al. [39] Utilized the DenseNet algorithm to classify bridge cracks.

Li et al. [40] developed a model called Skip–Squeeze-and-Excitation Networks (SSENets), which had two modules of Skip–Squeeze–Excitation (SSE) and Atrous Spatial Pyramid Pooling (ASPP). The utilization of skip-connections in the SSENets model addresses the vanishing gradient problem found in deep networks, a technique also employed in the ResNet algorithm. Subsequently, the feature map produced by the SSE module is fed into the ASPP module to capture multi-scale features. In the context of crack detection, cracks occupy only a small portion of the image, and their widths vary significantly. Traditional convolutional methods need to be improved for analyzing cracks with different widths at multiple scales, which hinders capturing their features effectively. The Atrous Spatial Pyramid Pooling (ASPP) module addresses this by using atrous convolutions with varying rates to extract multi-scale features from cracks, improving detection accuracy.

CNNs were one of the most frequently used algorithms for defect classification, but lightweight models like MobileNet became popular to enable real-time defect detection. Chen et al. [41] proposed a lightweight model named MobileNetV3-Large-CBAM, which utilized sliding window to locate and classify cracks. The researchers used two open-source data sets, the first containing 6532 images after data augmentation and the second containing 15,068 images. The model combined with Focal Loss handled an imbalanced dataset and achieved 95.90% overall accuracy using the first dataset. They also compared this method with other cutting-edge CNNs, demonstrating the superior performance of their model.

Researchers have expanded their focus to detect multiple defects in addition to cracks. For instance, in 2019, Hüthwohl et al. [42] introduced a three-stage multi-classifier powered by Inception V3 to identify concrete bridge defects. The study recognized eight defect categories: spallation, cracks, rust staining, efflorescence, scaling, abrasion/wear, exposed bars, and general defects. They considered that some types of defects occur along with other types. For example, Exposed bars happens with spalling, scaling, and efflorescence. Rust staining occurs with exposed bars. As a result, they proposed a hierarchical three-stage classifier. Again, in 2019, Kruachottikul et al. [43] developed a novel approach using image processing and deep learning to facilitate bridge inspection. They utilized CNN to classify cracks, Spallation, Erosion, and stains. The model achieved 89% accuracy on a dataset containing 3926 images of bridges across Thailand. Also, Mundt et al. [44] created an open-source dataset named COncrete DEfect BRidge IMage (CODEBRIM) to classify cracks, Spallation, exposed bars, efflorescence (calcium leaching), and corrosion (stains). The data set contains 1590 images, utilized in several research studies to train classification algorithms. At last, two meta-learning approaches, MetaQNN and efficient neural architecture search (ENAS), were employed to classify the defects mentioned in the dataset. Cardellicchio, et al. [45] compared five different deep learning algorithms, namely InceptionV3, ResNet50V2, DenseNet121, MobileNetV3, and NASNetMobile, for the classification of corroded steel reinforcement, cracks, deteriorated concrete, honeycombs, moisture spots, pavement degradation, and shrinkage cracks. ResNet50V2 outperformed other models in detecting the majority of these damages.

In 2023, Trach [46] proposed a method based on convolutional neural networks to classify images of concrete bridge elements into four classes: defect-free, crack, spallation, and popout. The model was trained and evaluated on a dataset of real bridge inspection images obtained from 15 years of inspections in various regions of Ukraine. After comparing the performance of eight different CNN architectures, Trach determined that the MobileNet model was the most effective in terms of accuracy, loss function, and training time.

Abubakr et al. [47] compared Xception & Vanilla models based on convolutional neural networks for the classification of bridge defects. They trained the algorithms on the concrete defect bridge image (CODEBRIM) dataset to detect cracks, corrosion, efflorescence, spallation, and exposed bars on the bridge surface. The result illustrated the superiority of Xception over Vanilla in terms of accuracy.

Since training deep learning algorithms requires considerable data, scholars utilized transfer learning due to the limited concrete bridge defect images. Transfer learning is a significant concept in machine learning, where a trained model on a dataset is used as the starting point of the second model. A study by Zhu et al. [5] utilized CNN and transfer learning to accurately classify spallation, exposed bars, cracks, and pockmarks on concrete bridge surfaces, achieving a testing set accuracy of 97.8%. In 2022, Bukhsh et al. [48] created a dataset of 3588 images of cracks, spallation, exposed bars, and corrosion stains on Slovenian concrete bridges. They then implemented three separate training methods on the dataset using the VGG16 model: without transfer learning, pretraining VGG16 on the ImageNet dataset, and pretraining VGG16 on the CODEBRIM dataset. The second approach yielded the highest accuracy, likely attributed to the vast number of images in the ImageNet dataset. Zoubir et al. [49] constructed a dataset with more than 6900 images of bridge cracks, spallation, and efflorescence. Then, a VGG16 with three different levels of transfer learning on the dataset was trained to identify and localize bridge defects.

In 2021, Kruachottikul et al. [50] introduced an innovative visual monitoring system that utilized deep learning to detect various types of damage in concrete bridge substructures. The system captured images of the substructure, used a Modified ResNet-50 algorithm to identify defects, and then classified them accordingly. Lastly, they utilized an ANN model for defect severity prediction. The ANN model took 12 input variables related to bridge components and defect types to predict the severity of defects in severe and non-severe classes.

Aliyari et al. [51] captured several images from the Skodsberg Bridge and constructed two distinct datasets. One contained heterogeneous and noisy images, and the other one contained images with less noise. They pre-trained VGG16, ResNet50, Resnet152v2, Xception, InceptionV3, DenseNet210, NASNetLarge, and EfficientNetB4 on ImageNet and then on the Skodsberg Bridge datasets and SDNET2018 after applying some preprocessing. SDNET2018 is a benchmark dataset for concrete crack detection introduced by Dorafshan et al. [52]. Among the models, EfficientNetB4 and DenseNet210 achieved the highest accuracy of 96%. Following closely are ResNet50 and VGG16, both with an accuracy of 95%. However, VGG16, ResNet50, and Xception outperformed the other models in the UAV-based bridge inspection datasets. They also determined the optimal amount of transfer learning, which other researchers had not previously explored.

Table 3 presents a summary of papers that were reviewed in this section. It presents the utilized algorithm, the types of identified defects, preprocessing actions, and the employed dataset.

Table 3. Summary of reviewed papers on bridge defect detection using classification algorithms.

Author	Algorithm	Type of Defect(s)	Preprocessing	Dataset Details		Performance (%)
				Dataset Name	Number of Images	
Chen [38]	A Variant of the VGG structure	Crack	Labeling Data augmentation Cropping	Own data collection using a UAV and a handheld DSLR camera	2000	Accuracy = 99.1%
Hüthwohl et al. [42]	Inception V3	Cracks	Labeling	Own data collection and authority image sets	38,408	Accuracy = 97.4%
		Efflorescence				Accuracy = 96.9%
		Scaling				Accuracy = 94.6%
		Spallation				Accuracy = 94.3%
		General defects				Accuracy = 95.4%
		No defect				Accuracy = 97.2%
		Spallation scaling				Accuracy = 95.2%
		Exposed bars				Accuracy = 87%
		Corrosion				Accuracy = 94.9%
Zhu et al. [5]	CNN	Spallation Exposed bars Crack Pockmark	Brightness adjustment, saturation adjustment and flip adjustment labeling	Own data collection using a digital camera (Canon EOS 700 D, 18 megapixels)	1458	Accuracy = 97.8%
Mundt et al. [44]	Efficient Neural Architecture Search (ENAS)	Crack Spallation Exposed bars Efflorescence Corrosion (stains)	Labeling	COncrete DEfect Bridge IMage: CODEBRIM	1590	Accuracy = 70.78%
	MetaQNN					Accuracy = 72.19%
Zoubir et al. [49]	VGG16	Efflorescence Spallation Crack	Data augmentation (flipping, rotating) Labeling	Own data collection using two 20-MP consumer digital cameras with 5 mm focal length	6952	Accuracy = 97.13%
Bukhsh et al. [48]	CNN	Crack Spallation Exposed bars Corrosion stain	Labeling	BiNet	3588	Accuracy = 80 ± 1%
	VGG16					Accuracy = 80 ± 2%
	Pretrained VGG16-ImageNet					Accuracy = 88 ± 1%
	Pretrained VGG16-CODEBRIM					Accuracy = 81 ± 1%
Kruachottikul et al. [50]	ResNet-50	Crack Erosion Honeycombing Spallation Scaling	Data augmentation (image rotations, adding noises, and flipping images), labeling	Own data collection using a mobile application	3618	Accuracy = 81%

Table 3. *Cont.*

Author	Algorithm	Type of Defect(s)	Preprocessing	Dataset Details: Dataset Name	Dataset Details: Number of Images	Performance (%)
Xu et al. [36]	End-to-end CNN	Crack	Filtering, cropping, flipping, resizing	Own data collection using the Phantom 4 Pro's CMOS surface array camera	6069	Accuracy = 96.37%
Zhang et al. [37]	1D-CNN-LSTM	Crack	Images were transformed into the frequency domain Cropping Labeling	SDNET2018	16,789	Accuracy = 99.25%
Cardellicchio et al. [45]	DenseNet121 + transfer learning, InceptionV3 + transfer learning, ResNet50V2 + transfer learning, MobileNetV3 + transfer learning, NASNetMobile + transfer learning	Corroded/oxidized steel reinforcement Cracks Deteriorated concrete Honeycombs Moisture spots Pavement degradation shrinkage cracks	Labeling Data enrichment and augmentation (horizontal and vertical flipping)	Own data collection	2436	Best Accuracy = 85.19% for crack 85.4% for corroded and oxidized steel reinforcement 91.78% for deteriorated concrete 90.96% for honeycombs 76.43% for moisture spots 93.44% for pavement degradation 96.13% for shrinkage cracks
Trach [46]	CNN (MobileNet architecture)	Crack Spallation (this class includes two groups of defects—(scaling) and (spallation)) Popout	Manually crop 256 by 256 pixel rotating part of the images by 90○ to overcome the unbalanced dataset	Data collected from real inspection reports of bridge structures, which were taken for over 15 years in different regions of Ukraine	5200	Accuracy = 94.61%
Chen et al. [41]	MobileNetV3-Large-CBAM	Crack	Data augmentation (flipping, mirroring, cropping, and rotating data) Labeling	Bridge crack dataset [53]	6532	Accuracy = 95.9%
				Open-source datasets and images collected from the web	15,068	Accuracy = 99.66% for various material cracks test set and 99.69 for huge-width cracks test set
Zhang et al. [54]	Hybrid probabilistic deep convolutional neural network	Crack	Labeling	Own data collection	40,000	Accuracy = 99.09%
				Own data collection	2600	Accuracy = 92.49%
Abubakr et al. [47]	Xception	Cracks, corrosion, efflorescence, spallation, and exposed bars	Labeling	CODEBRIM	1590	Accuracy = 94.95%
	Vanilla					Accuracy = 85.71%
Alfaz et al. [39]	Dense Convolutional Network (DenseNet)	Crack	Cleaning, resizing, and manual categorization	Crack dataset, gathered by Xu et al. [36]	6069	Accuracy = 99.83%
Aliyari et al. [51]	CNNs	crack	Rotation, zooming in, cropping, flipping	SDNET	19,023	Accuracy (DenseNet210) = 97%
			Rotation, vertical and horizontal flipping, and zoom-in	Heterogeneous dataset: own data collection	308	Accuracy (VGG16, Resnet152v2) = 69%
			Rotation, vertical and horizontal flipping, and zoom-in	Homogeneous dataset: own data collection	400	Accuracy (Xception) = 74%
Li et al. [40]	A convolutional neural network named Skip-Squeeze-and-Excitation Networks (SSENets)	Crack	Filtering, cropping, and flipping Labeling	Crack dataset, gathered by Xu et al. [36]	6069	Accuracy = 97.77% Precision = 95.45%

4.2.2. Object Detection

Conventional computer vision classification algorithms can only assign a single label to an image. However, in the case of bridge defect detection, an image may contain multiple damages [8]. To address this challenge, object detection algorithms are commonly employed to identify defects using bounding boxes. Furthermore, object detection algorithms offer an approximate location of the defect, as opposed to classification algorithms. Among object detection algorithms, YOLO is a particularly effective tool for detecting concrete damages in images. Researchers like Murao et al. [55], Yu et al. [56], Li et al. [57], Deng et al. [58], and Ji [59] have developed methodologies for detecting bridge cracks by employing images and YOLO algorithms. For example, Murao et al. [55] employed YOLOv2 on a set of UAV-captured images, but the model accuracy was only 60% due to insufficient training images. This paper highlights the importance of employing high-quality datasets. Similarly, Deng et al. [58] modified YOLOv2 to detect bridge cracks and handwriting on bridge surfaces.

Ruggieri et al. [60] also emphasized the importance of datasets and trained YOLOv5 for detecting Cracks, Corroded steel reinforcement, Deteriorated concrete, Honeycombs, Moisture spots, and Pavement degradation on concrete bridge surfaces employing 2685 images. They declared that their dataset was challenging due to visual similarities among some classes, which could result in misclassifying one defect class as another. Moreover, overlapping two defect classes could lead to the failure of the algorithm. Zhang et al. [61] proposed a method for highway concrete bridge defect detection using YOLOv3. They trained the model on a dataset containing 2206 images from cracks, spallation, exposed bars, and pop-out. Furthermore, in 2023, Liu et al. [62] employed YOLOv5 to detect cracks in concrete bridges and also utilized deep convolutional generative adversarial network (DCGAN) to produce artificial images to enhance dataset image number and reduce the cost of data acquisition. Yamane et al. [63] proposed a novel method for detecting exposed bars from images using YOLOv5 and then reflecting it on a 3D bridge model. SfM (Structure from Motion) was employed for 3D model creation. SfM is a technique that creates 3D models from 2D images. Using SfM, a 3D model of the bridge and defects was constructed, which enhanced the integration of damage detection with BIM. In another research employing the YOLOv3 model, Teng et al. [64] successfully identified cracks and exposed bars in concrete bridges. Their study included a comparative analysis of YOLOv2, YOLOv3, and Faster RCNN. In addition, they implemented transfer learning and data augmentation techniques to enhance the performance of YOLOv3. The transfer learning contained retraining the SqueezeNet model, which extracts features in their YOLO model, using bridge defect images to enhance its capabilities. Consequently, the improved YOLOv3 model exhibited superior accuracy compared to the other models.

Other object detection algorithms that can help identify bridge damages include R-CNN (Region-based Convolutional Neural Networks) and Faster R-CNN. A study by Yu et al. [65] utilized Faster R-CNN to develop an object detector capable of detecting cracks, spallation, and exposed bars, achieving an mAP_{50} of 84.56%. Similarly, Li et al. [66] utilized Faster R-CNN, training it with a dataset of 637 images captured by UAVs. Their model was tested on a cable bridge in China to detect three types of cracks (transverse, longitudinal, and alligator cracks), achieving a precision of 92.03%. Lin et al. [67] generated a UAV flight plan by considering various input parameters such as a 3D map, georeferenced inspection areas, target capture resolution, cameras' relative orientation, required visibility, completeness, and safety. Subsequently, they generated 3D point clouds from the captured images. Then, they employed a feature pyramid network (FPN) in a Faster R-CNN system to identify bridge damages, including cracks, spallation, efflorescence, corrosion stains, and exposed bars, in the images. The results were projected back into the 3D point cloud. In 2023, using UAV images, Gan et al. [68] proposed a method for detecting, quantifying, and visualizing cracks on the bottom of concrete bridges. The process involved utilizing deep learning techniques such as Faster R-CNN and BIM. The authors captured 637 images of the bridge using UAV and augmented the dataset by applying flipping and other methods. Subsequently, they trained the model on the augmented dataset and utilized spline curves and crack box girder families to create realistic crack shapes and map them to the bridge model. Finally, they employed a comprehensive judgment method to evaluate the health status of a region of the bridge with cracks. The authors used crack width as a criterion to identify the structural health grade referred to technical specifications for highway maintenance and bridge evaluation [69]. The Faster R-CNN model achieved an accuracy and recall of 92.03% and 96.54%, respectively.

Image quality can highly affect algorithm performance. As a result, Hong et al. [70] initially processed the images using a super-resolution (SR) model to improve image quality. Subsequently, they labeled dataset images based on damage type and bridge member. Then, they developed n models for each member-specific dataset to identify efflorescence, water leaks, concrete sealing, concrete spallation, cracks, and RC corrosion. Where n denotes the number of bridge members, they trained and evaluated Mask R-CNN, YOLO, and Blendmask and ultimately selected the most effective model. Their rationale for this approach was the observation that the shapes of the same type of damage differed across different bridge members, prompting them to develop distinct models for each member. They achieved precise detection results by training optimized detection models for each member. Specifically, the Blendmask model demonstrated an 11.31% improvement, while the Mask-RCNN model showed an 8.893% enhancement in accuracy performance. Notably, the Blendmask outperformed the mask-RCNN overall, with a mean average precision (mAP) of 92.87 compared to 90.374. However, for specific cases like rail and road pavement models, mask-RCNN performed better. In another study, Deng et al. [71] trained Faster R-CNN on a set of images to detect cracks and handwriting on bridges and compared it to YOLOv2. The results indicated that Faster R-CNN achieved higher detection accuracy for concrete crack detection, while YOLOv2 had higher speed.

Recent advancements in defect detection for concrete bridges incorporated attention mechanisms into deep learning models. A research [72] proposed an enhanced YOLO11 model that integrated attention mechanisms to better focus on relevant details in bridge defect images. This attention mechanism not only improved model performance but also enabled real-time implementation on constrained devices, making it a practical tool for on-site inspections.

Table 4 presents a summary of papers that were reviewed in this section.

Table 4. Summary of reviewed papers on bridge defect detection using object detection algorithms.

Author	Algorithm	Type of Defect(s)	Preprocessing	Dataset Details		Performance (%)
				Dataset Name	Number of Images	
Murao et al. [55]	YOLOv2	Crack	Labeling, resized, inverted, rotated, HSV (Hue, Saturation, Value)	Own data collection using a UAV equipped with a camera and various image data collected from the Internet	164 for system 1	
		Concrete joints with line forms other than cracks:, etc., Crack Branch-crack chalk Branch-chalk			166 for system 2	Precision = 27.95%
		Concrete joints with line forms other than cracks: etc., Crack Branch-crack chalk Branch-chalk			253 for system 3	Precision = 41.63%
Yu et al. [56]	YOLOv4-FPM	Crack	Cropping, Labeling	Own data collection using digital SLR camera (Canon EOS M10) and UAV	2768	mAP = 97.6%
Li et al. [57]	Lite-YOLO-v4	Crack	Mosaic data augmentation Labeling	Data collected from a bridge underwater inspection company and the Internet, including Crack 500 dataset	8780	mAP = 77.07% Precision = 93.97% Recall = 47.98%
Deng et al. [58]	YOLOv2	Cracks and handwriting	Horizontal flipping labeling	Data collected by inspectors during annual visual inspections	3010	mAP = 77%
Ji [59]	YOLO Autobot	Crack	labeling	Crack dataset provided by RoboFlow [73]	3917	mAP = 82.6%
Deng et al. [71]	Faster R-CNN	Cracks and handwriting	labeling horizontal flipping cropping	Images from inspection records of concrete bridges taken by consumer-grade cameras with complex background information	5009	mAP = 82%
Ruggieri et al. [60]	YOLOv5 (different versions)	Crack, corroded steel reinforcement, deteriorated concrete, honeycombs, moisture spots, pavement degradation	Labeling	Collecting figures from the observations of some existing bridges in Italy	2685	For YOLOv5m6: mAP = 20.66% Precision = 43.63% Recall = 24.24%

Table 4. *Cont.*

Author	Algorithm	Type of Defect(s)	Preprocessing	Dataset Details		Performance (%)
				Dataset Name	Number of Images	
Zhang et al. [61]	Modified YOLOv3	Crack Spallation Exposed bars Pop-out	Downsampling Labeling Data augmentation: (scaling and cropping, flipping, manipulating the images by applying the motion blur, changing brightness, or adding salt and pepper noise)	Data acquired from the Hong Kong Highways Department + Own data collection	2206	mAP = 79.9%
Liu et al. [62]	YOLOv5	Crack	Labeling	Own data collection: The original dataset	180 sets of training samples and 10 sets of validation samples	
				The extended dataset produced by DCGAN	180 sets of training samples and 10 sets of validation samples)	
Yamane et al. [63]	YOLOv5	Exposed bars	Labeling	Data were taken during the inspection of bridges managed by the Kanto Regional Development Bureau of Japan's Ministry of Land, Infrastructure, Transport and Tourism from 2004 to 2018 using Skydio 2, a small UAV developed by Skydio.	1000	AP = 64%
Teng et al. [64]	Improved YOLOv3	Cracks Exposed bars	Data augmentation (color jitter augmentation, flipping, scaling) Labeling	Data from concrete bridges (along the Erenhot–Guangzhou and Guangzhou–Kunming Expressways)	1660	mAP= 91%
Hong et al. [70]	(BDODC-F): Mask R-CNN Blendmask	Efflorescence Spallation Crack Corrosion Water leak Concrete scaling	Enhancing the image quality Labeling	Own data collection	100 for Aggregated model training	Blendmask accuracy = 92.675% Mask-RCNN accuracy = 98.679%

Table 4. *Cont.*

Author	Algorithm	Type of Defect(s)	Preprocessing	Dataset Details		Performance (%)
				Dataset Name	Number of Images	
Yu et al. [65]	Modified Faster R-CNN	Crack Spallation Exposed bars	Resizing Labeling	Images taken by a Canon EOS 5DS R camera from the periodic inspection of bridges by the CCCC First Highway Consultants Co., Ltd.	1000	mAP = 84.56%
Gan et al. [68]	Faster R-CNN	Crack	Labeling	Own data collection using DJI M210-RTK	637	Precision = 92.03% Recall = 92.26%
Lin et al. [67]	Faster-RCNN	Crack	Labeling Data augmentation (flipping, rotation, shearing, and changing brightness and contrast)	Own data collection using UAV	742	Average precision = 49.2%
		Spallation				Average precision = 84.6%
		Efflorescence				Average precision = 57.6%
		Corrosion stains				Average precision = 74.1%
		Exposed bars				Average precision = 84.5%
Ngo et al. [74]	CNN	Crack	Changes in brightness and rotation to increase the training image's quality	Own data collection and the existing datasets (from Kaggle's image, SDNet2018, and …)	51,000	Accuracy = 95.19%
Lu et al. [75]	MSCNet	Crack	Labeling	SDNET dataset and CCIC dataset	30,000	Accuracy = 92.7% Precision = 93.5% Recall = 94.2% F1-score = 93.8%
Ruggieri et al. [72]	Yolo11x+ attention mechanisms	Crack Corroded steel bar Deteriorated concrete Honeycomb Moisture spot Shrinkage Pavement degradation	Labeling	Dataset proposed by [45]	6580	mAP = 59.38% Precision = 82.87% Recall = 52.61% F1-score = 64.36%

4.2.3. Image Segmentation

Initially, researchers segmented bridge defects using image processing methods, such as threshold-based segmentation [76–78]. Recently, deep learning models have transformed this field, achieving remarkable performance improvements [79]. The pioneering deep learning algorithm for image segmentation was the Fully Convolutional Network (FCN) [21]. Since then, multiple traditional deep learning segmentation algorithms have been developed, including U-Net, SegNet, PSPNet, DeepLab, and more [80]. The outputs of segmentation algorithms and the object detection algorithm are compared in Figure 8. In 2019, Rubio et al. [81] applied FCN to segment delamination and exposed bars on concrete bridge decks. Their dataset consisted of 734 images of bridges located in Japan. In addition, they employed transfer learning by training VGG-16 on the PASCAL VOC 2012 dataset. The model achieved a mean accuracy of 89.7% for delamination and 78.4% for exposed bars detection. In 2022, Lopez Droguett et al. [82] proposed a DenseNet algorithm with 13 layers for crack segmentation on concrete bridges. Additionally, they provided two datasets for crack detection. The achieved Intersection over Union (IoU) for this model was 94.51%. Merkle et al. [83] segmented crack using U-Net, achieving a precision of 56.1%. The precision was not very high due to the limited amount of data used for training (76 images). By combining photogrammetry and SfM, they reconstructed a metric 3D model of the bridge so that each image pixel could be projected onto the bridge surface with real-world coordinates. The crack masks were post-processed to extract the medial axis and crack contours, which were then back-projected onto the 3D CAD/SfM model. Crack length was computed as the Euclidean distance between the farthest 3D points along each crack centreline, and for crack width at each 3D medial-axis point, they searched for the nearest 3D contour point and computed the distance between them. The local crack width was then defined as

$$W = 2 \times d_{axis-contour} + GSD \tag{7}$$

where GSD (ground sampling distance) is the local real-world size of one pixel (e.g., 0.2 mm/pixel), derived from the SfM reconstruction and camera parameters. In other words, GSD tells how many millimeters on the bridge correspond to one image pixel at that location. In another study [84], U-Net was applied to multi-class segmentation of concrete defects, and different loss functions were tried to check performance. The study demonstrated that U-Net, when trained with IoU loss functions, achieved the best results for detecting defects in concrete structures.

In 2023, Fukuoka and Fujiu [85] introduced a segmentation method for detecting bridge defects using a transfer learning algorithm called SegFormer to identify delamination and exposed bars. Additionally, they trained SegNet, a convolutional neural network, using the same dataset with images of various dimensions to compare the model results. Their findings revealed that larger input image dimensions resulted in improved model performance. SegFormer exhibited slightly lower precision when detecting delamination than the SegNet model; however, it demonstrated significantly better recall. Conversely, in exposed bars detection, the SegFormer model outperformed the SegNet model in all aspects.

Li et al. [86] presented a neural network for bridge crack detection that used an encoder–decoder structure and an MPM module. The MPM module helped the network to detect long and thin objects like cracks. The authors also collected and processed 2240 original photos of bridge cracks, which were expanded to a bridge crack data set of 10,000 images after image processing, and used it to train and test their model. They compared their model with four other image segmentation networks and showed that their model performs better in detecting bridge cracks. However, they acknowledged that their method does not calculate the width, length, area, and other parameters of the cracks, which are important for assessing the bridge condition.

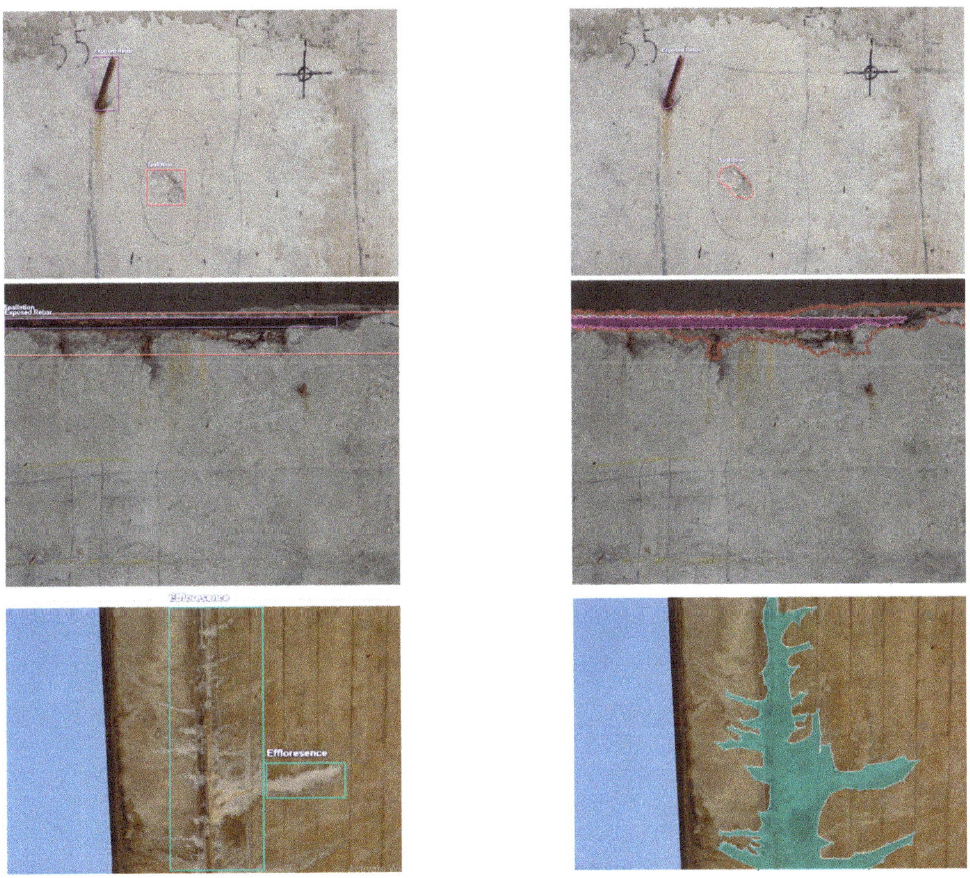

Figure 8. Object detection bounding boxes and segmentation masks.

Yamane et al. [87] proposed integrating and recording damage detected from multiple images into a 3D model using deep learning and structure from motion. The authors used Mask R-CNN with PointRend to detect corrosion damage from bridge images. They trained and evaluated the model on a dataset of 1966 images and demonstrated its high accuracy and precision in detecting damage. They employed SfM to construct a 3D model of the bridge from 2316 images captured by a UAV.

Some scholars modified the traditional deep learning segmentation algorithm architecture to improve their performance; for instance, Xu et al. [88] modified DeepLabv3+ with a MobileNetV2 backbone, modified ASPP module, and piecewise loss function to segment cracks on bridges. They also trained the model on two additional datasets (one containing concrete spallation with exposed bars and the other with cable corrosion) to segment spallation, exposed bars, and cable corrosion. Bae et al. [89] introduced "SrcNet", a deep neural network based on CNN designed to enhance automatic crack detection in raw digital images captured by drones from large civil infrastructure. The network addressed the blurriness issue in these images caused by motion. To address this issue, SrcNet used deep learning to generate high-resolution images from raw input. Then, it employed VGG-16 to detect cracks at the pixel level automatically. Deng et al. [90] introduced a novel deep-learning method for detecting defects in concrete bridges, including exposed bars and delamination. Their developed network, LinkASPPNet, integrated a ResNet encoder with an enhanced ASPP module to capture multi-scale information. The paper also introduced an improved loss function to handle dataset imbalance. The model was trained using a dataset of 732 labeled images, which three experts annotated. Its performance was then compared to U-Net and LinkNet, and it outperformed both models, achieving F1 scores of 93.46% and accuracy of 92.91%, respectively. There are also further studies on bridge defect segmentation using deep-learning algorithms [10,91–94].

Creating a dataset of images that show cracks in pixel-level labels is time-consuming, and it takes about 1.5 min to annotate a crack image with 256 $\times$ 256 resolution at the pixel level. This means annotating thousands of images with pixel-level labels is challenging. To deal with this problem, Jin et al. [95] suggested a GAN-based method that simultaneously generated synthesized crack images and their corresponding labels. To do that, a dataset containing crack images and their pixel-wise labels was prepared first. Then, the Deep Convolutional GAN model was adopted to generate synthesized crack annotations, while the Pixel2Pixel model was used to create the corresponding synthesized crack images. They used 7800 pixel-wise labeled crack images and produced 23,400 synthesized crack annotation images. Finally, they trained PCR-Net on a mixed dataset consisting of both real and synthesized crack images. They stated that the proposed method for establishing the synthesized crack image dataset could be helpful in training deep neural networks. Also, Subedi et al. [96] presented a novel method for segmenting bridge components and damages using synthetic images and deep-learning models. Their method consisted of a new architecture combining a RegNet encoder with a Nested U-Net decoder, which they claimed to outperform other state-of-the-art architectures (like DeepLabV3+, U-Net, PSPNet, LinkNet, RefineNet) in component and damage segmentation tasks. They also employed a Lovasz–Softmax loss function to address the class imbalance problem and a weighted softmax ensembling to enhance the performance. In another study, Ayele et al. [97] used Mask-RCNN to segment bridge cracks. To overcome the problem of pixel-wise labeling, they used an active learning approach for labeling images captured by UAV from Skodsberg Bridge in Norway, which led to 90% less manually labeled data. They also measured length and width as Euclidean distances between contour points in the binary mask, then converted those pixel distances to physical crack length/width using the known ground sampling (pixel size) of the orthomosaic. Wang et al. [98] used a refinement network embedded with an attention mechanism (RefineNet-AM) on a synthetic (Tokaido) dataset to detect bridge components and damages after an earthquake. The attention mechanism, inspired by the human ability to pay more attention to important areas of images, increased the weight of images' important regions. They demonstrated that the proposed RefineNet-AM could achieve satisfactory accuracy and robustness on the test dataset and outperform a baseline model (U-Net) on several metrics.

Some researchers combined the defect information from deep learning algorithms with 3D models and BIM. Montes et al. [99] developed a unique 3D dataset based on a real bridge structure using a low-cost LiDAR (Light Detection and Ranging)-enabled imaging device. The dataset contained point clouds and RGB images of bridge components and damages, with semantic annotations. They proposed a deep learning method for semantic segmentation of bridge components and damages using an enhanced 3D graph neural network. A 3D Graph Neural Network (3D GNN) is a neural network created to process data structured as graphs in three-dimensional space. First, a colored point cloud image was input into SubNet1 of enhanced 3D GNN. This image was used to create a directed graph, where each pixel was considered a node and was connected to its K-nearest neighbors in 3D space. Pixel features were then calculated using color information and a transformed CNN, which serves as the initial node information, resulting in a component prediction map. In the second step, initial information and weights were re-assigned to each node based on the results of the first sub-network and the constructed graph network. Color information and node types were combined, and initial convolution was performed through CNN, serving as the initial node information for the second network. The final output was a damage type prediction map, effectively utilizing texture and geometric information throughout the process. Artus et al. [100] conducted a study on applying Building Information Modeling to model bridge defects. Their research established a systematic approach for automated collection and transfer of defect data and creating a defect information model for data exchange. The study used Industry Foundation Classes (IFC) for this purpose. Additionally, they utilized TernausNet to segment spalled areas on the captured images of the bridge and an SfM-based pipeline to generate a metric point cloud of the bridge components. The segmented spall pixels were then back-projected onto this 3D point cloud so that the defect geometry (area and volume of the spall) was obtained directly in real-world units and stored as voiding features in the BIM model. Finally, by using an existing BIM model in the form of an IFC file and point clouds created from the photos of the damaged structure, they developed a geometric-semantic as-damaged BIM model, including the damage data. Borin and Cavazzini [101] introduced an approach that integrates deep learning, imaging, and BIM to evaluate the state of concrete bridges. They employed Mask R-CNN to identify and segment spallation. The model was trained on a dataset containing 575 manually annotated images and tested on a bridge in northern Italy. This study used images to generate point clouds, forming the foundation for constructing the BIM. The BIM then functioned as a platform for organizing and visualizing information regarding the bridge's condition.

In some studies, measurements of the defects have also been carried out in addition to identifying them. For example, a study conducted by Jang et al. [102] aimed to detect cracks in high-rise bridge piers using a climbing robot and deep learning algorithms. The robot captured high-quality images while climbing the piers, which were then used to train a modified SegNet to segment cracks. To determine the length and width of the cracks, the images were converted to grayscale, and a median filter was applied. Background information was removed by subtracting the filtered image from the grayscale image, and the Euclidean distance transform (EDT) and skeletonization were used to calculate the crack length and width. Using a pinhole-camera scale factor:

$$s = d_w l / P f \tag{8}$$

where d_w is the working distance, l the physical sensor size, P the image resolution, and f the focal length, they obtain a metric resolution of 0.086 mm per pixel. Crack length was then obtained by summing the lengths of the skeleton pixels and multiplying by this scale, while crack width at each skeleton point was derived from the EDT radius (distance to the crack boundary) multiplied by the same scale. In this way, both crack length and width were expressed in real-world units (millimeters) rather than pixels.

In 2020, McLaughlin et al. [103] automated the inspection and quantification of damaged areas in concrete bridges using robotics and deep learning. Their study involved using a robot equipped with a camera and LiDAR to capture images of bridges. The approach integrated SLAM to combine LiDAR data and defect information in images into a labeled 3D map of the bridge structure. Additionally, they employed a CNN architecture to automatically identify spallation and delamination in both regular and infrared images. The researchers also proposed a technique to measure defect areas on the 3D map. For automatic damage assessment, they utilized MobileNetV2 for feature extraction and DeepLab V3 for pixel-level segmentation. This CNN architecture achieved high accuracy in detecting structural defects.

Detecting post-earthquake bridge defects through deep learning has proven to be a valuable application. In a study by Ye et al. [104], the multi-task high-resolution net (MT HRNet) model was utilized to identify concrete bridge defects such as cracks, spallation, and exposed bars and evaluate bridge safety levels. It is a neural network designed to handle multiple tasks simultaneously while maintaining high-resolution representations throughout the process by avoiding downsampling. They trained the model on the Tokaido dataset and achieved an accuracy of 0.9661 in detecting the bridge components and an accuracy of 0.9944 in detecting bridge damages. Based on the findings, the authors determined that columns are the most crucial component of a bridge. By considering three key indices, including the damage type of the column (shear and torsional damage and bending damage), spallation area, and the width of the cracks (in pixels), the post-earthquake bridge safety risk level was assessed.

Recently, researchers have started exploring transformer-based models for concrete defect segmentation, moving beyond traditional CNN architectures. In 2025, a study [105] used the ALF-ViT model, a Vision Transformer-based approach for crack detection in low-light concrete images, which showed good segmentation performance in challenging illumination conditions, showing promising potential for real-world inspection scenarios.

Table 5 summarizes the reviewed papers on bridge defect detection using segmentation algorithms.

Table 5. Summary of reviewed papers on bridge defect detection using segmentation algorithms.

Author	Algorithm	Type of Defect(s)	Aim	Quantified Attribute	Preprocessing	Dataset Details		Performance (%)
						Dataset Name	Number of Images	
Rubio et al. [81]	Fully convolutional networks	Delamination	Detection		Labeling Flipping Changing in illumination and slight affine transformations	Data collected from inspection records of bridges in Niigata Prefecture, Japan	734	Mean accuracy = 89.7%
		Exposed bars						Mean accuracy = 78.4%
Tian et al. [77]	Otsu binarization method (threshold segmentation)	Crack Other damages, such as concrete cavities and damage	Detection and quantification (in physical units)	Width Length Area	Grayscale processing Gray adjustment Adaptive filter function Wiener2 for noise removal			Image processing method
Vivekananthan et al. [78]	Image processing-based segmentation (gray-level discrimination approach)	Crack	Detection and quantification	Area and orientation	Gray level discrimination	Own data collection using CMOS cluster camera	2068	Accuracy = 95%
Adhikari et al. [76]	Image processing-based segmentation (threshold operation:(a) maximum entropy; (b) Otsu; and mean of the intensity in the given image Erode and Dilate)	Crack	Detection and quantification (in physical units)	Width Length	Image enhancement using point processing, histogram equalization, and mask processing	Own data collection using SONY-DSC T5 digital camera		Image processing method
Lopez Droguett et al. [82]	DenseNet-13	Crack	Detection		Labeling	Own data collection: CRACKV2	409,432	mIoU = 92.13%
Merkle et al. [83]	U-NET	Crack	Detection and quantification (in physical units)	Width Length	Labeling	Own data collection	126	Precision = 56.1% Recall = 59.9% IoU = 40.8%
Fukuoka and Fujiu [85]	SegFormer	Delamination Exposed bars	Detection		Data augmentation (flipping, scaling, rotation) Labeling	Data collected from the Japanese bridge inspection report: dataset A	17,810	Precision: Delamination detection = 80.8% Rebar-exposure detection = 70.8%
						Data collected from the Japanese bridge inspection report: dataset B	17,810	Precision: Delamination detection = 77.1% Rebar-exposure detection = 74.7%
						Data collected from the Japanese bridge inspection report: dataset C	17,810	Precision: delamination detection = 76.9% rebar-exposure detection = 76.5%
Li et al. [86]	A neural network based on the encoder–decoder	Crack	Detection		90°, 180° clockwise rotation and horizontal inversion Labeling	Own data collection using a telephoto camera on the equipment cart to take pictures of the bridge cracks	2240 which were expanded to 10,000 images	mIoU = 84.5% Precision = 98.3% Recall = 97.3%
Yamane et al. [87]	Mask-R-CNN	Corrosion	Detection		Resizing Labeling	Own data collection	1966	Accuracy = 94% Precision = 80%

Table 5. *Cont.*

Author	Algorithm	Type of Defect(s)	Aim	Quantified Attribute	Preprocessing	Dataset Details		Performance (%)
						Dataset Name	Number of Images	
Xu et al. [88]	Modified DeepLabv3+	Crack	Detection		Data augmentation (horizontal and vertical flipping, random rotation, and random translation) Labeling	Own data collection captured from several actual bridges in multiple scenes, scales, and resolutions	4303	mAP = 88.8% mIoU = 77.6%
Bae et al. [89]	End-to-end deep super-resolution crack network (SrcNet)	Crack	Detection		Rotating, flipping, labeling	Own data collection and web scraping	4055	Precision on Jang-Duck bridge = 81.18% Recall on Jang-Duck bridge = 92.65%
Deng et al. [90]	LinkASPPNet	Delamination, exposed bars	Detection		Labeling	Images from inspection records of Japanese concrete bridges	732	Mean precision= 73.59% Mean IoU = 61.95%
Munawar et al. [92]	CycleGAN	Corrosion	Detection		Adjusting the image brightness and size Data augmentation (flipping, rotation, cropping) Labeling	Own data collection using the UAV model DJI-M200 Also, images were extracted from public datasets	1300	mIoU = 87.8% Precision = 84.9% Recall = 81.8%
Li et al. [106]	Fully convolutional network and a Naive Bayes data fusion (NB-FCN) model	Crack-only Handwriting Peel off Water stain Repair trace	Detection and Quantification (in physical units)	Crack Length and width	Rotating Flipping Labeling Resizing	Own data collection using an image acquisition device named Bridge Substructure Detection (BSD-10)	7200	Accuracy = 97.96% Precision = 81.73% Recall = 78.97% F1-score = 79.95%
Fu et al. [91]	DeepLabv3+	Crack	Detection		Flipping, rotation, scaling Labeling	Own data collection using digital equipment and from the Internet in various environments.	5000	mIoU = 82.37%
Qiao et al. [93]	EMA-DenseNet	Cracks, exposed bars	Detection		Pixel-level labeling Rotation Cropping	Yang et al. [107] dataset	800	Validation mIoU = 87.42%
						Own data collection from bridges in Xuzhou (Zhejiang Province, China)	1800 crack images and 2500 rebar images	Validation mIoU = 79.87% Test mIoU= 80.4%
Jin et al. [95]	PCR-Net-based model	Crack	Detection		Labeling	Synthesized crack image dataset named Bridge Crack Library 2.0	26,600	Accuracy = 98.9% mIoU = 61.49% Precision = 81.63% Recall = 71.36%
Subedi et al. [96]	Nested Reg-Unet	Concrete damage and exposed bars	Detection		Data augmentations (flipping, random brightness. and contrast change) Labeling	Tokaido Dataset [108]	7079	mIoU = 84.19% Precision = 90.98% Recall = 91.04%
Ayele et al. [97]	Mask R-CNN	Crack	Detection and quantification (in physical units)	Cracks length and width	Labeling	Own data collection using a UAV		Accuracy = 90%
Wang et al. [98]	RefineNet with Attention Mechanism (RefineNet-AM)	Concrete damage and exposed bars	Detection			Tokaido Dataset [108]	8599	mIoU = 67.3% Precision = 84.4% Recall = 73.8%

Table 5. *Cont.*

Author	Algorithm	Type of Defect(s)	Aim	Quantified Attribute	Preprocessing	Dataset Details		Performance (%)
						Dataset Name	Number of Images	
Montes et al. [99]	3D GNN	Corrosion, spallation, cracks, and leaking water	Detection		Labeling	Own data collection using a lidar		Accuracy = 93.914% mIoU = 33.98%
Artus et al. [100]	TernausNet16	Spallation	Detection and quantification (in physical units)			Spallation and crack (CSSC) database [109]	715	Accuracy = 91.96% mIoU = 83.26% Precision = 87.55% Recall = 81.36%
Borin and Cavazzini [101]	Mask R-CNN	Spallation	Detection and visualization		Labeling	Own data collection	575	
Jang et al. [102]	Modified SegNet	Intact Crack Marker	Detection and quantification (in physical units)	Crack length and width	Contrast enhancement, Labeling	Own data collection using a ring-type climbing robot at Jang-Duck Bridge in Gangneung City. South Korea	1021	Precision = 90.92% Recall = 97.47%
McLaughlin et al. [103]	DeepLab V3	Spallation Delaminations	Detection and quantification (in physical units)	Area	Flip horizontal Flip vertical, Rotation Width and height shift Labeling	Own data collection	496 infrared images 600 visual spectrum images	Validation mIoU = 71.4% Validation mIoU = 82.7%
Ye et al. [104]	Multi-task high-resolution net (MT HRNet)	Concrete damage (crack and spallation) and exposed bars	Detection and quantification	Spallation area and the width of the cracks (in pixels)		Tokaido dataset [108]	13,956	Accuracy = 99.44% mIoU = 80.47% Precision = 94.66% Recall = 83.62%
Shen et al. [105]	Adaptive learning filters vision transformer (ALF-ViT),	Crack	Detection		Enhancement	CrackForest data set [110], Crack500 data set [111] and own data collection using mobile devices	1339	mIoU = 73.3%
Na and Kim [84]	U-Net	Crack, spallation and delamination, water leakage, exposed bars, and paint peeling	Detection		Data augmentation Labeling	Own data collection using a UAV	14,155	Accuracy = 95.7% Precision = 91.2% Recall = 91.4% F1-Score = 91.3%

4.2.4. Hybrid Methods

Segmentation algorithms were often integrated with classification and object detection algorithms to enhance the speed and accuracy of detection as depicted in Figure 9; for instance, Peng et al. [112] proposed a method based on drones and machine vision for detecting cracks and measuring their width. This approach utilizes the R-FCN for crack detection and localization, along with the Haar-AdaBoost algorithm for segmentation. Additionally, they incorporated a three-point laser rangefinder to accurately measure the distance to the object, enabling crack width calculation. Flah et al. [113] developed a model based on image processing and deep learning algorithms to detect cracks on concrete surfaces such as buildings and bridges. The method utilized CNN for crack detection, followed by edge detection and thresholding techniques to accurately measure the crack's length, width, and orientation angle. Crack dimensions were first obtained in pixels and then converted to real-world units by asking the user to input the physical area of the imaged surface; using the image resolution, the algorithm scaled pixel distances to metric values, achieving small quantification errors for crack length, width, and orientation. Kim et al. [114] trained well-known CNNs, including AlexNet, VGG-16, and ResNet-50, to detect cracks in blurry images. The images were cropped into square patches, and the CNNs classified the patches into two classes: crack and non-crack. After identifying the region of a connected crack during the detection stage, it was visualized in a blob image. A blob image typically refers to a Binary Large Object (BLOB) in the context of image processing. In a binary image, a blob is a group of connected pixels that share specific properties, such as color or intensity. Blobs with low confidence were filtered out. The remaining refined blobs were then transformed into lines using a morphological-thinning algorithm, a process used in image processing to simplify the structure of a binary image while preserving its essential shape. To convert the measured length of these cracks from pixels to real-world units, the study used a pixel-to-centimeter conversion rate. Finally, the lengths of these lines were measured.

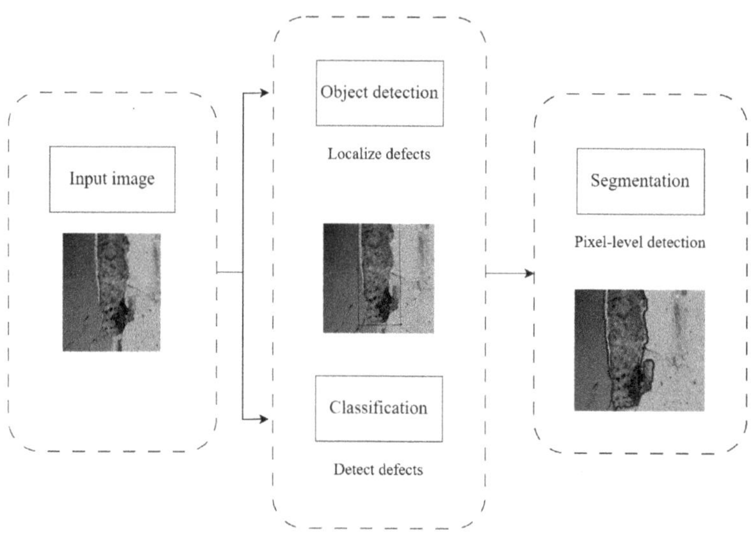

Figure 9. Hybrid methods for bridge defect detection workflow.

Over time, the advancements in deep learning segmentation algorithms have overtaken traditional segmentation methods. In 2019, Liang [115] introduced an innovative image-based technique for inspecting concrete bridge columns after incidents using deep learning and Bayesian optimization. This method involved three key steps; firstly, a CNN was utilized to categorize images into damage-free and damaged. Secondly, the Faster R-CNN model was employed to identify bridge columns as critical load-bearing components using object detection. Finally, damages on the columns were determined at the pixel level. This method enabled engineers and decision-makers to promptly evaluate the bridge condition following an incident and make informed decisions about repairs or replacements. Also, Mirzazade et al. [116] utilized the SegNet model to detect bridge components from the background. They then trained the Inception v3 model to classify images with and without damage. Following this, they employed the U-Net to segment the bridge joint openings. For damage quantification, they used 3D point cloud data generated through Structure-from-Motion (SfM) and compared it with laser scanning data. The dimensions of the detected damage were then calculated directly from the 3D point clouds in real-world units (cm). Additionally, other researchers like Kun et al. [117], also combined classification and segmentation algorithms to detect bridge defects.

In some articles, the defective region was initially identified using object detection algorithms, followed by defect detection at the pixel level using segmentation algorithms; for example, Kim et al. [118] utilized a UAV to capture images of a bridge to create the bridge 3D model and train the R-CNN algorithm to detect crack regions. They then used image processing algorithms to measure the length and thickness of the detected cracks and placed a marker on the bridge to calibrate the image scale, enabling accurate conversion of crack dimensions from pixels to millimetres. The researchers observed that environmental conditions, such as lighting and shadows, could affect the accuracy of crack measurements. As a solution, they suggested using proper lighting equipment during the photography process for future research. Also, Ni et al. [119] developed a deep learning framework for accurately and quickly detecting and measuring concrete cracks from images. They used DCGANs to synthesize crack images from real ones and augment the dataset for training the YOLOv5s model. They also applied the Otsu and edge detection methods to segment the crack images and used the medial axis algorithm to estimate the crack length and width. In their laboratory tests, each crack image was captured together with a checkerboard calibration target with known square size, so that the pixel-based skeleton and distance map could be converted into physical crack dimensions; using this pixel-to-millimetre scale.

Jiang et al. [120] wanted to identify bridge cracks in over 100.000 high-resolution images (5120 × 5120 pixels). Since using segmentation algorithms for this process would have taken too long, they initially used YOLOv4 to identify images with cracks and their approximate area. They then applied HDCB-Net, a deep learning algorithm, to overcome the problem of downsampling loss information of blurred cracks containing three downsampling stages for crack pixel-level identification. Similarly, Tran et al. [121], Kao et al. [122], and Ma et al. [123] utilized object detection followed by a segmentation algorithm to detect concrete bridge cracks. Tran et al. [121] compared five advanced networks for object detection and concluded that YOLOv7 was the most effective one for high-quality images. Additionally, they enhanced two networks for object segmentation, U-Net and pix2pix, by adjusting parameters such as network depth, activation function, loss function, and data augmentation. The modified U-Net model outperformed the original U-Net and pix2pix models. Their approach achieved a 92.38% accuracy in measuring the length of cracks and demonstrated a 0.87 R^2 value and 91% average accuracy in estimating the width and type of cracks. Furthermore, they addressed the challenge of dealing with huge images (18.000 × 10.000 pixels corresponding to 3.6 m × 2.0 m, i.e., about 0.2 mm per pixel) by splitting them into smaller parts. Kao et al. [122] employed YOLOv4 to locate the cracks and used a combination of thresholding (Sauvola local limit method) and edge detection (canny and morphological method) to estimate crack width. They collected 1463 images and used an additional 3006 images from the SDNET 2018 dataset to train their YOLOv4 model, which achieved a 92% accuracy in crack detection. This study also compared two different crack width measurement methods (the planar marker and the total station) for bridges, revealing a slight difference of 0.005 mm between them. However, the planar marker method was deemed unsuitable for bridges over water or roads. On the other hand, the total station measurement method offered more advantages and was a better alternative. It was easier to use and had no limitations associated with the planar marker method, making it the preferred approach for measuring bridge crack sizes. Ma et al. [123] assessed the effectiveness of different deep learning frameworks (DLFs) and convolutional neural networks in identifying bridge cracks from images. The researchers chose three images of bridge cracks with varying orientation characteristics: oblique, horizontal, and vertical. They used these images to measure the performance of different DLFs and CNNs. Ultimately, they compared the performance of three object detection models (Faster R-CNN, SSD, and YOLO-v5(x)) in detecting bridge cracks using three DLFs (PyTorch, TensorFlow2, and Keras). They also assessed two CNN models (U-Net and PSPNet) for segmenting bridge cracks using the same three DLFs. The study concluded that Faster R-CNN displayed the best performance for object detection and U-Net for segmentation. However, the optimal model choice depended on the DLF employed. Inam et al. [124] developed a method for identifying and localizing concrete bridge cracks using YOLOv5. They evaluated three versions of YOLOv5, including (s.m. and l). YOLOv5m outperformed the other versions with mAP of 99.3%. They used a dataset of 2270 images, consisting of 1370 images of Pakistani bridges, and the SDNET2018 dataset. Afterward, they employed U-Net to segment the cracks and estimate their length, width, and area in pixel units, which had an accuracy of 93.4%. In light of their findings, they recommended utilizing LIDAR or estimating the camera resolution and distance from the surface to measure the cracks in millimetres in the future. Also, Zakaria et al. [125] designed a method for detecting cracks and spallation on concrete bridges using portable devices such as smartphones and tablets. YOLOv5s was employed to accurately locate the defects, while U-Net measured the defects, including crack width and spallation area. In their framework, U-Net first produces a pixel-level segmentation of the damaged region, and the defect size is calculated in pixel coordinates; these pixel measurements are then converted to physical dimensions on the bridge surface using a homography-based transformation between image pixels and global coordinates. Similarly, Meng et al. [126] utilized a robot to capture images of bridge defects. The robot had advanced sensors that captured high-quality images, with optical antishake and thermal antifog capabilities. It used the BeiDou Navigation Satellite System for positioning and navigation. Its high-strength mechanical arm could extend up to 6.5 m horizontally and 80 m vertically, powered by a silent DC rotary motor. They then employed YOLOv3 for defect detection and DeepLab for defect segmentation.

Using CR-YOLO and PSPNet algorithms, Zhang et al. [80] aimed to detect and segment cracks on bridge surfaces. To train their model, they created a dataset of 800 bridge crack images, which they augmented to 5000 images by adding publicly available images and applying image enhancement techniques. They also evaluated their model on an edge device and found it suitable for real-time crack detection and segmentation. Table 6 presents the summary of the analysed hybrid methods.

Table 6. Summary of reviewed papers on bridge defect detection using Hybrid Methods.

Author	Task (Segmentation/Object Detection/ Classification)	Algorithm	Type of Defect (s)	Aim	Quantified Attribute	Preprocessing	Dataset Details		Performance (%)
							Dataset Name	Number of Data	
Peng et al. [112]	Object detection + Segmentation	R-FCN + Haar-AdaBoost + A local threshold segmentation	Crack	Detection and quantification (in physical units)	Width	Labeling Data augmentation	Own data collection using a UAV	3540	Average IoU = 90% for crack segmentation Precision = over 95% for detection
Flah et al. [113]	Classification + Segmentation	CNN + Edge detection and thresholding (Otsu method)	Crack	Detection and quantification (in physical units)	Length, width, angle of crack	Manual image selection Labeling Filter utilization to remove the non-uniform background intensity	Concrete crack image classification database [127]	6000	Classification accuracy = 98.25%
Kim et al. [114]	Classification + Segmentation	CNN (AlexNet, VGG16, ResNet152) + Morphological segmentation	Crack	Detection and quantification (in physical units)	Length	Labeling	Data collected by the Korea Expressway Corporation's bridge monitoring system	192	Precision: AlexNet = 87.74% VGG-16 = 88.76% ResNet 152 = 87.59%
Liang [115]	Classification	Pre-trained VGG16	Bridge damage	Detection		Labeling Data augmentation	Collected from different studies and Google images	492	Accuracy = 98.98%
	Semantic segmentation	Fully deep CNN				Pixel-wise labeling		436	Accuracy = 93.14% Weighted IoU = 87.65%
Mirzazade et al. [116]	Classification + Localization + Segmentation	Inception v3 for damage area detection+ U-Net and SegNet for pixel-wise defect segmentation	Damage classification + joints segmentation	Detection and quantification (in physical units)	Joint length and width	Cropping Labeling	Own data collection using a UAV	140 images cropped to 8344 slices and 238 images for segmentation	Inception v3 validation accuracy = 96.2% SegNet mIoU = 49.98% U-Net mIoU = 47.102
Kun et al. [117]	Classification + Segmentation	Deep bridge crack classification (DBCC)-Net: (A CNN-based neural network for crack patch classification) segmentation: DDRNet	Crack	Detection		Labeling	Own data collection using an I-800 UAV	385	DBCC-Net F1-score = 72.6% IoU = 83.4%
Kim et al. [118]	Object detection + Segmentation	R-CNN+ image processing	Crack	Detection, localization, and quantification (in physical units)	Thickness and length	Labeling	Images were taken using human resources and the UAV	384	Crack quantification relative error = 1–2%,

Table 6. *Cont.*

Author	Task (Segmentation/Object Detection/Classification)	Algorithm	Type of Defect (s)	Aim	Quantified Attribute	Preprocessing	Dataset Details		Performance (%)
							Dataset Name	Number of Data	
Yu et al. [128]	Object detection + Segmentation	YOLOv5+ image processing segmentation	Crack	Detection and quantification (in physical units)	Length and width	Cropping Extract the binary image Median filtering and graying Mask filter and ratio filter Labeling Data augmentation (mosaic, random rotation, random cropping, Gaussian noise, and manual exposure)	Own data collection using a Canon EOS 5DS R camera.	487 processed by offline data augmentation to obtain 3453 images	For YOLOv5: mAP = 98.7% Precision = 92% Recall = 97.5% For crack quantification: Absolute error is within 0.05 mm
Ni et al. [119]	Object detection + Segmentation	YOLOv5s+ Ostu method and the medial axis algorithm	Crack	Detection and quantification (in physical units)	Crack length and width	Denoising, filtering, and image enhancement (including image flipping, mirroring, rotation, translation, shearing) Dataset Augmentation through DCGANs Labeling	The concrete crack dataset is divided into two sections. The authors created one portion, while the other was obtained from publicly available data sources [52]	4000	For YOLOv5 Precision = 83.3% Recall = 95.3% mAP = 94%
Jiang et al. [120]	Object detection + Segmentation	YOLOv4 + HDCB-Net	Crack	Detection		Labeling	Own data collection: Blurred Crack, which contains five sub-datasets	150 632	Highest mAP of YOLOv4 = 80.1% on the BridgeXQ48
							Own data collection: Bridge Nonblurred	1536	HDCB-Net precision = 61.72%
Tran et al. [121]	Object detection + Segmentation	Detecting cracks: YOLOv7 Segmentation: optimized U-Net	Crack	Detection and quantification (in physical units)	Crack length and width	Labeling	Own data collection from bridge decks in South Korea using a 3D-mounted vehicle camera	1441 for the crack detection algorithm 800 to train the crack segmentation algorithm	mAP@0.5 = 74.5% for YOLOv7 IoU= 64% for optimized U-Net
Kao et al. [122]	Object detection + Segmentation	YOLOv4 + Thresholding (Sauvola local thresholding method) and edge detection for segmentation (Canny edge detection and morphological edge detection)	Crack	Detection and quantification (in physical units)	Crack width	Labeling for object detection For segmentation: Cropping to the Bounding Box Converting into grayscale image Image Binarization	Own data collection using Smartphones and UAVs + Open-source data on cracks (SDNET 2018 dataset [52])	1463 + 3006 = 4469	mAP = 92% for YOLOv4

Table 6. *Cont.*

Author	Task (Segmentation/Object Detection/ Classification)	Algorithm	Type of Defect (s)	Aim	Quantified Attribute	Preprocessing	Dataset Details		Performance (%)
							Dataset Name	Number of Data	
Ma et al. [123]	Object detection + Segmentation	Object detection: Faster R-CNN SSD (YOLO)-v5(x) Segmentation: U-Net PSPNet	Crack	Detection		Labeling	Open-source dataset [53]	2068	F1-score: Faster R-CNN = 76% SSD = 67% YOLOv5 = 67% Accuracy: U-Net = 98.37% PSPNet = 97.86%
Inam et al. [124]	Object detection + Segmentation	YOLOv5 s, m, and l U-Net	Crack	Detection	Width, height, area	Data resizing and augmentation (rotation and cropping) Labeling	Own data collection dataset from Pakistan bridges + SDNET2018 dataset [52]	1370 2270 after data augmentation	YOLOv5 s, m, and l mAP = 97.8%, 99.3%, and 99.1% U-Net accuracy on validation set of SDNET2018 = 93.4%
Zakaria et al. [125]	Object detection + Segmentation	Object detection: Yolov5s Segmentation: U-Net	Crack and spallation	Detection, localization, and quantification (in physical units)	Maximum crack width or area of spallation	bounding box labeling data augmentation	Own data collection + also from datasets published by other researchers, including CODEBRM [44] and SDNET2018 [52]	1600 + 180	mAP = 51% for Yolov5s mIoU = 74% for U-Net
Meng et al. [126]	Object detection + Segmentation	Object detection: Improved YOLOv3 Crack Segmentation algorithm: DeepLab	cracks, spallation, exposed bars, and efflorescence	Detection and quantification	Crack length and width	Retinex image denoising algorithm Labeling	Own data collection	20,033	For improved YOLOv3: mAP = 86.3% precision = 92.9% Recall = 86.9% F1-score = 90% For DeepLab: Accuracy = 85%
Zhang et al. [80]	Object detection + Segmentation	Object detection: CR-YOLO Segmentation: PSPNet	Cracks	Detection		Data enhancement (geometric transformation, optical transformation, and noise addition)	Own data collection using a digital camera + portion of an open-sourced dataset [36]	5000	For CR-YOLO: Precision = 90.88% Recall = 88.69% PSPNet: Precision = 87.03 Recall = 85.45%

5. Discussion and Comparison of Algorithms

This section compares deep learning algorithms for image-based bridge defect detection and measurement. Table 7 reviews the existing literature that has evaluated these algorithms. As mentioned earlier, image classification, segmentation, and object detection algorithms, or their combination can be used for the bridge damage detection. However, employing segmentation algorithms is inevitable if defect measurement is also needed.

Table 7. Algorithms comparison.

Reference	Task	Algorithm	Performance (%)	
Mirzazade et al. [116]	Segmentation	SegNet	Mean Accuracy in large-scale object segmentation = 97.299 Mean Accuracy in small-scale object segmentation = 50	
		U-Net	Mean Accuracy in large-scale object segmentation = 84.144 Mean Accuracy in small-scale object segmentation = 79.5	
Liang [115]	Classification	VGG-16	Testing Accuracy = 98.98	
		AlexNet	Testing Accuracy = 93.88	
Tran et al. [121]	Object Detection	YOLOv7	MAP@0.5 = 74.8	Speed (Time to analyze a 1024 × 1024-pixel image) (s) = 0.022
		Faster RCNN-ResNet101	MAP@0.5 = 72.8	Speed (s) = 0.18
		RetinaNet-ResNet101	MAP@0.5 = 72.7	Speed (s) = 0.065
		RetinaNet-ResNet50	MAP@0.5 = 72	Speed (s) = 0.055
		Faster RCNN-ResNet50	MAP@0.5 = 71.2	Speed (s) = 0.164
Trach [46]	Classification	MobileNet	Accuracy = 94.61	Mean times (training time of one epoch) (s) = 63
		ResNet 50	Accuracy = 93.1	Mean times (s) = 79
		VGG16	Accuracy = 92.73	Mean times (s) = 101
		DenseNet201	Accuracy = 91.82	Mean times (s) = 120
		Inception V3	Accuracy = 89.54	Mean times (s) = 63
Cardellicchio et al. [45]	Classification	DenseNet121	Accuracy = 85.19 for crack, 81.92 for Corroded and oxidized steel reinforcement, 91.10 for Deteriorated concrete, 90.37 for Honeycombs, 73.94 for Moisture spots, 89.89 for Pavement degradation, 96.05 for Shrinkage cracks	
		InceptionV3	Accuracy = 79.7 for crack, 85.03 for Corroded and oxidized steel reinforcement, 91.78 for Deteriorated concrete, 90.37 for Honeycombs, 73.26 for Moisture spots, 93.44 for Pavement degradation, 96.04 for Shrinkage cracks	
		ResNet50V2	Accuracy = 83.33 for crack, 81.85 for Corroded and oxidized steel reinforcement, 91.78 for Deteriorated concrete, 90.96 for Honeycombs, 74.77 for Moisture spots, 89.88 for Pavement degradation, 96.12 for Shrinkage cracks	
		MobileNetV3	Accuracy = 84.30 for crack, 83.26 for Corroded and oxidized steel reinforcement, 90.96 for Deteriorated concrete, 89.56 for Honeycombs, 76.43 for Moisture spots, 92.25 for Pavement degradation, 96.13 for Shrinkage cracks	
		NASNetMobile	Accuracy = 81.93 for crack, 85.4 for Corroded and oxidized steel reinforcement, 91.78 for Deteriorated concrete, 90.44 for Honeycombs, 76.43 for Moisture spots, 89.88 for Pavement degradation, 96.05 for Shrinkage cracks	

Table 7. *Cont.*

Reference	Task	Algorithm	Performance (%)				
Ma et al. [123]	Object Detection	Faster R-CNN	F1-score = 76	Precision = 80.53	Recall = 71.37		
		YOLOv5	F1-score = 67	Precision = 87.5	Recall = 54.9		
		SSD	F1-score = 67	Precision = 82.76	Recall = 56.47		
Ni et al. [119]	Object Detection	YOLOv5l	mAP = 99.6	Inference time (s = frame) = 0.083			
		Faster R-CNN	mAP = 98.3	Inference time (s = frame) = 0.5			
		YOLOv3	mAP = 98.3	Inference time (s = frame) = 0.103			
		YOLOv4	mAP = 82.8	Inference time (s = frame) = 0.022			
		SSD	mAP = 71.8	Inference time (s = frame) = 0.022			
Qiao et al. [93]	Segmentation	DeepLab v3+	mIoU (%) = 86.5	FPS (f/s) = 12.8			
		FCN	mIoU (%) = 85.77	FPS (f/s) = 15.6			
		SegNet	mIoU (%) = 85.35	FPS (f/s) = 18.5			
Jin et al. [95]	Segmentation	DeepLab	mIoU (%) = 54.23	Precision = 67.56	Recall = 76.18	Accuracy = 98.6	F1-score = 71.61
		U-Net	mIoU (%) = 52.65	Precision = 75.93	Recall = 65.39	Accuracy = 98.75	F1-score = 70.26
		FCN	mIoU (%) = 51.22	Precision = 62.35	Recall = 77.1	Accuracy = 98.72	F1-score = 68.91
Subedi et al. [96]	Segmentation	DeepLabV3+	Precision = 89.24	Recall = 87.27	F1-score = 88.24	mIoU (%) = 80.03	Inference time (ms) = 32
		U-Net	Precision = 88.83	Recall = 87.9	F1-score = 88.36	mIoU (%) = 80.19	Inference time (ms) = 32
		LinkNet	Precision = 88.8	Recall = 87.86	F1-score = 88.33	mIoU (%) = 80.15	Inference time (ms) = 31
		RefineNet	Precision = 85.64	Recall = 65.65	F1-score = 85.64	mIoU (%) = 76.5	Inference time (ms) = 31
		PSPNet	Precision = 85.42	Recall = 82.87	F1-score = 84.13	mIoU (%) = 74.42	Inference time (ms) = 12

Among classification algorithms, the reviewed studies consistently indicate that deeper CNN backbones such as VGG, ResNet, and DenseNet achieve higher accuracy than shallower or handcrafted-feature approaches, whereas lightweight architectures (e.g., MobileNet) tend to offer faster inference. Trach [46] conducted a comprehensive study on classifying cracks, spallation, and pop-outs. In this investigation, various classification algorithms were compared, concluding that ResNet50 outperforms VGG16 and, in turn, VGG16 surpasses DenseNet121 in terms of accuracy. Additionally, the research highlighted the superior speed of MobileNet and InceptionV3 compared to other algorithms. Furthermore, Cardellicchio et al. [45] trained five convolutional neural networks, DenseNet121, InceptionV3, ResNet50V2, MobileNetV3, and NASNetMobile, to classify images into seven distinct defect classes. The findings revealed that ResNet performed better in detecting most defect categories. Subsequently, InceptionV3 and DenseNet121 demonstrated comparable accuracy levels. InceptionV3 is often preferred for identifying defects of various sizes because it uses different kernel sizes. This algorithm can accurately detect both small and large defects while benefiting from efficient processing speed because it is not very deep, and multiple kernels of different sizes are implemented within the same layer [129]. Taken together, these results suggest that ResNet- and DenseNet-type backbones are generally preferred when accuracy is prioritized, whereas MobileNet- and NASNet-type models are attractive for resource-constrained or near-real-time applications.

Object detection algorithms are classified into one- and two-stage object detectors. Object detection algorithms first find the objects and then classify them with a bounding box. One-stage object detectors perform both stages in one step, making them faster. In contrast, two-stage algorithms first propose approximate object regions likely to contain the objects, then classify the regions in the next stage, making them more accurate [130].

Several object detection algorithms have been developed, including R-CNN, Faster R-CNN, YOLO, Single Shot MultiBox Detector (SSD), RetinaNet, and EfficientDet. R-CNN and Faster R-CNN are two-stage object detectors and are generally known for their accuracy, while YOLO, SSD, RetinaNet, and EfficientDet are faster because they are one-stage object detectors [131]. One-stage object detectors, such as early versions of the YOLO algorithm and SSD, might be less suitable for small or irregularly shaped objects, such as thin cracks, due to potential limitations in accuracy [132,133].

Ma et al. [123] conducted a comparison of various object detection algorithms. They determined that the Faster R-CNN model demonstrated the highest recall and F1-score compared to other object detection CNNs such as SSD and YOLO-v5(x). Yet, according to Tran et al. [121] and Ni et al. [119] recent versions of the YOLO model have demonstrated superior accuracy and computational speed performance compared to Faster R-CNN. Overall, the reviewed evidence indicates that modern one-stage detectors (YOLO family) provide a favorable accuracy–speed trade-off for UAV-based and real-time inspection, while two-stage detectors remain competitive and can still be advantageous when detection of small or overlapping defects is prioritized and higher computational cost is acceptable.

Several algorithms have been widely studied in image segmentation, including DeepLabv3+, U-Net, SegNet, and Fully Convolutional Networks. In a study by Qiao et al. [93], which focused on segmenting cracks and exposed bars, DeepLabv3+, SegNet, and FCN were evaluated. Among these, DeepLabv3+ demonstrated superior performance. Another comparative study conducted by Jin et al. [95] found that U-Net outperformed both DeepLab and FCN regarding precision and accuracy, while DeepLab exhibited better Mean Intersection over Union (MeanIoU) than U-Net. Lastly, Subedi et al. [96] addressed concrete damage and exposed bars using Nested Reg-UNet, comparing it with six other state-of-the-art segmentation algorithms. In this case, DeepLabv3+ surpassed U-Net in terms of precision and MeanIoU. Overall, these studies suggest that DeepLabv3+ tends to provide higher mIoU and more robust segmentation on complex, multi-defect datasets, while U-Net and its variants can yield competitive or superior precision on simpler or more homogeneous datasets. Also, Mirzazade et al. [116] concluded that U-Net outperforms SegNet in small object segmentation, while SegNet performs better in large-scale object segmentation. Furthermore, among the various segmentation algorithms, FCNs stand out for their computational efficiency. By utilizing locally connected layers and avoiding dense layers, FCNs achieve faster training with fewer parameters.

The comparative analysis and evaluation considered critical metrics such as precision, recall, mean average precision (mAP), mean intersection over union (mIoU), and total inference time for comparison. Figure 10 categorically arranged the algorithms, considering their performance in terms of speed and accuracy.

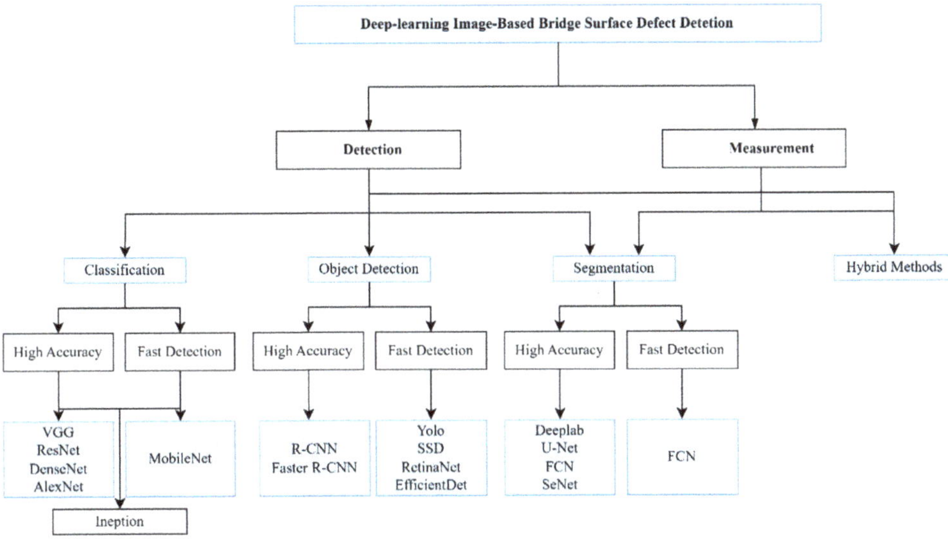

Figure 10. Algorithms performance comparison.

The comparison of deep learning models for 2D image-based bridge defect detection also reveals several important trends that highlight the trade-offs between performance, dataset size, and model architecture.

In the context of single-class object detection, YOLOv5 and its variants consistently show strong performance, even with moderate dataset sizes. As shown in Figure 11, YOLOv5m achieves a 99.3% mAP with 2270 images, demonstrating that one-stage detectors can offer excellent performance with a balanced speed-accuracy trade-off. This is significant when compared to YOLOv4, which, although using a larger dataset of 4469 images, achieves 92% mAP (also in Figure 11), showing that YOLOv5m is highly effective, even with moderate dataset sizes

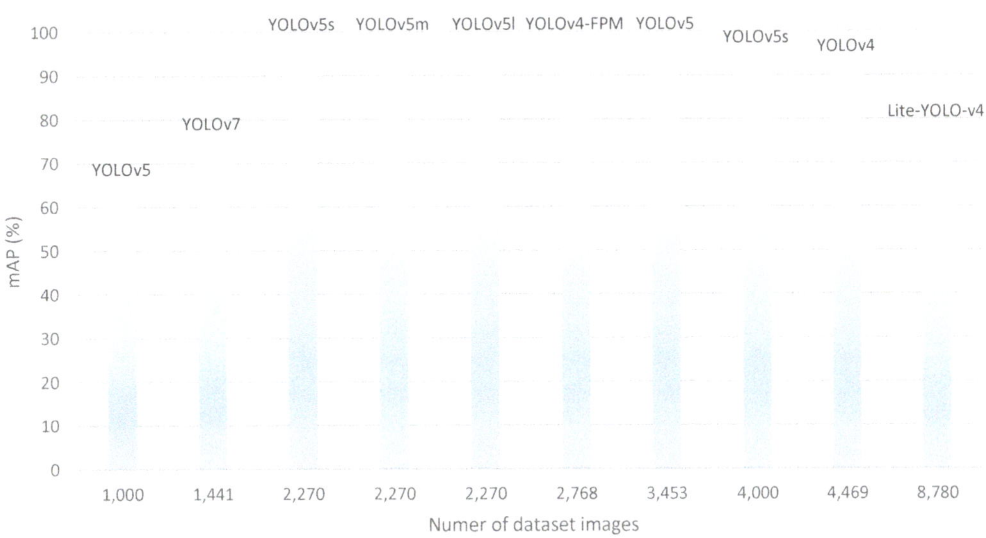

Figure 11. Performance of object detection algorithms for single-class problems across different dataset sizes.

As shown in Figure 12, Faster R-CNN outperforms YOLO even with smaller datasets, indicating that two-stage detectors are often more accurate for multi-class detection tasks. For instance, Faster R-CNN achieves 90% mAP with 1660 images, while YOLOv3 achieves 88% mAP with the same number of images. This suggests that Faster R-CNN tends to provide better performance even with smaller datasets when the task involves multiple object classes. Furthermore, YOLOv3+ with Transfer Learning (TL) and Data Augmentation (DA) achieves the highest mAP in the table with 91% at 1660 images, showing that data augmentation and transfer learning significantly enhance YOLO's performance, narrowing the gap with two-stage detectors.

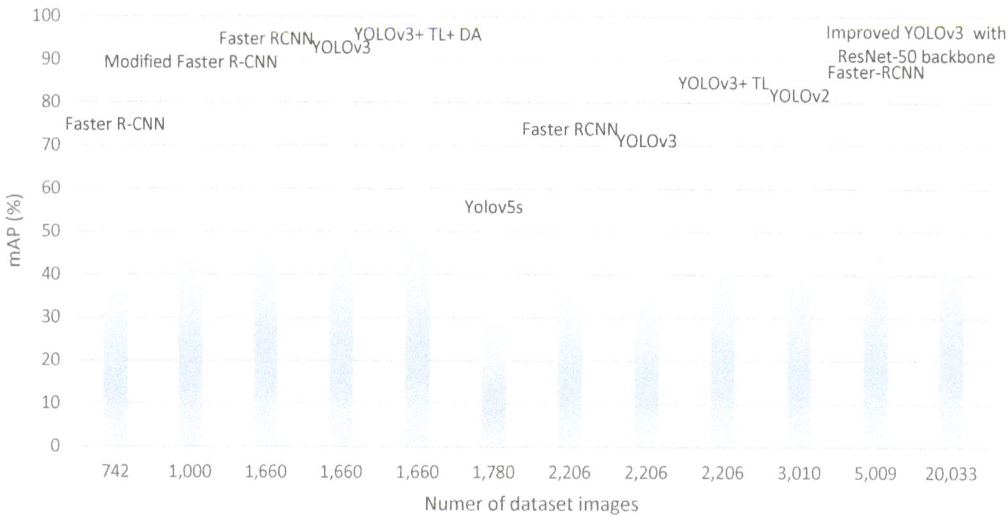

Figure 12. Performance of object detection algorithms for multi-class problems across different dataset sizes.

On the other hand, YOLOv5s (with 1780 images) shows a significantly lower mAP of 51%, indicating that lightweight models may struggle in multi-class detection, especially when the dataset is not large enough.

Additionally, models like YOLOv2 and YOLOv3 achieve mAP values of around 70% on larger dataset, suggesting that YOLO versions perform reasonably well on larger datasets but are still not as good as Faster R-CNN in multi-class detection scenarios.

In single-class segmentation, as shown in Figure 13, DeepLabv3+ outperforms other models, achieving 82.37% IoU with 5000 images.

In multi-class segmentation, as shown in Figure 14, U-Net achieves the highest mIoU of 80.19% with 7079 images, slightly outperforming DeepLabv3+, which achieves 80.03% mIoU with the same dataset size. Models like SegNet and FCN, both using 4300 images, show lower mIoU values of 75.03% and 74.75%, respectively.

Models with deeper backbones or more complex decoder structures (e.g., ResNet/DenseNet classifiers, DeepLabv3+ segmenters, and two-stage detectors such as Faster R-CNN) typically achieve higher mAP or mIoU but require longer inference times, whereas lightweight CNNs (e.g., MobileNet, NASNet) and one-stage detectors from the YOLO family provide faster inference at the cost of a modest reduction in accuracy. Therefore, the algorithm selection relies on the particular use case and application domain. Moreover, algorithms can be adjusted to enhance accuracy or speed.

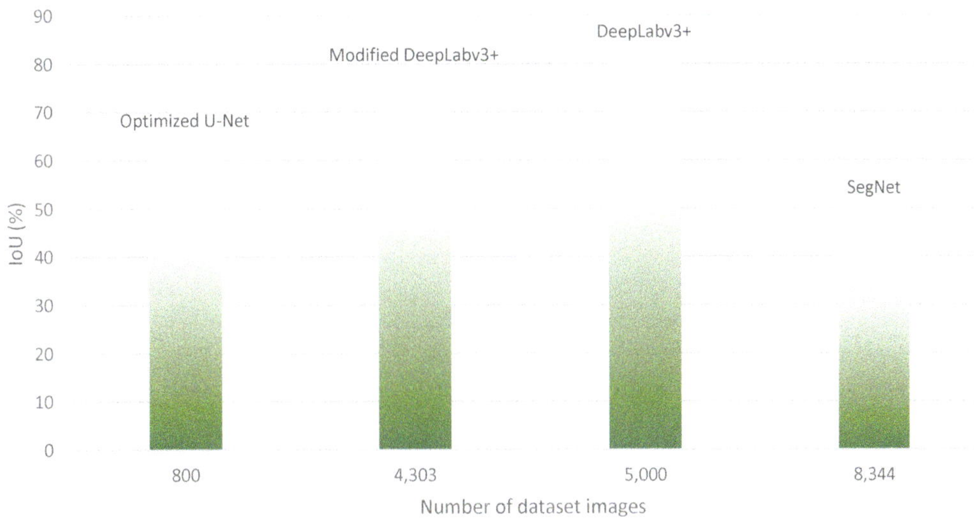

Figure 13. Performance of segmentation algorithms for single-class problems across different dataset sizes.

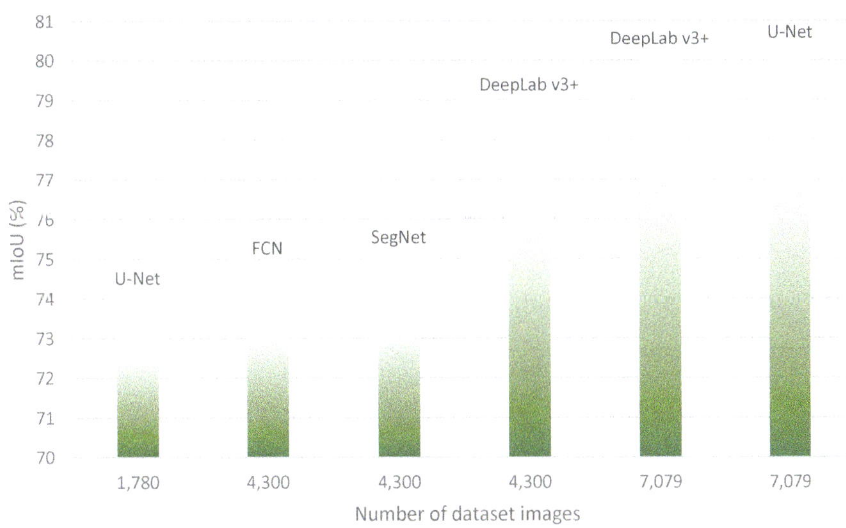

Figure 14. Performance of segmentation algorithms for multi-class problems across different dataset sizes.

6. Conclusions

The increasing application of AI across diverse industries has encouraged bridge managers to adopt advanced AI methodologies in bridge inspection processes. This paper systematically reviewed recent advancements in deep learning-based detection of concrete bridge surface defects, highlighting the limitations of traditional image-processing methods and outlining the transition towards machine learning and deep learning approaches. Hence, several key findings from this systematic review were derived.

Initially, classification algorithms were predominantly utilized for defect detection; however, their limitations prompted the adoption of object detection algorithms, which can handle multiple defect types within single images more effectively. Among various deep learning techniques, object detection and segmentation algorithms have emerged as the most effective methods for concrete bridge defect detection. Object detection algorithms are particularly suited when identification is the primary requirement. Faster R-CCNs (Region-based Convolutional Neural Networks) are preferable for scenarios prioritizing accuracy, whereas YOLO (You Only Look Once) or SSD (Single-Shot Detector) models are employed for faster detection. Recent versions of YOLO demonstrate superior performance in terms of both accuracy and computational speed.

For precise quantification and measurement of defects, segmentation algorithms, either alone or combined with other deep learning methods, are essential as they provide pixel-level accuracy. Among segmentation models, U-Net and DeepLab have shown higher accuracy, while FCNs (Fully Convolutional Networks) offer advantages in computational efficiency. A key limitation identified with segmentation methods is the representation of defects in pixel units rather than real-world measurements. This issue has been addressed by incorporating known-sized reference targets within images or by calculating real-world dimensions based on camera distances. Within the reviewed literature, YOLO was the most frequently employed deep learning algorithm for bridge defect detection tasks.

Beyond algorithmic performance, the reviewed studies also showed how deep learning methods have begun to be embedded in smart surveillance workflows for concrete bridges. Several works integrated CNN- and transformer-based models into UAV platforms or climbing robots for data acquisition and automated inspection [85,87,100,113]. Other studies combined image-based defect detection with photogrammetry, SfM, LiDAR, and SLAM to localize defects on 3D models or within BIM-like representations of bridges [85,87,99,103]. In post-earthquake scenarios, multi-task networks have been used to detect components and damages and to derive risk levels or safety indices for bridges [98,104]. However, a number of practical challenges remain for real-world deployment, including variable field lighting, view obstruction from vehicles and vegetation, adverse weather conditions, camera calibration and scale conversion for defect quantification, and the substantial effort required for data collection, storage, and pixel-level annotation. Real-time or near-real-time inference on edge devices is still not routinely achieved, and human–AI interaction typically remains at the stage where inspectors manually validate model outputs rather than working with well-designed decision-support interfaces. To move from paper performance to routine deployment by agencies, future research needs to focus on robust domain adaptation, uncertainty-aware predictions, active learning, efficient annotation strategies, and tighter integration with BIM/BMS and asset-management workflows that explicitly support inspectors in interpreting and acting on automated defect detections.

Considering the insights gathered, several gaps and potential areas for future research have been identified. Integration of deep learning-based defect detection with advanced 3D visualization techniques such as SfM (Structure from Motion) and SLAM (Simultaneous Localization and Mapping) remains relatively unexplored. Future research could further develop methods to measure bridge condition indices and visualize these effectively using color-coded BIM systems. A significant limitation identified is the lack of readily available, comprehensive datasets containing annotated images of diverse bridge defects, particularly for segmentation tasks. Developing and openly sharing such datasets would significantly benefit the research community by reducing the extensive manual annotation currently required. While various deep learning algorithms reviewed here have demonstrated strong performance, further refinement and modification of deep learning architectures remain necessary to optimize performance, particularly when dealing with lower-quality images or complex backgrounds. Also, advancing research on computationally efficient deep learning models is vital to enable real-time defect detection, thereby enhancing their usability in portable devices such as smartphones and other edge computing devices.

As we've seen in some recent studies, generative models and synthetic data are promising solutions to the problem of data scarcity in bridge defect detection. Since gathering and labeling real-world data is both time-consuming and resource-intensive, focusing more on these methods in the future could significantly streamline the process. Additionally, the use of large-scale vision models, such as transformers, is an exciting emerging trend. These models have the potential to push the boundaries of defect detection accuracy and automation, making them an important area for future research in bridge inspection.

Author Contributions: Conceptualization, N.L.K., E.S., N.S. and S.B.; methodology, N.L.K., N.S. and S.B.; formal analysis, N.L.K., E.S., N.S. and S.B.; data curation, N.L.K., N.S. and S.B.; investigation, N.L.K.; writing—original draft preparation, N.L.K.; writing—review and editing, N.L.K., E.S., N.S. and S.B. All authors have read and agreed to the published version of the manuscript.

Funding: This research received no external funding.

Data Availability Statement: Some or all data that support the findings of this study are available from the corresponding author upon reasonable request.

Conflicts of Interest: The authors declare no conflicts of interest.

Appendix A

Table A1. Risk of Bias.

Studies	Study Design	Data Handling	Reporting Transparency
Chen [38]	low	low	Some concerns
Hüthwohl et al. [42]	low	low	low
Zhu et al. [5]	low	low	low
Mundt et al. [44]	low	low	low
Zoubir et al. [49]	low	low	low
Bukhsh et al. [48]	low	low	low
Kruachottikul et al. [50]	low	Some concerns	low
Xu et al. [36]	low	low	low
Zhang et al. [37]	low	low	low
Cardellicchio et al. [45]	low	Some concerns	low
Trach [46]	low	Some concerns	low
Chen et al. [41]	low	low	low
Zhang et al. [54]	low	low	low
Abubakr et al. [47]	low	low	low
Alfaz et al. [39]	low	Some concerns	low
Aliyari et al. [51]	low	low	low
Li et al. [40]	low	Some concerns	low
Murao et al. [55]	Some concerns	Some concerns	Some concerns
Yu et al. [56]	low	low	low
Li et al. [57]	low	low	low
Deng et al. [58]	low	Some concerns	low
Ji [59]	low	Some concerns	low
Deng et al. [71]	low	Some concerns	low
Ruggieri et al. [60]	low	low	low
Zhang et al. [61]	low	low	low
Liu et al. [62]	low	Some concerns	low
Yamane et al. [63]	low	low	low
Teng et al. [64]	low	low	low
Hong et al. [70]	low	low	low

Table A1. *Cont.*

Studies	Study Design	Data Handling	Reporting Transparency
Yu et al. [65]	low	low	low
Gan et al. [68]	low	low	Some concerns
Lin et al. [67]	low	low	low
Ngo et al. [74]	low	low	Some concerns
Lu et al. [75]	low	low	Some concerns
Ruggieri et al. [72]	low	low	low
Rubio et al. [81]	low	low	low
Lopez Droguett et al. [82]	low	Some concerns	low
Merkle et al. [83]	low	low	low
Fukuoka and Fujiu [85]	low	low	low
Li et al. [86]	low	Some concerns	low
Yamane et al. [87]	low	low	Some concerns
Xu et al. [88]	low	low	low
Bae et al. [89]	low	low	Some concerns
Deng et al. [90]	low	low	low
Munawar et al. [92]	low	Some concerns	low
Li et al. [106]	low	low	Some concerns
Fu et al. [91]	low	Some concerns	low
Qiao et al. [93]	low	Some concerns	low
Jin et al. [95]	low	low	low
Subedi et al. [96]	low	low	low
Ayele et al. [97]	low	Some concerns	Some concerns
Wang et al. [98]	low	low	low
Montes et al. [99]	low	Some concerns	low
Artus et al. [100]	low	Some concerns	low
Borin and Cavazzini [101]	low	Some concerns	Some concerns
Jang et al. [102]	low	low	low
McLaughlin et al. [103]	low	low	low
Ye et al. [104]	low	low	low
Shen et al. [105]	low	low	low
Na and Kim [84]	low	low	Some concerns
Peng et al. [112]	low	Some concerns	low
Flah et al. [113]	low	low	low
Kim et al. [114]	low	low	low
Liang [115]	low	low	low
Mirzazade et al. [116]	low	Some concerns	low
Kun et al. [117]	low	Some concerns	low
Kim et al. [118]	low	Some concerns	Some concerns
Yu et al. [128]	low	low	low
Ni et al. [119]	low	low	low
Jiang et al. [120]	low	low	low
Tran et al. [121]	low	Some concerns	low
Kao et al. [122]	low	Some concerns	low
Ma et al. [123]	low	low	Some concerns
Inam et al. [124]	low	Some concerns	low
Zakaria et al. [125]	low	low	low
Meng et al. [126]	low	low	low
Zhang et al. [80]	low	Some concerns	Some concerns

References

1. Munawar, H.S.; Hammad, A.W.; Haddad, A.; Soares, C.A.P.; Waller, S.T. Image-based crack detection methods: A review. *Infrastructures* **2021**, *6*, 115. [CrossRef]
2. Iron, A.; Institute, S. *Steel Bridge Construction: Myths & Realities*; American Iron and Steel Institute: Washington, DC, USA, 2007.

3. Koch, C.; Georgieva, K.; Kasireddy, V.; Akinci, B.; Fieguth, P. A review on computer vision based defect detection and condition assessment of concrete and asphalt civil infrastructure. *Adv. Eng. Inform.* **2015**, *29*, 196–210. [CrossRef]
4. Oh, J.-K.; Jang, G.; Oh, S.; Lee, J.H.; Yi, B.-J.; Moon, Y.S.; Lee, J.S.; Choi, Y. Bridge inspection robot system with machine vision. *Autom. Constr.* **2009**, *18*, 929–941. [CrossRef]
5. Zhu, J.; Zhang, C.; Qi, H.; Lu, Z. Vision-based defects detection for bridges using transfer learning and convolutional neural networks. *Struct. Infrastruct. Eng.* **2020**, *16*, 1037–1049. [CrossRef]
6. Spencer, B.F., Jr.; Hoskere, V.; Narazaki, Y. Advances in computer vision-based civil infrastructure inspection and monitoring. *Engineering* **2019**, *5*, 199–222. [CrossRef]
7. Khan, M.A.-M.; Kee, S.-H.; Pathan, A.-S.K.; Nahid, A.-A. Image Processing Techniques for Concrete Crack Detection: A Scientometrics Literature Review. *Remote Sens.* **2023**, *15*, 2400. [CrossRef]
8. Luo, K.; Kong, X.; Zhang, J.; Hu, J.; Li, J.; Tang, H. Computer vision-based bridge inspection and monitoring: A review. *Sensors* **2023**, *23*, 7863. [CrossRef] [PubMed]
9. Zhang, Y.; Yuen, K.-V. Review of artificial intelligence-based bridge damage detection. *Adv. Mech. Eng.* **2022**, *14*, 1–21. [CrossRef]
10. Alzubaidi, L.; Zhang, J.; Humaidi, A.J.; Al-Dujaili, A.; Duan, Y.; Al-Shamma, O.; Santamaría, J.; Fadhel, M.A.; Al-Amidie, M.; Farhan, L. Review of deep learning: Concepts, CNN architectures, challenges, applications, future directions. *J. Big Data* **2021**, *8*, 53. [CrossRef] [PubMed]
11. Kulkarni, S.; Harnoorkar, S.; Pintelas, P. Comparative analysis of CNN architectures. *Int. Res. J. Eng. Technol.* **2020**, *7*, 1459–1464.
12. Neu, D.A.; Lahann, J.; Fettke, P. A systematic literature review on state-of-the-art deep learning methods for process prediction. *Artif. Intell. Rev.* **2022**, *55*, 801–827. [CrossRef]
13. Gu, J.; Wang, Z.; Kuen, J.; Ma, L.; Shahroudy, A.; Shuai, B.; Liu, T.; Wang, X.; Wang, G.; Cai, J. Recent advances in convolutional neural networks. *Pattern Recognit.* **2018**, *77*, 354–377. [CrossRef]
14. Krizhevsky, A.; Sutskever, I.; Hinton, G.E. Imagenet classification with deep convolutional neural networks. *Adv. Neural Inf. Process. Syst.* **2012**, *25*, 84–90. [CrossRef]
15. Simonyan, K.; Zisserman, A. Very deep convolutional networks for large-scale image recognition. *arXiv* **2014**, arXiv:1409.1556. [CrossRef]
16. He, K.; Zhang, X.; Ren, S.; Sun, J. Deep residual learning for image recognition. In Proceedings of the IEEE Conference on Computer Vision and Pattern Recognition, Las Vegas, NV, USA, 27–30 June 2016; pp. 770–778. [CrossRef]
17. Huang, G.; Liu, Z.; Van Der Maaten, L.; Weinberger, K.Q. Densely connected convolutional networks. In Proceedings of the IEEE Conference on Computer Vision and Pattern Recognition, Honolulu, HI, USA, 21–26 July 2017; pp. 4700–4708. [CrossRef]
18. Howard, A.G.; Zhu, M.; Chen, B.; Kalenichenko, D.; Wang, W.; Weyand, T.; Andreetto, M.; Adam, H. Mobilenets: Efficient convolutional neural networks for mobile vision applications. *arXiv* **2017**, arXiv:1704.04861. [CrossRef]
19. Chollet, F. Xception: Deep learning with depthwise separable convolutions. In Proceedings of the IEEE Conference on Computer Vision and Pattern Recognition, Honolulu, HI, USA, 21–26 July 2017; pp. 1251–1258.
20. Gurita, A.; Mocanu, I.G. Image segmentation using encoder-decoder with deformable convolutions. *Sensors* **2021**, *21*, 1570. [CrossRef]
21. Long, J.; Shelhamer, E.; Darrell, T. Fully convolutional networks for semantic segmentation. In Proceedings of the IEEE Conference on Computer Vision and Pattern Recognition, Boston, MA, USA, 7–12 June 2015; pp. 3431–3440. [CrossRef]
22. Ronneberger, O.; Fischer, P.; Brox, T. U-net: Convolutional networks for biomedical image segmentation. In Proceedings of the Medical Image Computing and Computer-Assisted Intervention–MICCAI 2015: 18th International Conference, Munich, Germany, 5–9 October 2015; part III 18. pp. 234–241. [CrossRef]
23. Chen, L.-C.; Papandreou, G.; Kokkinos, I.; Murphy, K.; Yuille, A.L. Semantic image segmentation with deep convolutional nets and fully connected crfs. *arXiv* **2014**, arXiv:1412.7062. [CrossRef]
24. He, K.; Gkioxari, G.; Dollár, P.; Girshick, R. Mask r-cnn. In Proceedings of the IEEE International Conference on Computer Vision, Venice, Italy, 22–29 October 2017; pp. 2961–2969. [CrossRef]
25. Girshick, R.; Donahue, J.; Darrell, T.; Malik, J. Rich feature hierarchies for accurate object detection and semantic segmentation. In Proceedings of the IEEE Conference on Computer Vision and Pattern Recognition, Columbus, OH, USA, 24–28 June 2014; pp. 580–587. [CrossRef]
26. Girshick, R. Fast r-cnn. *arXiv* **2015**, arXiv:1504.08083. [CrossRef]
27. Ren, S. Faster r-cnn: Towards real-time object detection with region proposal networks. *arXiv* **2015**, arXiv:1506.01497. [CrossRef] [PubMed]
28. Redmon, J. You only look once: Unified, real-time object detection. In Proceedings of the IEEE Conference on Computer Vision and Pattern Recognition, Las Vegas, NV, USA, 27–30 June 2016. [CrossRef]
29. Otsu, N. A threshold selection method from gray-level histograms. *Automatica* **1975**, *11*, 23–27. [CrossRef]
30. Borrmann, A.; König, M.; Koch, C.; Beetz, J. *Building Information Modeling: Why? What? How?* Springer: Berlin/Heidelberg, Germany, 2018. [CrossRef]

31. Hüthwohl, P.; Borrmann, A.; Sacks, R. Integrating RC bridge defect information into BIM models. *J. Comput. Civ. Eng.* **2018**, *32*, 04018013. [CrossRef]
32. Sacks, R.; Eastman, C.; Lee, G.; Teicholz, P. *BIM Handbook: A Guide to Building Information Modeling for Owners, Designers, Engineers, Contractors, and Facility Managers*; John Wiley & Sons: Hoboken, NJ, USA, 2018. [CrossRef]
33. Kim, B.; Cho, S. Automated vision-based detection of cracks on concrete surfaces using a deep learning technique. *Sensors* **2018**, *18*, 3452. [CrossRef]
34. Prasanna, P.; Dana, K.; Gucunski, N.; Basily, B. Computer-vision based crack detection and analysis. In Proceedings of the Sensors and Smart Structures Technologies for Civil, Mechanical, and Aerospace Systems 2012, San Diego, CA, USA, 12–15 March 2012; pp. 1143–1148. [CrossRef]
35. Zhou, S.; Canchila, C.; Song, W. Deep learning-based crack segmentation for civil infrastructure: Data types, architectures, and benchmarked performance. *Autom. Constr.* **2023**, *146*, 104678. [CrossRef]
36. Xu, H.; Su, X.; Wang, Y.; Cai, H.; Cui, K.; Chen, X. Automatic bridge crack detection using a convolutional neural network. *Appl. Sci.* **2019**, *9*, 2867. [CrossRef]
37. Zhang, Q.; Barri, K.; Babanajad, S.K.; Alavi, A.H. Real-time detection of cracks on concrete bridge decks using deep learning in the frequency domain. *Engineering* **2021**, *7*, 1786–1796. [CrossRef]
38. Chen, R. Migration learning-based bridge structure damage detection algorithm. *Sci. Program.* **2021**, *2021*, 1102521. [CrossRef]
39. Alfaz, N.; Hasnat, A.; Khan, A.M.R.N.; Sayom, N.S.; Bhowmik, A. Bridge crack detection using dense convolutional network (densenet). In Proceedings of the 2nd International Conference on Computing Advancements, Dhaka, Bangladesh, 10–12 March 2022; pp. 509–515. [CrossRef]
40. Li, H.; Xu, H.; Tian, X.; Wang, Y.; Cai, H.; Cui, K.; Chen, X. Bridge crack detection based on SSENets. *Appl. Sci.* **2020**, *10*, 4230. [CrossRef]
41. Chen, L.; Yao, H.; Fu, J.; Ng, C.T. The classification and localization of crack using lightweight convolutional neural network with CBAM. *Eng. Struct.* **2023**, *275*, 115291. [CrossRef]
42. Hüthwohl, P.; Lu, R.; Brilakis, I. Multi-classifier for reinforced concrete bridge defects. *Autom. Constr.* **2019**, *105*, 102824. [CrossRef]
43. Kruachottikul, P.; Cooharojananone, N.; Phanomchoeng, G.; Chavarnakul, T.; Kovitanggoon, K.; Trakulwaranont, D.; Atchariyachanvanich, K. Bridge sub structure defect inspection assistance by using deep learning. In Proceedings of the 2019 IEEE 10th International Conference on Awareness Science and Technology (iCAST), Morioka, Japan, 23–25 October 2019; pp. 1–6. [CrossRef]
44. Mundt, M.; Majumder, S.; Murali, S.; Panetsos, P.; Ramesh, V. Meta-learning convolutional neural architectures for multi-target concrete defect classification with the concrete defect bridge image dataset. In Proceedings of the IEEE/CVF Conference on Computer Vision and Pattern Recognition, Long Beach, CA, USA, 16–20 June 2019; pp. 11196–11205. [CrossRef]
45. Cardellicchio, A.; Ruggieri, S.; Nettis, A.; Patruno, C.; Uva, G.; Renò, V. Deep learning approaches for image-based detection and classification of structural defects in bridges. In Proceedings of the International Conference on Image Analysis and Processing, Lecce, Italy, 23–27 May 2022; pp. 269–279. [CrossRef]
46. Trach, R. A Model Classifying Four Classes of Defects in Reinforced Concrete Bridge Elements Using Convolutional Neural Networks. *Infrastructures* **2023**, *8*, 123. [CrossRef]
47. Abubakr, M.; Rady, M.; Badran, K.; Mahfouz, S.Y. Application of deep learning in damage classification of reinforced concrete bridges. *Ain Shams Eng. J.* **2024**, *15*, 102297. [CrossRef]
48. Bukhsh, Z.A.; Anžlin, A.; Stipanović, I. BiNet: Bridge Visual Inspection Dataset and Approach for Damage Detection. In Proceedings of the 1st Conference of the European Association on Quality Control of Bridges and Structures: EUROSTRUCT 2021, Padua, Italy, 29 August–1 September 2021; pp. 1027–1034. [CrossRef]
49. Zoubir, H.; Rguig, M.; El Aroussi, M.; Chehri, A.; Saadane, R.; Jeon, G. Concrete Bridge defects identification and localization based on classification deep convolutional neural networks and transfer learning. *Remote Sens.* **2022**, *14*, 4882. [CrossRef]
50. Kruachottikul, P.; Cooharojananone, N.; Phanomchoeng, G.; Chavarnakul, T.; Kovitanggoon, K.; Trakulwaranont, D. Deep learning-based visual defect-inspection system for reinforced concrete bridge substructure: A case of Thailand's department of highways. *J. Civ. Struct. Health Monit.* **2021**, *11*, 949–965. [CrossRef]
51. Aliyari, M.; Droguett, E.L.; Ayele, Y.Z. UAV-based bridge inspection via transfer learning. *Sustainability* **2021**, *13*, 11359. [CrossRef]
52. Dorafshan, S.; Thomas, R.J.; Maguire, M. SDNET2018: An annotated image dataset for non-contact concrete crack detection using deep convolutional neural networks. *Data Brief* **2018**, *21*, 1664–1668. [CrossRef]
53. Li, L.-F.; Ma, W.-F.; Li, L.; Lu, C. Research on detection algorithm for bridge cracks based on deep learning. *Acta Autom. Sin.* **2019**, *45*, 1727–1742.
54. Zhang, Y.; Ni, Y.-Q.; Jia, X.; Wang, Y.-W. Identification of concrete surface damage based on probabilistic deep learning of images. *Autom. Constr.* **2023**, *156*, 105141. [CrossRef]

55. Murao, S.; Nomura, Y.; Furuta, H.; Kim, C.-W. Concrete crack detection using uav and deep learning. In Proceedings of the 13th International Conference on Applications of Statistics and Probability in Civil Engineering (ICASP), Seoul, Republic of Korea, 26–20 May 2019; pp. 1–8.
56. Yu, Z.; Shen, Y.; Shen, C. A real-time detection approach for bridge cracks based on YOLOv4-FPM. *Autom. Constr.* **2021**, *122*, 103514. [CrossRef]
57. Li, X.; Sun, H.; Song, T.; Zhang, T.; Meng, Q. A method of underwater bridge structure damage detection method based on a lightweight deep convolutional network. *IET Image Process.* **2022**, *16*, 3893–3909. [CrossRef]
58. Deng, J.; Lu, Y.; Lee, V.C.-S. Imaging-based crack detection on concrete surfaces using You Only Look Once network. *Struct. Health Monit.* **2021**, *20*, 484–499. [CrossRef]
59. Ji, H. Development of an Autonomous Column-Climbing Robotic System for Real-time Detection and Mapping of Surface Cracks on Bridges. In Proceedings of the 2023 IEEE IAS Global Conference on Emerging Technologies (GlobConET), London, UK, 19–21 May 2023; pp. 1–6. [CrossRef]
60. Ruggieri, S.; Cardellicchio, A.; Nettis, A.; Renò, V.; Uva, G. Using machine learning approaches to perform defect detection of existing bridges. *Procedia Struct. Integr.* **2023**, *44*, 2028–2035. [CrossRef]
61. Zhang, C.; Chang, C.c.; Jamshidi, M. Concrete bridge surface damage detection using a single-stage detector. *Comput.-Aided Civ. Infrastruct. Eng.* **2020**, *35*, 389–409. [CrossRef]
62. Liu, Y.; Gao, W.; Zhao, T.; Wang, Z.; Wang, Z. A Rapid Bridge Crack Detection Method Based on Deep Learning. *Appl. Sci.* **2023**, *13*, 9878. [CrossRef]
63. Yamane, T.; Chun, P.-J.; Honda, R. Detecting and localising damage based on image recognition and structure from motion, and reflecting it in a 3D bridge model. *Struct. Infrastruct. Eng.* **2022**, *20*, 594–606. [CrossRef]
64. Teng, S.; Liu, Z.; Li, X. Improved YOLOv3-based bridge surface defect detection by combining High-and low-resolution feature images. *Buildings* **2022**, *12*, 1225. [CrossRef]
65. Yu, L.; He, S.; Liu, X.; Ma, M.; Xiang, S. Engineering-oriented bridge multiple-damage detection with damage integrity using modified faster region-based convolutional neural network. *Multimed. Tools Appl.* **2022**, *81*, 18279–18304. [CrossRef]
66. Li, R.; Yu, J.; Li, F.; Yang, R.; Wang, Y.; Peng, Z. Automatic bridge crack detection using Unmanned aerial vehicle and Faster R-CNN. *Constr. Build. Mater.* **2023**, *362*, 129659. [CrossRef]
67. Lin, J.J.; Ibrahim, A.; Sarwade, S.; Golparvar-Fard, M. Bridge inspection with aerial robots: Automating the entire pipeline of visual data capture, 3D mapping, defect detection, analysis, and reporting. *J. Comput. Civ. Eng.* **2021**, *35*, 04020064. [CrossRef]
68. Gan, L.; Liu, H.; Yan, Y.; Chen, A. Bridge bottom crack detection and modeling based on faster R-CNN and BIM. *IET Image Process.* **2023**, *18*, 664–677. [CrossRef]
69. Zhejiang-Highway-Administration. JTG H10-2009: Technical Specification for Highway Maintenance. Available online: https://www.chinesestandard.net/PDF/BOOK.aspx/JTGH10-2009 (accessed on 5 November 2024).
70. Hong, S.-S.; Hwang, C.-H.; Chung, S.-W.; Kim, B.-K. A deep-learning-based bridge damaged object automatic detection model using a bridge member model combination framework. *Appl. Sci.* **2022**, *12*, 12868. [CrossRef]
71. Deng, J.; Lu, Y.; Lee, V.C.S. Concrete crack detection with handwriting script interferences using faster region-based convolutional neural network. *Comput.-Aided Civ. Infrastruct. Eng.* **2020**, *35*, 373–388. [CrossRef]
72. Ruggieri, S.; Cardellicchio, A.; Nettis, A.; Renò, V.; Uva, G. Using attention for improving defect detection in existing RC bridges. *IEEE Access* **2025**, *13*, 18994–19015. [CrossRef]
73. Dwyer, B.; Nelson, J.; Hansen, T.; Solawetz, J. *Roboflow*, Version 1.0; Roboflow: Des Moines, IA, USA, 2022.
74. Ngo, L.; Xuan, C.L.; Luong, H.M.; Thanh, B.N.; Ngoc, D.B. Designing image processing tools for testing concrete bridges by a drone based on deep learning. *J. Inf. Telecommun.* **2023**, *7*, 227–240. [CrossRef]
75. Lu, G.; He, X.; Wang, Q.; Shao, F.; Wang, J.; Zhao, X. MSCNet: A Framework With a Texture Enhancement Mechanism and Feature Aggregation for Crack Detection. *IEEE Access* **2022**, *10*, 26127–26139. [CrossRef]
76. Adhikari, R.; Moselhi, O.; Bagchi, A. Image-based retrieval of concrete crack properties. In Proceedings of the ISARC International Symposium on Automation and Robotics in Construction, Eindhoven, The Netherlands, 26–29 June 2012; p. 1.
77. Tian, Y.; Zhang, X.; Chen, H.; Wang, Y.; Wu, H. A Bridge Damage Visualization Technique Based on Image Processing Technology and the IFC Standard. *Sustainability* **2023**, *15*, 8769. [CrossRef]
78. Vivekananthan, V.; Vignesh, R.; Vasanthaseelan, S.; Joel, E.; Kumar, K.S. Concrete bridge crack detection by image processing technique by using the improved OTSU method. *Mater. Today Proc.* **2023**, *74*, 1002–1007. [CrossRef]
79. Minaee, S.; Boykov, Y.; Porikli, F.; Plaza, A.; Kehtarnavaz, N.; Terzopoulos, D. Image segmentation using deep learning: A survey. *IEEE Trans. Pattern Anal. Mach. Intell.* **2021**, *44*, 3523–3542. [CrossRef] [PubMed]
80. Zhang, J.; Qian, S.; Tan, C. Automated bridge surface crack detection and segmentation using computer vision-based deep learning model. *Eng. Appl. Artif. Intell.* **2022**, *115*, 105225. [CrossRef]
81. Rubio, J.J.; Kashiwa, T.; Laiteerapong, T.; Deng, W.; Nagai, K.; Escalera, S.; Nakayama, K.; Matsuo, Y.; Prendinger, H. Multi-class structural damage segmentation using fully convolutional networks. *Comput. Ind.* **2019**, *112*, 103121. [CrossRef]

82. Lopez Droguett, E.; Tapia, J.; Yanez, C.; Boroschek, R. Semantic segmentation model for crack images from concrete bridges for mobile devices. *Proc. Inst. Mech. Eng. Part O J. Risk Reliab.* **2022**, *236*, 570–583. [CrossRef]
83. Merkle, D.; Solass, J.; Schmitt, A.; Rosin, J.; Reiterer, A.; Stolz, A. Semi-automatic 3D crack map generation and width evaluation for structural monitoring of reinforced concrete structures. *J. Inf. Technol. Constr.* **2023**, *28*, 774–805. [CrossRef]
84. Na, Y.-H.; Kim, D.-K. Deep Learning Strategy for UAV-Based Multi-Class Damage Detection on Railway Bridges Using U-Net with Different Loss Functions. *Appl. Sci.* **2025**, *15*, 8719. [CrossRef]
85. Fukuoka, T.; Fujiu, M. Detection of Bridge Damages by Image Processing Using the Deep Learning Transformer Model. *Buildings* **2023**, *13*, 788. [CrossRef]
86. Li, G.; Fang, Z.; Mohammed, A.M.; Liu, T.; Deng, Z. Automated Bridge Crack Detection Based on Improving Encoder–Decoder Network and Strip Pooling. *J. Infrastruct. Syst.* **2023**, *29*, 04023004. [CrossRef]
87. Yamane, T.; Chun, P.j.; Dang, J.; Honda, R. Recording of bridge damage areas by 3D integration of multiple images and reduction of the variability in detected results. *Comput.-Aided Civ. Infrastruct. Eng.* **2023**, *38*, 2391–2407. [CrossRef]
88. Xu, Y.; Fan, Y.; Li, H. Lightweight semantic segmentation of complex structural damage recognition for actual bridges. *Struct. Health Monit.* **2023**, *22*, 3250–3269. [CrossRef]
89. Bae, H.; Jang, K.; An, Y.-K. Deep super resolution crack network (SrcNet) for improving computer vision–based automated crack detectability in in situ bridges. *Struct. Health Monit.* **2021**, *20*, 1428–1442. [CrossRef]
90. Deng, W.; Mou, Y.; Kashiwa, T.; Escalera, S.; Nagai, K.; Nakayama, K.; Matsuo, Y.; Prendinger, H. Vision based pixel-level bridge structural damage detection using a link ASPP network. *Autom. Constr.* **2020**, *110*, 102973. [CrossRef]
91. Fu, H.; Meng, D.; Li, W.; Wang, Y. Bridge crack semantic segmentation based on improved Deeplabv3+. *J. Mar. Sci. Eng.* **2021**, *9*, 671. [CrossRef]
92. Munawar, H.S.; Ullah, F.; Shahzad, D.; Heravi, A.; Qayyum, S.; Akram, J. Civil infrastructure damage and corrosion detection: An application of machine learning. *Buildings* **2022**, *12*, 156. [CrossRef]
93. Qiao, W.; Ma, B.; Liu, Q.; Wu, X.; Li, G. Computer vision-based bridge damage detection using deep convolutional networks with expectation maximum attention module. *Sensors* **2021**, *21*, 824. [CrossRef]
94. Siddhartha, V.R. Bridge Crack Detection Using Horse Herd Optimization Algorithm. In Proceedings of the 4th International Conference on Smart Electronics and Communication (ICOSEC), Trichy, India, 20–22 September 2023. [CrossRef]
95. Jin, T.; Ye, X.; Li, Z. Establishment and evaluation of conditional GAN-based image dataset for semantic segmentation of structural cracks. *Eng. Struct.* **2023**, *285*, 116058. [CrossRef]
96. Subedi, A.; Tang, W.; Mondal, T.G.; Wu, R.-T.; Jahanshahi, M.R. Ensemble-based deep learning for autonomous bridge component and damage segmentation leveraging Nested Reg-UNet. *Smart Struct. Syst.* **2023**, *31*, 335–349. [CrossRef]
97. Ayele, Y.Z.; Aliyari, M.; Griffiths, D.; Droguett, E.L. Automatic crack segmentation for UAV-assisted bridge inspection. *Energies* **2020**, *13*, 6250. [CrossRef]
98. Wang, J.; Lei, Y.; Yang, X.; Zhang, F. A refinement network embedded with attention mechanism for computer vision based post-earthquake inspections of railway viaduct. *Eng. Struct.* **2023**, *279*, 115572. [CrossRef]
99. Montes, K.; Zhang, M.; Liu, J.; Hajmousa, L.; Chen, Z.; Dang, J. Integrated 3D Structural Element and Damage Identification: Dataset and Benchmarking. In Proceedings of the International Conference on Experimental Vibration Analysis for Civil Engineering Structures, Milan, Italy, 30 August–1 September 2023; pp. 712–720.
100. Artus, M.; Alabassy, M.S.H.; Koch, C. A BIM Based Framework for Damage Segmentation, Modeling, and Visualization Using IFC. *Appl. Sci.* **2022**, *12*, 2772. [CrossRef]
101. Borin, P.; Cavazzini, F. Condition assessment of RC bridges. Integrating machine learning, photogrammetry and BIM. *Int. Arch. Photogramm. Remote Sens. Spat. Inf. Sci.* **2019**, *42*, 201–208. [CrossRef]
102. Jang, K.; An, Y.K.; Kim, B.; Cho, S. Automated crack evaluation of a high-rise bridge pier using a ring-type climbing robot. *Comput.-Aided Civ. Infrastruct. Eng.* **2021**, *36*, 14–29. [CrossRef]
103. McLaughlin, E.; Charron, N.; Narasimhan, S. Automated defect quantification in concrete bridges using robotics and deep learning. *J. Comput. Civ. Eng.* **2020**, *34*, 04020029. [CrossRef]
104. Ye, X.W.; Ma, S.Y.; Liu, Z.X.; Ding, Y.; Li, Z.X.; Jin, T. Post-earthquake damage recognition and condition assessment of bridges using UAV integrated with deep learning approach. *Struct. Control Health Monit.* **2022**, *29*, e3128. [CrossRef]
105. Shen, Q.; Xiao, B.; Mi, H.; Yu, J.; Xiao, L. Adaptive Learning Filters–Embedded Vision Transformer for Pixel-Level Segmentation of Low-Light Concrete Cracks. *J. Perform. Constr. Facil.* **2025**, *39*, 04025007. [CrossRef]
106. Li, G.; Liu, Q.; Zhao, S.; Qiao, W.; Ren, X. Automatic crack recognition for concrete bridges using a fully convolutional neural network and naive Bayes data fusion based on a visual detection system. *Meas. Sci. Technol.* **2020**, *31*, 075403. [CrossRef]
107. Yang, X.; Li, H.; Yu, Y.; Luo, X.; Huang, T.; Yang, X. Automatic pixel-level crack detection and measurement using fully convolutional network. *Comput.-Aided Civ. Infrastruct. Eng.* **2018**, *33*, 1090–1109. [CrossRef]
108. Narazaki, Y.; Hoskere, V.; Yoshida, K.; Spencer, B.F.; Fujino, Y. Synthetic environments for vision-based structural condition assessment of Japanese high-speed railway viaducts. *Mech. Syst. Signal Process.* **2021**, *160*, 107850. [CrossRef]

109. Yang, L.; Li, B.; Li, W.; Liu, Z.; Yang, G.; Xiao, J. A robotic system towards concrete structure spalling and crack database. In Proceedings of the 2017 IEEE International Conference on Robotics and Biomimetics (ROBIO), Macau, China, 5–8 December 2017; pp. 1276–1281.
110. Shi, Y.; Cui, L.; Qi, Z.; Meng, F.; Chen, Z. Automatic road crack detection using random structured forests. *IEEE Trans. Intell. Transp. Syst.* **2016**, *17*, 3434–3445. [CrossRef]
111. Yang, F.; Zhang, L.; Yu, S.; Prokhorov, D.; Mei, X.; Ling, H. Feature pyramid and hierarchical boosting network for pavement crack detection. *IEEE Trans. Intell. Transp. Syst.* **2019**, *21*, 1525–1535. [CrossRef]
112. Peng, X.; Zhong, X.; Zhao, C.; Chen, A.; Zhang, T. A UAV-based machine vision method for bridge crack recognition and width quantification through hybrid feature learning. *Constr. Build. Mater.* **2021**, *299*, 123896. [CrossRef]
113. Flah, M.; Suleiman, A.R.; Nehdi, M.L. Classification and quantification of cracks in concrete structures using deep learning image-based techniques. *Cem. Concr. Compos.* **2020**, *114*, 103781. [CrossRef]
114. Kim, J.-Y.; Park, M.-W.; Huynh, N.T.; Shim, C.; Park, J.-W. Detection and Length Measurement of Cracks Captured in Low Definitions Using Convolutional Neural Networks. *Sensors* **2023**, *23*, 3990. [CrossRef]
115. Liang, X. Image-based post-disaster inspection of reinforced concrete bridge systems using deep learning with Bayesian optimization. *Comput.-Aided Civ. Infrastruct. Eng.* **2019**, *34*, 415–430. [CrossRef]
116. Mirzazade, A.; Popescu, C.; Blanksvärd, T.; Täljsten, B. Workflow for off-site bridge inspection using automatic damage detection-case study of the pahtajokk bridge. *Remote Sens.* **2021**, *13*, 2665. [CrossRef]
117. Kun, J.; Zhenhai, Z.; Jiale, Y.; Jianwu, D. A deep learning-based method for pixel-level crack detection on concrete bridges. *IET Image Process.* **2022**, *16*, 2609–2622. [CrossRef]
118. Kim, I.-H.; Jeon, H.; Baek, S.-C.; Hong, W.-H.; Jung, H.-J. Application of crack identification techniques for an aging concrete bridge inspection using an unmanned aerial vehicle. *Sensors* **2018**, *18*, 1881. [CrossRef]
119. Ni, Y.; Mao, J.; Wang, H.; Xi, Z.; Xu, Y. Toward High-Precision Crack Detection in Concrete Bridges Using Deep Learning. *J. Perform. Constr. Facil.* **2023**, *37*, 04023017. [CrossRef]
120. Jiang, W.; Liu, M.; Peng, Y.; Wu, L.; Wang, Y. HDCB-Net: A neural network with the hybrid dilated convolution for pixel-level crack detection on concrete bridges. *IEEE Trans. Ind. Inform.* **2020**, *17*, 5485–5494. [CrossRef]
121. Tran, T.S.; Nguyen, S.D.; Lee, H.J.; Tran, V.P. Advanced crack detection and segmentation on bridge decks using deep learning. *Constr. Build. Mater.* **2023**, *400*, 132839. [CrossRef]
122. Kao, S.-P.; Chang, Y.-C.; Wang, F.-L. Combining the YOLOv4 deep learning model with UAV imagery processing technology in the extraction and quantization of cracks in bridges. *Sensors* **2023**, *23*, 2572. [CrossRef] [PubMed]
123. Ma, K.; Meng, X.; Hao, M.; Huang, G.; Hu, Q.; He, P. Research on the Efficiency of Bridge Crack Detection by Coupling Deep Learning Frameworks with Convolutional Neural Networks. *Sensors* **2023**, *23*, 7272. [CrossRef]
124. Inam, H.; Islam, N.U.; Akram, M.U.; Ullah, F. Smart and Automated Infrastructure Management: A Deep Learning Approach for Crack Detection in Bridge Images. *Sustainability* **2023**, *15*, 1866. [CrossRef]
125. Zakaria, M.; Karaaslan, E.; Catbas, F.N. Advanced bridge visual inspection using real-time machine learning in edge devices. *Adv. Bridge Eng.* **2022**, *3*, 27. [CrossRef]
126. Meng, Q.; Yang, J.; Zhang, Y.; Yang, Y.; Song, J.; Wang, J. A Robot System for Rapid and Intelligent Bridge Damage Inspection Based on Deep-Learning Algorithms. *J. Perform. Constr. Facil.* **2023**, *37*, 04023052. [CrossRef]
127. Ozgenel, F. Concrete crack images for classification. *Mendeley Data* **2019**. [CrossRef]
128. Yu, L.; He, S.; Liu, X.; Jiang, S.; Xiang, S. Intelligent crack detection and quantification in the concrete bridge: A deep learning-assisted image processing approach. *Adv. Civ. Eng.* **2022**, *2022*, 1813821. [CrossRef]
129. Feng, C.; Zhang, H.; Wang, S.; Li, Y.; Wang, H.; Yan, F. Structural damage detection using deep convolutional neural network and transfer learning. *KSCE J. Civ. Eng.* **2019**, *23*, 4493–4502. [CrossRef]
130. Carranza-García, M.; Torres-Mateo, J.; Lara-Benítez, P.; García-Gutiérrez, J. On the performance of one-stage and two-stage object detectors in autonomous vehicles using camera data. *Remote Sens.* **2020**, *13*, 89. [CrossRef]
131. Zou, Z.; Chen, K.; Shi, Z.; Guo, Y.; Ye, J. Object detection in 20 years: A survey. *Proc. IEEE* **2023**, *111*, 257–276. [CrossRef]
132. Srivastava, S.; Divekar, A.V.; Anilkumar, C.; Naik, I.; Kulkarni, V.; Pattabiraman, V. Comparative analysis of deep learning image detection algorithms. *J. Big Data* **2021**, *8*, 66. [CrossRef]
133. Wu, W.-H.; Lee, J.-C.; Wang, Y.-M. A study of defect detection techniques for metallographic images. *Sensors* **2020**, *20*, 5593. [CrossRef]

MDPI AG
Grosspeteranlage 5
4052 Basel
Switzerland
Tel.: +41 61 683 77 34

Infrastructures Editorial Office
E-mail: infrastructures@mdpi.com
www.mdpi.com/journal/infrastructures

www.ingramcontent.com/pod-product-compliance
Lightning Source LLC
Chambersburg PA
CBHW040822250926
51850CB00007BA/1237
* 9 7 8 3 7 2 5 8 7 9 3 5 9 *